Atlas of Plant Viruses

Volume II

Authors

R. I. B. Francki, Ph.D.
Department of Plant Pathology
Waite Agricultural Research Institute
University of Adelaide
Glen Osmond
South Australia

Robert G. Milne, Ph.D.
Institute of Applied Plant Virology
National Research Council
Turin
Italy

T. Hatta, Ph.D.
Aburahi Laboratories
Shionogi & Co., Ltd.
Shiga
Japan

CRC Press
Taylor & Francis Group
Boca Raton London New York

CRC Press is an imprint of the
Taylor & Francis Group, an **informa** business

First published 1985 by CRC Press
Taylor & Francis Group
6000 Broken Sound Parkway NW, Suite 300
Boca Raton, FL 33487-2742

Reissued 2018 by CRC Press

Library of Congress Cataloging in Publication Data

Francki, R. I. B., 1930-
Atlas of plant viruses.

Includes bibliographies and indexes.
1. Plant viruses--Atlases. I. Milne, Robert G.
II. Hatta, T. III. Title.
QR351.F73 1985 632'.8 84-6999
ISBN 0-8493-6501-5 (v. 1)
ISBN 0-8493-6502-3 (v. 2)

A Library of Congress record exists under LC control number: 84006999

Publisher's Note
The publisher has gone to great lengths to ensure the quality of this reprint but points out that some imperfections in the original copies may be apparent.

Disclaimer
The publisher has made every effort to trace copyright holders and welcomes correspondence from those they have been unable to contact.

ISBN 13: 978-1-315-89084-5 (hbk)
ISBN 13: 978-1-351-06994-6 (ebk)

Visit the Taylor & Francis Web site at http://www.taylorandfrancis.com and the
CRC Press Web site at http://www.crcpress.com

PREFACE

Electron microscopy has played an important role in the development of virology, contributing significantly to an understanding of the molecular biology and pathology of viruses and to the development of virus taxonomy and diagnostics. Nevertheless, our impression is that, while electron microscopes and their attendant techniques are generally capable of producing micrographs of high quality, the micrographs presented to journals (and accepted by them) or published in books, vary in quality from excellent to poor. No publication exists that assembles a comprehensive collection of plant virus electron micrographs of good quality, offers a consistent treatment, and backs the visual data with a consistent and comprehensive text.

It was with these considerations in mind that we decided to embark on preparing an atlas of plant viruses. Our aim has been to bring together, in a systematic way, the available information about the morphology and cytopathology of plant viruses in all the taxonomic groups recognized by the International Committee on Taxonomy of Viruses (ICTV). We have also tried to consider some viruses that are quite well known but still unclassified.

The project might arguably have been better tackled by an editor marshalling a group of authors to write chapters on viruses with which they were especially familiar. By this procedure each group of viruses would have been treated in the most authoritative way. In practice, however, such an approach has its pitfalls. Uniformity and coherence are difficult to attain in such a book, and the near-impossibility of extracting manuscripts from all the authors on time can mean that certain chapters (written by the promptest authors) are painfully out of date at the time of publication. Believing that uniformity and coherence are important in an atlas, we chose the rather amibitious task of preparing the book ourselves. In striving to be authoritative we have relied heavily on many colleagues for information and material.

Although this book is primarily about the structure of virus particles and infected cells, we see little virtue in studying structure without function. Hence we have referred to the results of biochemical experiments where relevant, so that the virus particles described appear as part of a replicating complex. Similarly, we have tried to portray infected cells as active rather than static structures.

In Chapter 1 a brief account is given of the development of studies on virus structure, nucleic acids, cytopathology and taxonomy from their beginnings. We hope that this will provide a useful background to the chapters that follow. The two authoritative guides to the taxonomy of plant viruses used are the Reports of the ICTV and the Commonwealth Mycological Institute and Association of Applied Biologists' *CMI/AAB Descriptions of Plant Viruses*, edited by B. D. Harrison and A. F. Murant. It seems important to standardize on a list of virus abbreviations or sigla that will come to be used unambiguously by all plant virologists. We have used the list of van Regenmortel published in his book *Serology and Immunochemistry of Plant Viruses* (Academic Press, 1982) where possible, and have extended it.

We have tried to present not only the current state of knowledge about the viruses discussed, but also the current state of uncertainty (or sheer ignorance) in particular areas. In several of the chapters, data are included on unclassified viruses with possible affinities to the group under discussion; in doing this, we hope to stimulate investigations that will clarify the positions of such viruses.

The bibliographies at the end of each chapter are not exhaustive, but they include references that we consider the most relevant, and sufficient for the reader who wishes to explore further.

Many of the electron micrographs in this book concern viruses investigated as a part of a research project supported by a Commonwealth Special Research Grant from the Australian Department of Primary Industry on the "Characterization of Plant Viruses in Australia".

Many others were prepared in the Italian National Research Council Plant Virus Institute in Turin, Italy (the Istituto di Fitovirologia Applicata del CNR). Other photographs, or virus preparations from which we prepared photographs, were kindly supplied by colleagues. For easier comparison, we have presented the particles of at least one example of each virus group negatively stained in the same material (uranyl acetate) and magnified a nominal (i.e., not always accurately calibrated) 300,000 times. This is not to deny the usefulness of other negative stains or preparative techniques. In selecting the micrographs of virus-infected cells, we have included large areas of the cells where possible, so as to give a better idea of the ultrastructural context.

The treatment of some virus groups is more comprehensive than that of others. This is, we hope, mostly a reflection of the current unevenness of the available data. Perhaps the *chiaroscuro* may encourage some readers to do experiments that would help a future atlas to be more comprehensive.

R. I. B. Francki
R. G. Milne
T. Hatta

AUTHORS

R. I. B. Francki is a Reader in Plant Pathology at the Waite Agricultural Research Institute of The University of Adelaide, South Australia.

Dr. Francki received his B.Sc., M.Sc., and Ph.D. degrees in Botany from the University of Auckland in 1955, 1958, and 1961, respectively.

He has published over 100 research papers and a number of reviews in the field of plant virology. He was Chairman of the Plant Virus Subcommittee of the International Committee on Taxonomy of Viruses from 1975 to 1981. He is also presently Secretary of the Virology Division of the International Union of Microbiological Societies.

Dr. Francki is an Associate Editor of the journal, *Virology*, and of the *Journal of General Virology* and is on the Advisory Boards of *Current Topics in Vector Research* and *Microbiological Sciences*. He has also served as an Associate Editor of *Intervirology* (from 1976 to 1981).

Dr. Francki is a member of a number of national and international scientific societies, including the Society for General Microbiology, the Association of Applied Biologists, the New York Academy of Science, the International Society for Plant Molecular Biology, American Society for Virology, and the American Phytopathological Society. In 1983 he was elected a Fellow of the American Phytopathological Society.

Dr. Francki has worked as a visiting researcher in a number of laboratories including the Department of Plant Biochemistry, University of California at Los Angeles in 1964 to 1965. He was a Fulbright Scholar at the Department of Agricultural Biochemistry, University of Arizona, in 1970 and a State Agricultural University Senior Research Fellow in the Department of Virology, Agricultural University, Wageningen, The Netherlands.

R. G. Milne, Ph.D., is a senior researcher with the Italian National Research Council, Institute of Applied Plant Virology, Turin.

He received his B.A. from Trinity College, Cambridge, in 1956, and his Ph.D. from Wye College, University of London, in 1960. From 1960 to 1964 he taught virology at the Botany School, Oxford University, and from 1964 to 1966 did postdoctoral research in electron microscopy at the Molecular Biology and Virus Laboratory, University of California, Berkeley. Here he was lucky to have contact with Wendell Stanley, Robley Williams, Heinz Fraenkel-Conrat, Arthur Knight, and "Geheimrat" Kleinschmidt. Returning to England, he joined the Plant Pathology Department, Rothamsted Experimental Station, Harpenden, from 1966 to 1970, and then did a year of medical electron microscopy at the Clinical Research Centre, Northwick Park, Harrow. In 1972 and again from 1974 to 1977 he was Visiting Fellow of the Italian National Research Council in the Institute of Applied Plant Virology in Turin. From 1972 to 1974 he was Associate Professor at the Institute of Molecular and Cell Biology of the University of Strasbourg. He assumed his present position in 1977.

Dr. Milne was born in Tanganyika, and when very small acquired the nickname *Bwana Madudu* ("He of the Creepy-Crawlies") due to his interest in large scorpions. As he grew bigger, the bugs that claimed his attention grew smaller. He is now author or co-author of about 80 research papers and reviews dealing mainly with thin sectioning, negative staining, and immunoelectron microscopy of plant viruses. The plant reo-like viruses have been a major interest, and lately a side interest in virus taxonomy and nomenclature has grown uncomfortably large. He has served on the editorial board of the *Journal of General Virology* and is currently a member of the Plant Virus Subcommittee of the International Committee on Taxonomy of Viruses.

T. Hatta, Ph.D., is a head of the Plant Pathology Division, Aburahi Laboratories, of Shionogi & Co., Ltd., Shiga, Japan.

Dr. Hatta received his B.Ag.Sc. (1968) and M.Ag.Sc. (1970) from the University of Osaka Prefecture, Japan, and the Ph.D. (1975) from the University of Auckland, Auckland, New Zealand, where he studied with S. Bullivant and R. E. F. Matthews.

He was Postdoctoral Fellow in the Department of Plant Pathology at the University of Adelaide from 1975 until he assumed his present position.

He is the author of a number of research and review papers, particularly in the field of electron microscopy of plant viruses.

ACKNOWLEDGMENTS

We are much indebted to many colleagues who provided illustrations used in this book; they are acknowledged individually in the text. We particularly thank Drs. G. P. Martelli, M. Russo, and V. Masenga for preparing a number of electron micrographs especially for this book. We also thank many colleagues who provided us with information by correspondence and who sent manuscripts prior to publication. We are especially indebted to the following colleagues who critically read and commented upon sections of the manuscript: Drs. A. Appiano, D. H. L. Bishop, G. Boccardo, M. Conti, C. P. de Jager, D. C. Gaidusek, A.-L. Haenni, B. D. Harrison, I. H. Holmes, R. Hull, T. Inouye, A. T. Jones, R. Koenig, D.-E. Lesemann, V. Lisa, R. Lister, O. Lovisolo, E. Luisoni, G. P. Martelli, R. E. F. Matthews, M. A. Mayo, G. I. Mink, D. Mossop, S. Pennazio, D. Peters, R. Redolfi, R. J. Shepherd, and T. Yamamoto.

We thank Mrs. L. Wichman for much of the art work and Messrs. C. J. Grivell and B. A. Palk for photographic reproductions. The help of Mr. C. J. Grivell was also invaluable in putting the manuscript together, as was that of Mrs. S. Suter and Mrs. B. Goldsmith in cheerfully typing and re-typing the manuscript.

Much of the work for this book was supported by a Commonwealth Special Grant from the Australian Department of Primary Industry. Research at the Waite Agricultural Research Institute was also supported by grants form the Australian Research Grants Scheme and the University of Adelaide. Work on the book done in Turin was supported by the Italian Consiglio Nazionale delle Richerche.

TABLE OF CONTENTS

Volume I

Volume II

Chapter 1

COMOVIRUS GROUP

Comoviruses derive their name from the type member, cowpea mosaic virus (CPMV).[1] Each Comovirus appears to be related to at least one other group member, the group forming a network of serologically related viruses.[2] The viruses have small isometric particles about 30 nm in diameter built from two species of polypeptides of M_r about 37×10^3 and 23×10^3. The particles sediment as three components which have identical protein coats but vary in their RNA content. They encapsidate two species of RNA of M_r about 2.4×10^6 and 1.4×10^6, which together constitute the viral genome.

Each Comovirus is transmitted to a narrow range of plants and most have been shown to be vectored by leaf-feeding beetles. Seed transmission usually occurs at a low but significant level.[3] Comoviruses can be transmitted by mechanical inoculation. They all induce characteristic cytopathological effects in the cytoplasm of infected cells.[2-4]

I. MEMBERS OF THE GROUP AND THEIR RELATIONSHIPS

CPMV, by far the best known member of the group, has been used extensively for studies on particle structure, mode of replication, and cytopathology. Most of the remaining Comoviruses have been studied rather superficially. However, all the available data, including those establishing extensive serological relationships among the members, make it fairly safe to assume that Comoviruses are all very similar.[2,3,5] Nevertheless, the serological data are not sufficiently precise to draw clear conclusions about the degrees of relationship among the members. In many cases it is difficult to decide whether some of the described viruses merit separate names or should be considered as strains of the same virus. This has already introduced some confusion in the literature; for example, the virus now known as cowpea severe mosaic virus (CPSMV) was originally referred to as the severe strain of CPMV. There are now good reasons for considering the two viruses as distinct on the basis of differences in host range, symptoms produced, antigenic properties, nucleotide sequences, and their inability to produce viable pseudorecombinants.[6-9] There is also confusion about the identity of viruses such as bean curly dwarf mosaic virus (BCDMV).[10] Although the authors concluded that BCDMV was sufficiently similar to quail pea mosaic virus (QPMV) to be considered as a strain of that virus, they referred to each isolate by a distinct name.[10]

All but four Comoviruses infect predominantly leguminous plants. Radish mosaic virus (RaMV), squash mosaic virus (SqMV), andean potato mottle virus (APMV), and Ullucus virus C (UVC) usually infect plants within the families Cruciferae, Cucurbitaceae, Solanaceae, and Basellaceae, respectively.[2,11] However, from the limited serological data available, one cannot conclude whether these host range differences are reflected in the antigenic properties of the viruses.[2,3] Much more work will be required to establish a satisfactory taxonomy of the group. This could be achieved with more extensive serological studies using uniform techniques throughout. Much useful taxonomic information might also be obtained by comparing nucleotide sequence homologies between the various Comoviruses and their strains. The ability to form viable pseudorecombinants between the isolates may also prove to be informative.[12] For the present, a tentative list of Comoviruses is presented in Table 1. However, as more information becomes available, some of the listed viruses may prove to be more appropriately considered as strains.

Although broad bean wilt virus (BBWV) has many chemical and physical properties in common with the Comoviruses,[28,29] its taxonomic position remains enigmatic. It has been assigned as a possible member of the Comovirus group,[1] but does not appear to be serologically related to any of the Comoviruses and is unlike them in being transmitted by aphids.

Table 1
MEMBERS OF THE COMOVIRUS GROUP AND SOME OF THEIR PROPERTIES

Virus	S_{20w} of particles			M_r of RNAs (× 10^6)		M_r of coat proteins (× 10^3)	Vectors	Ref.
	B	M	T	RNA 1	RNA 2			
Cowpea mosaic virus (CPMV)	115	95	58	2.0 (2.4)	1.4 (1.4)[a]	37 and 23	Beetles	13, 14
Andean potato mottle virus (APMV)	112	93	53	2.0	1.4	48 and 22	Unknown	15
Bean curly dwarf mosaic virus (BCDMV)		ND[b]		ND		ND	Beetles	10
Bean pod mottle virus (BPMV)	112	91	54	2.5	1.5	ND	Beetles	16
Bean rugose mosaic virus (BRMV)	113	97	59	2.3	1.4	38 and 23	Beetles	17
Broad bean stain virus (BBSV)	113	92	60	2.6	1.6	42 and 22	Weevils	18, 19
Broad bean true mosaic virus (BBTMV) (= echtes ackerbohnenmosaik virus)	114	93	—	2.7	1.8	41 and 22	Weevils	18
Cowpea severe mosaic virus (CPSMV)	115	95	58	ND		ND	Beetles	7
Glycine mosaic virus (GMV)	130	103	60	ND		42 and 22	Unknown	20
Pea green mottle virus (PGMV)		ND		ND		ND	Unknown	21, 22
Pea mild mosaic virus (PMMV)		ND		2.5	1.4	ND	Unknown	23
Quail pea mosaic virus (QPMV)	112	93	57	ND		40 and 22	Beetles	24
Radish mosaic virus (RaMV)	116	97	57	2.2	1.3	ND	Beetles	25
Red clover mottle virus (RCMV)	127	101	60	ND		ND	Unknown	26
Squash mosaic virus (SqMV)	118	95	57	2.4	1.6	ND	Beetles	27
Ullucus virus C (UVC)	116	95	51	2.2	1.45	45 and 21	Unknown	11

[a] Values in brackets are those obtained when determined under denaturing conditions.
[b] ND, not determined.

It has been suggested that BBWV and the serologically distantly related lamium mild mosaic virus (LMMV) (Table 2) should be considered as members of a new plant virus group.[29,30]

II. VIRUS STRUCTURE AND COMPOSITION

A. Particle Structure

Electron micrographs of Comovirus particles (Figure 1) negatively stained with uranyl acetate and those of BBWV stained with potassium phosphotungstate (Figure 2) show that they are polyhedra about 30 nm in diameter. Crowther and colleagues[41] reported the particle diameter of CPMV to be only 20 to 24 nm, and estimates of the diameters of other Comoviruses by other workers range from 24 to 30 nm. No doubt some of this variation is due to differences in preparing the particles for electron microscopy and errors in the calibration of the microscopes used. When stained in uranyl acetate and compared directly to virus particles in a number of different taxa, broad bean stain virus (BBSV) and broad bean true mosaic virus (BBTMV) particles were found to be 30 nm in diameter.[42]

Table 2
SOME PROPERTIES OF BROAD BEAN WILT AND LAMIUM MILD MOSAIC VIRUSES

Virus	S_{20w} of particles			M_r of RNAs ($\times 10^6$)		M_r of coat proteins ($\times 10^3$)		Ref.
	B	M	T	RNA 1	RNA 2			
Broad bean wilt virus (BBWV)								28—40
Serotype I	116	95	56	2.0	1.5	42	26	28, 29, 36
Petunia ringspot strain (PR strain)								
Type strain	Same as PR strain			2.0	1.5	42	26	
Nasturtium *(Tropaeolum)* ringspot strain	Same as PR strain			2.0	1.5	42	26	28, 29, 36
Plantago I isolate		ND[a]		ND		ND		36
Parsley virus 3 isolate		ND		ND		ND		33
Serotype II								
Plantago II isolate		ND		ND		ND		36
Some lettuce isolates		ND		ND		ND		36, 37
Helleborus isolate		ND		ND		ND		38
Argentine isolates		ND		ND		ND		39
Lamium mild mosaic virus (LMMV)	Sedimentation characteristics very like BBWV			2.2	1.6	45	28.5	30, 40

[a] ND, not determined.

Preparations of Comoviruses as well as those of BBWV and LMMV usually sediment as three components referred to as bottom (B), middle (M), and top (T) components, respectively (Table 1). All the particles have indistinguishable protein shells, but the T component is devoid of RNA, whereas the B and M components contain one molecule each of RNA 1 and RNA 2, respectively.[2]

The Comovirus protein shell is built from equimolar amounts of two distinct proteins of M_r about 37 $\times$ 10^3 and 23 $\times$ 10^3 (37K and 23K proteins, respectively).[13] Early studies on CPMV structure were complicated by the detection of two electrophoretic components in most virus preparations. However, it was shown that this was caused by an *in situ* degradation of the smaller coat protein subunit by proteolytic cleavage in the host cells and after isolation of the virus.[43] The cleavage involves the loss of a peptide of M_r about 2.5 $\times$ 10^3. From the size of the two intact proteins and the molecular weight of the CPMV protein shell, it has been calculated that each particle consists of 60 molecules of each of the two proteins.[13,41] Three-dimensional image analysis from electron micrographs of CPMV revealed that the particles have 5:3:2 symmetry. A model has been proposed for the CPMV particle consisting of 12 pentamers of the 37K protein at the fivefold positions and 20 trimers of the 23K protein at the threefold positions (Figure 3).[45]

B. The Viral Genome

The CPMV genome is bipartite.[2,45] It is composed of two single-stranded, positive-sense RNA molecules, both of whose complete sequences have been determined.[46,47] RNA 1 from B component capsids consists of 5889 nucleotides,[47] and RNA 2 from the M component of 3481 nucleotides; that is, excluding their poly A tails of 150 to 200 residues at their 3′ termini.[48] Although the sequences of the two RNAs are quite distinct, they are similar adjacent to these tails.[49] At their 5′ termini, both RNAs have a covalently linked protein (VPg) of M_r about 4 $\times$ 10^3 and stretches of about 40 nucleotides with similar se-

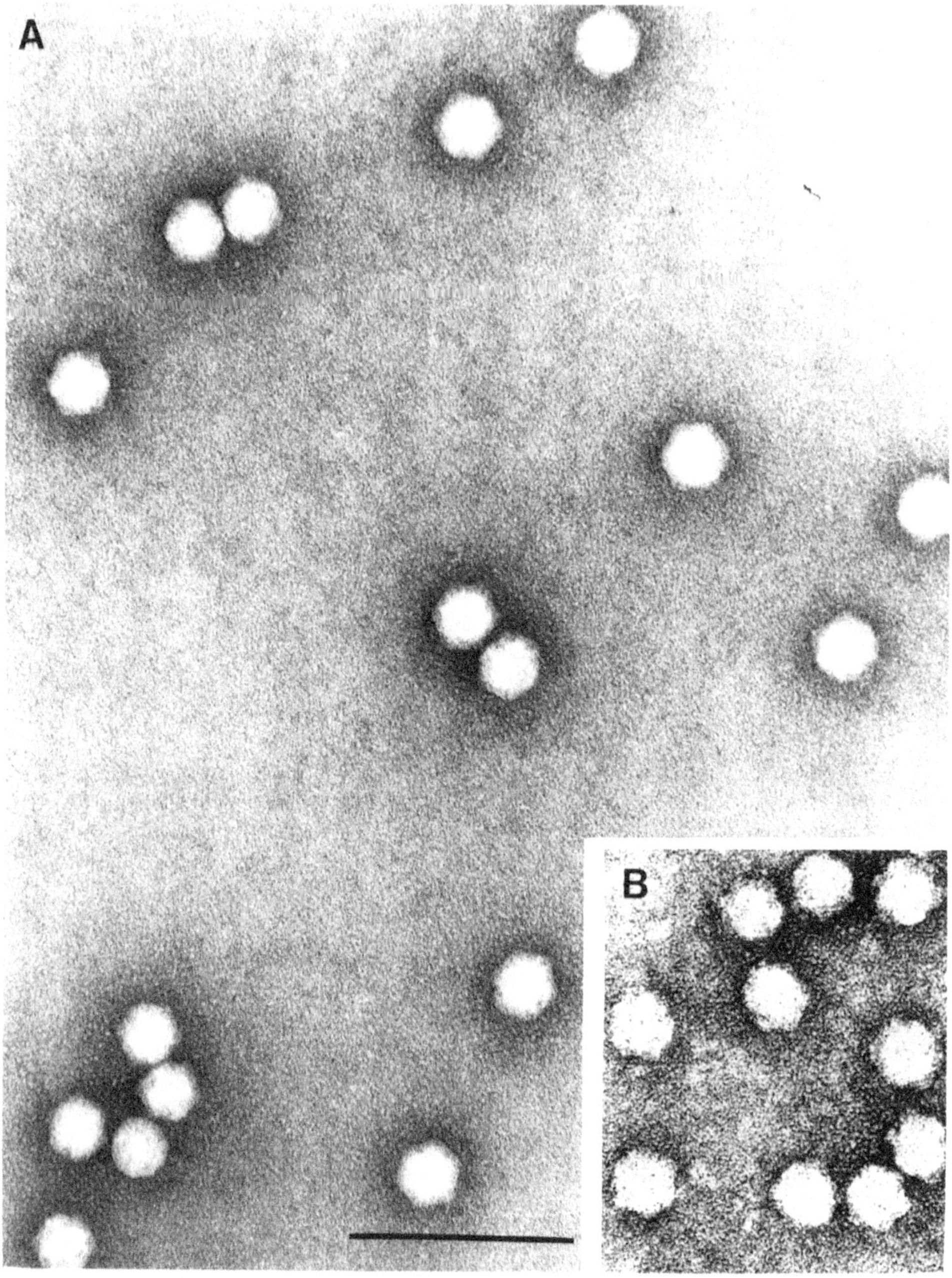

FIGURE 1. A, Purified particles of broad bean true mosaic virus (BBTMV); B, particles from a sap extract of plants infected with radish mosaic virus (RaMV). Both preparations stained in uranyl acetate and photographed at the same magnification. Bar = 100 nm.

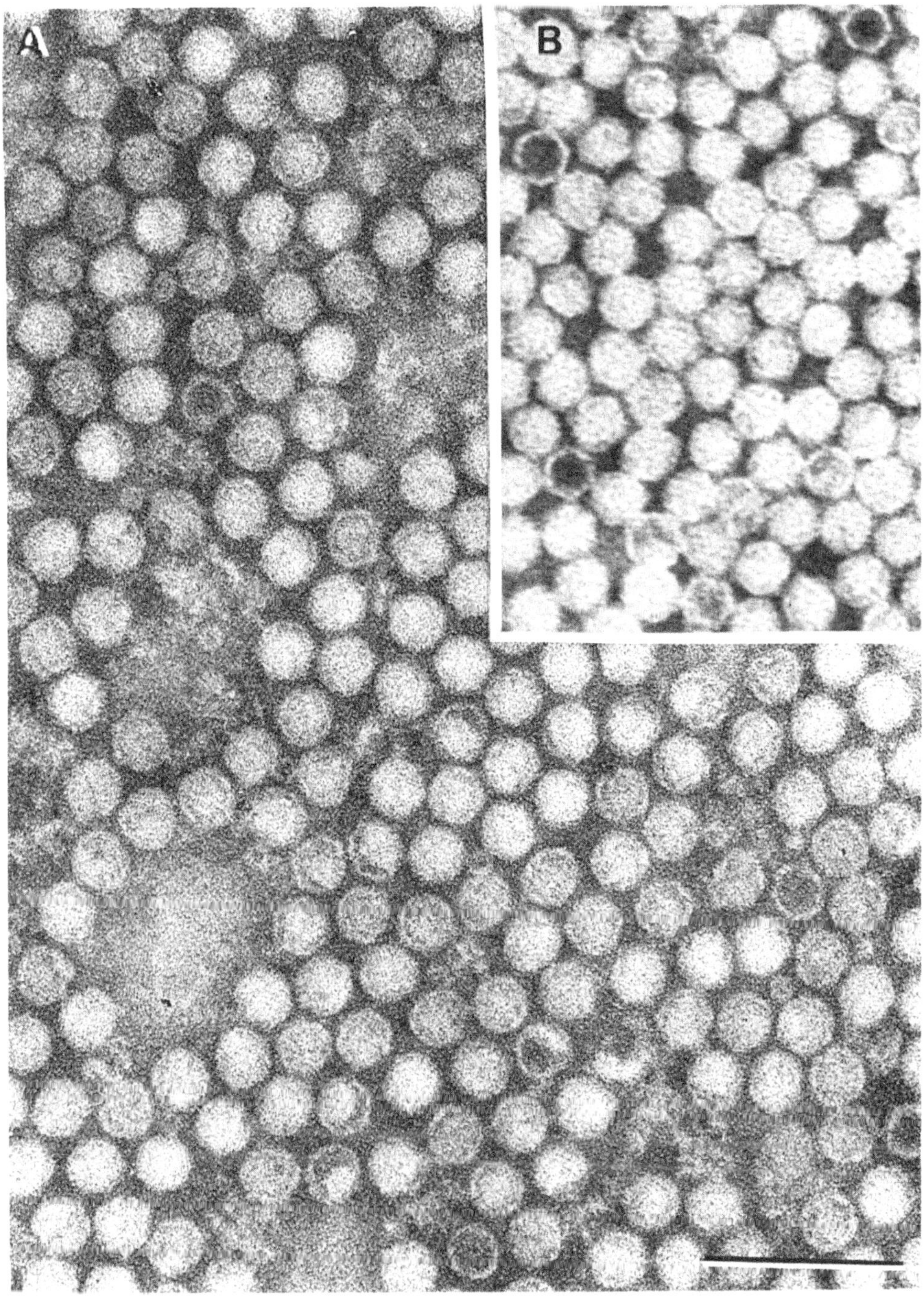

FIGURE 2. A, Purified particles of broad bean wilt virus (BBWV) in neutral potassium phosphotungstate; B, purified particles of broad bean stain virus (BBSV), a recognized Comovirus, for comparison (photographed at the same magnification), also in phosphotungstate. Bars = 100 nm. (Figure 2B courtesy of Dr. S. M. Moghal, Waite Agricultural Research Institute, University of Adelaide.)

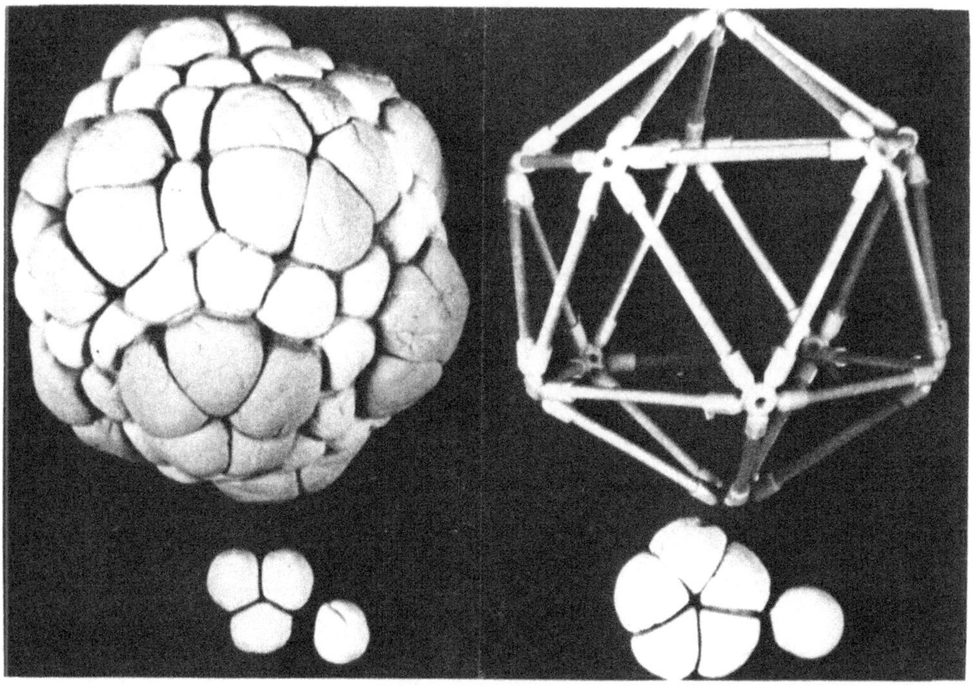

FIGURE 3. Models of the structure of the Comovirus particle. On the right is a simple icosahedral skeleton, and on the left an indication of how such a structure is realized in terms of protein. The subunits of coat protein I (M_r ~37 × 10^3) and coat protein II (M_r ~23 × 10^3) are shown on the right and left, respectively. (Courtesy of Dr. G. Bruening, Department of Biochemistry and Biophysics, University of California, Davis.)

quences.[50-54] The RNAs of other Comoviruses which have been investigated also appear to be polyadenylated and contain VPgs.[55] The VPgs are not essential for infectivity.[50,51]

Numerous studies have been directed towards gaining an understanding of the genetic functions of the two Comovirus RNA components by the construction and analysis of pseudorecombinants. For this, heterologous mixtures of M and B particle components from closely related spontaneously or chemically induced mutants as well as naturally occurring virus strains have been used.[56-67] A number of genetic determinants have been located on RNA 2 including control of symptom production,[57,59,62,66] lack of replication at elevated temperatures,[63] electrophoretic mobility,[64] and serological specificity.[58,65] Although work with strains of RaMV indicated that both RNA 1 and RNA 2 were involved in determining the serological properties, the pseudorecombinants were constructed from rather impure preparations. It now seems certain from in vitro translation studies using Comovirus RNAs as messengers that both the coat protein polypeptides and hence the serological specificity are determined by RNA 2.[65] However, there seems to be disagreement as to the exact locations of the genes of RNA 2 of SqMV[67] and CPMV.[68] The determinants located on RNA 1 include the production of certain symptoms and the ratio of M to B components.[59,62] Oxelfelt and Abdelmoeti[66] have reported that the production of T component of red clover mottle virus (RCMV) is determined by the interaction of both the genomic RNAs.

Although Comoviruses require both RNAs 1 and 2 for infection, it has been shown that inoculation of leaf protoplasts with B particles alone results in the synthesis of some virus-specific polypeptides, but neither of the coat proteins is apparently produced.[69] When similar protoplasts were inoculated with M particles only, no virus-specific polypeptides were detected. Coat protein was detected only in protoplasts which were inoculated with both M and B components.[69-71] These studies demonstrate the limited independence of RNA 1 and the complete dependence of RNA 2 on RNA 1.

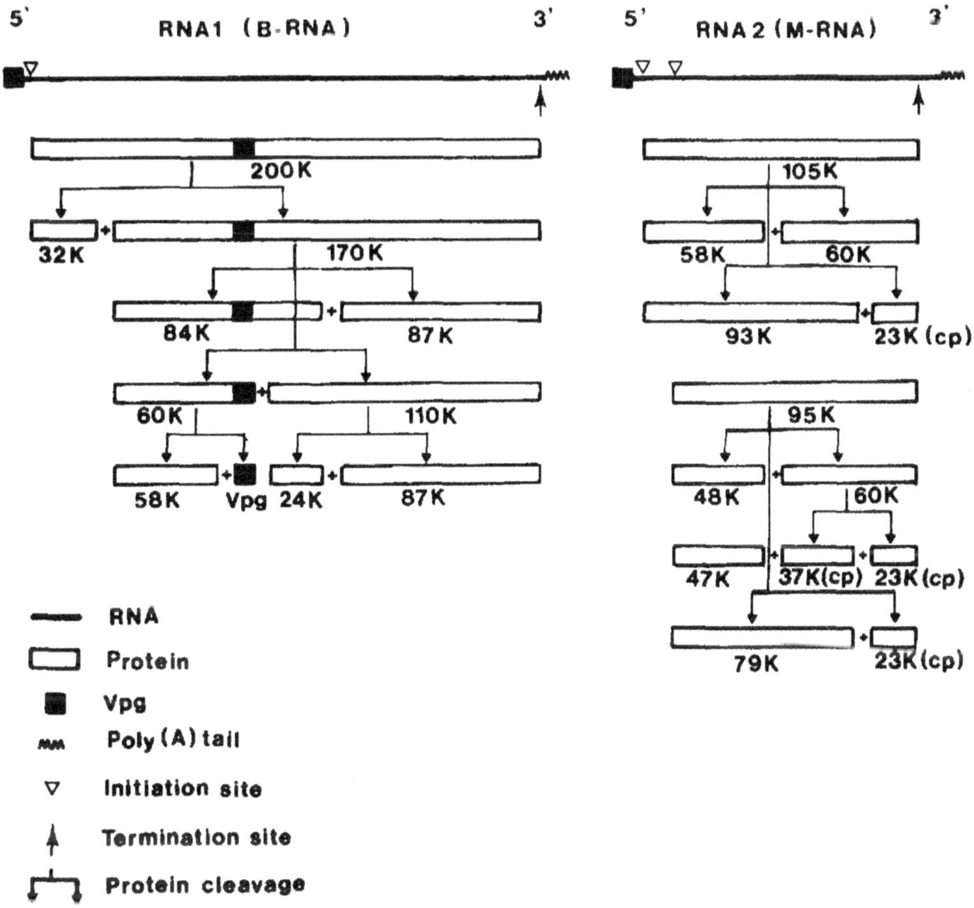

FIGURE 4. Physical map of the Comovirus genome showing structural features and translational strategies of the two RNAs based on data from CPMV.

In vivo studies with protoplasts have revealed that at least seven virus-specific polypeptides are coded for by RNA 1.[70] Definite functions could not be assigned to these proteins but at least one should be involved in RNA replication[69] and another should be responsible for the proteolytic in vitro cleavage of the RNA 2-coded primary translation product.[68,72,73] It has also been shown that RNA 1 codes for the VPg which is covalently linked to each of the viral RNAs.[53,74,75] RNA 2 has been shown to code for both the coat proteins,[68] but the functions of the remaining virus-specified proteins remain obscure. A model of the CPMV genome based on data from van Kammen's group[68,74-76] is presented in Figure 4.

III. CYTOPATHOLOGY

Plant cells infected with a number of Comoviruses have been investigated, including CPMV,[77-81] bean pod mottle virus (BPMV),[80,82-84] BBSV,[85] BBTMV,[85-87] glycine mosaic virus (GMV),[20] RaMV,[88-91] and RCMV.[92] From these studies it appears that the cytological changes associated with infection by these viruses, and hence probably by all Comoviruses, are very similar.[93] The most prominent changes, and probably the most significant in terms of virus replication, are the large vesiculate inclusions seen in the cytoplasm (Figures 5 through 8). They consist of labyrinths of membrane-bound vesicles often occupying a large volume of the cytoplasm and commonly surrounded by mitochondria and chloroplasts (Figure

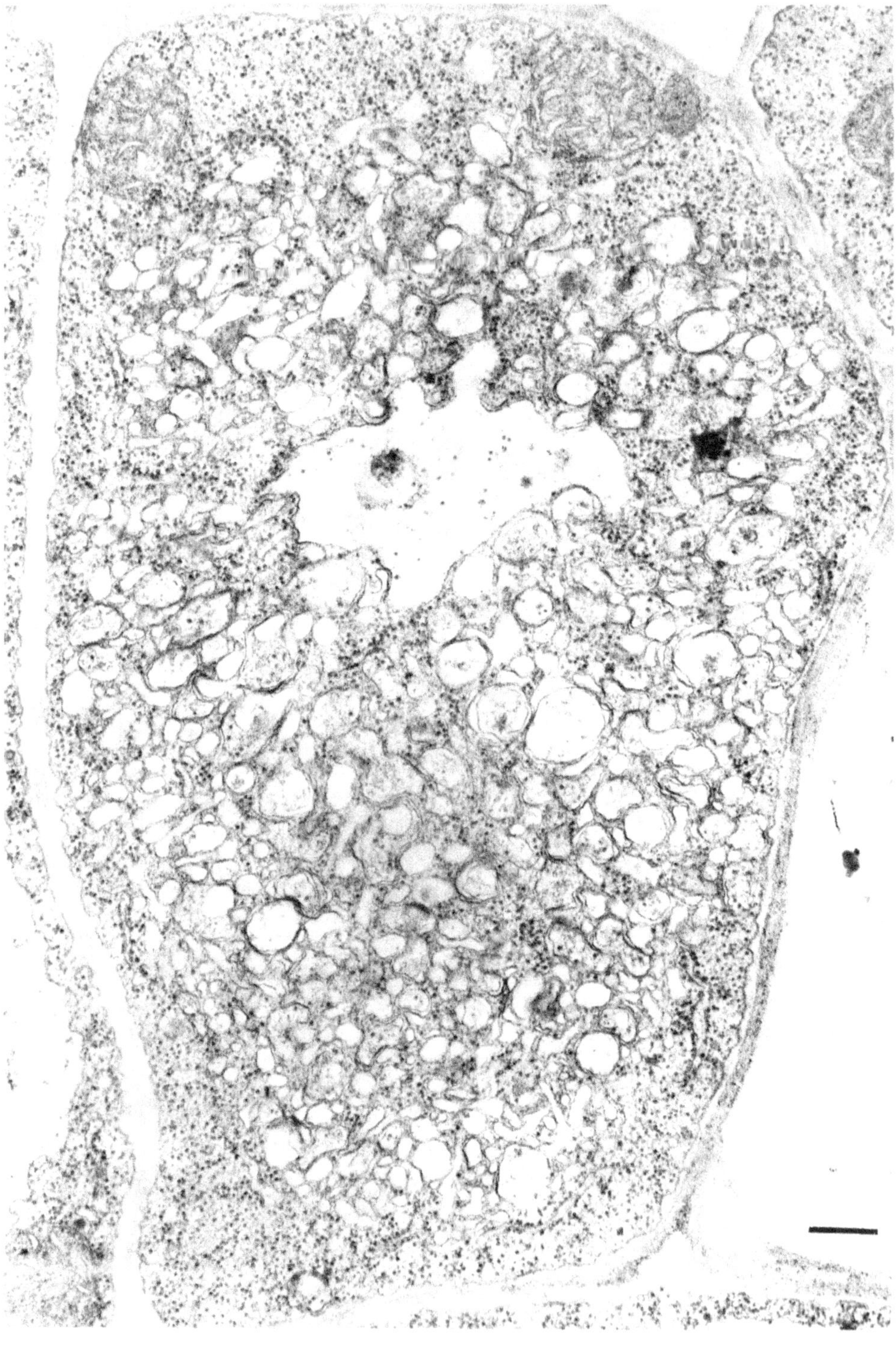

FIGURE 5. Thin section of a mesophyll cell from a leaf of *Eruca sativa* infected with radish mosaic virus (RaMV). The cell is packed with membranous vesicles characteristic of Comovirus infection. Bar = 500 nm.

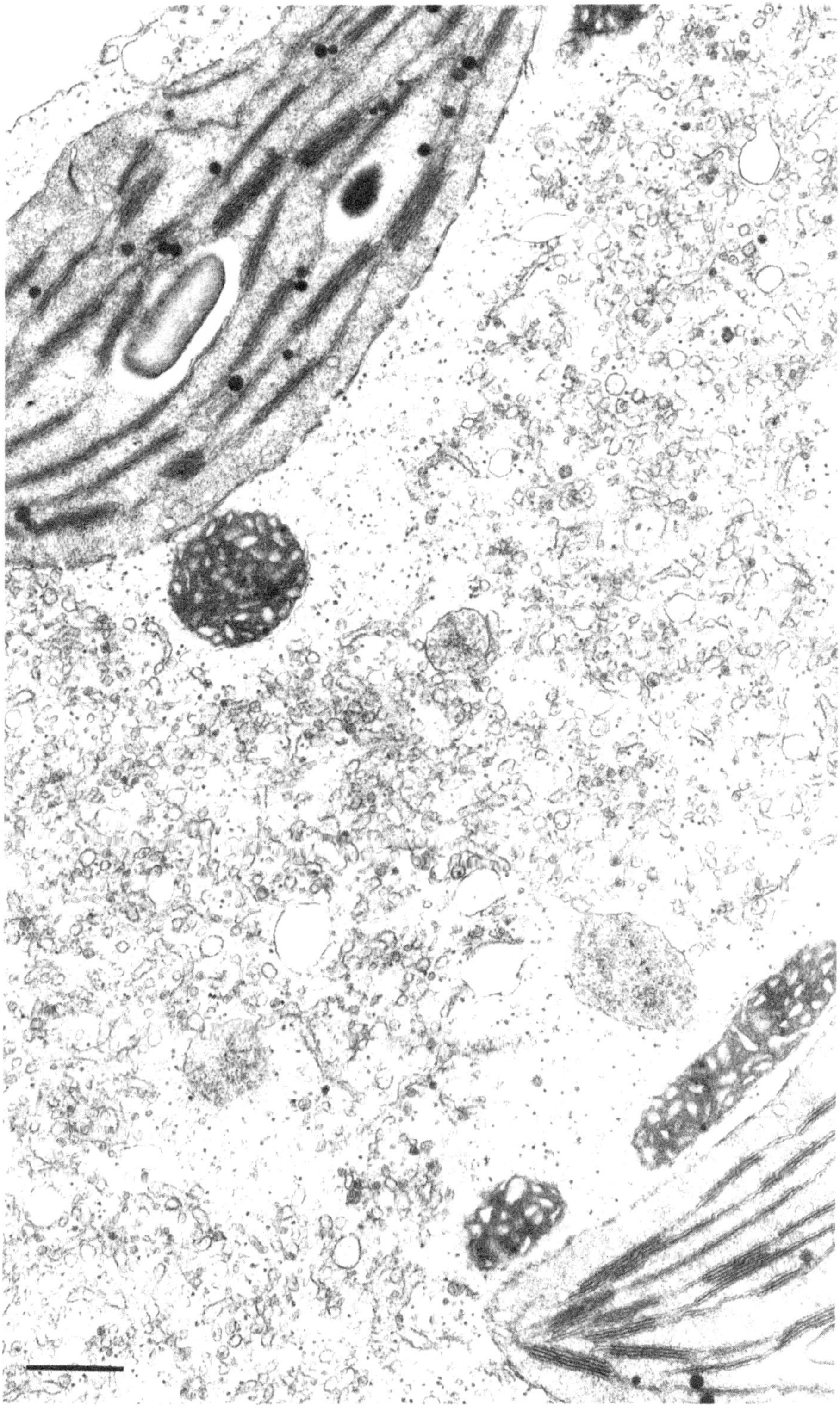

FIGURE 6. Thin sections of *Vicia faba* (broad bean) leaf cell infected with broad bean true mosaic virus (BBTMV) showing scattered virus particles and proliferation of membranes. (Tissue cleared of ribosomes by RNase treatment.) Bar = 500 nm.

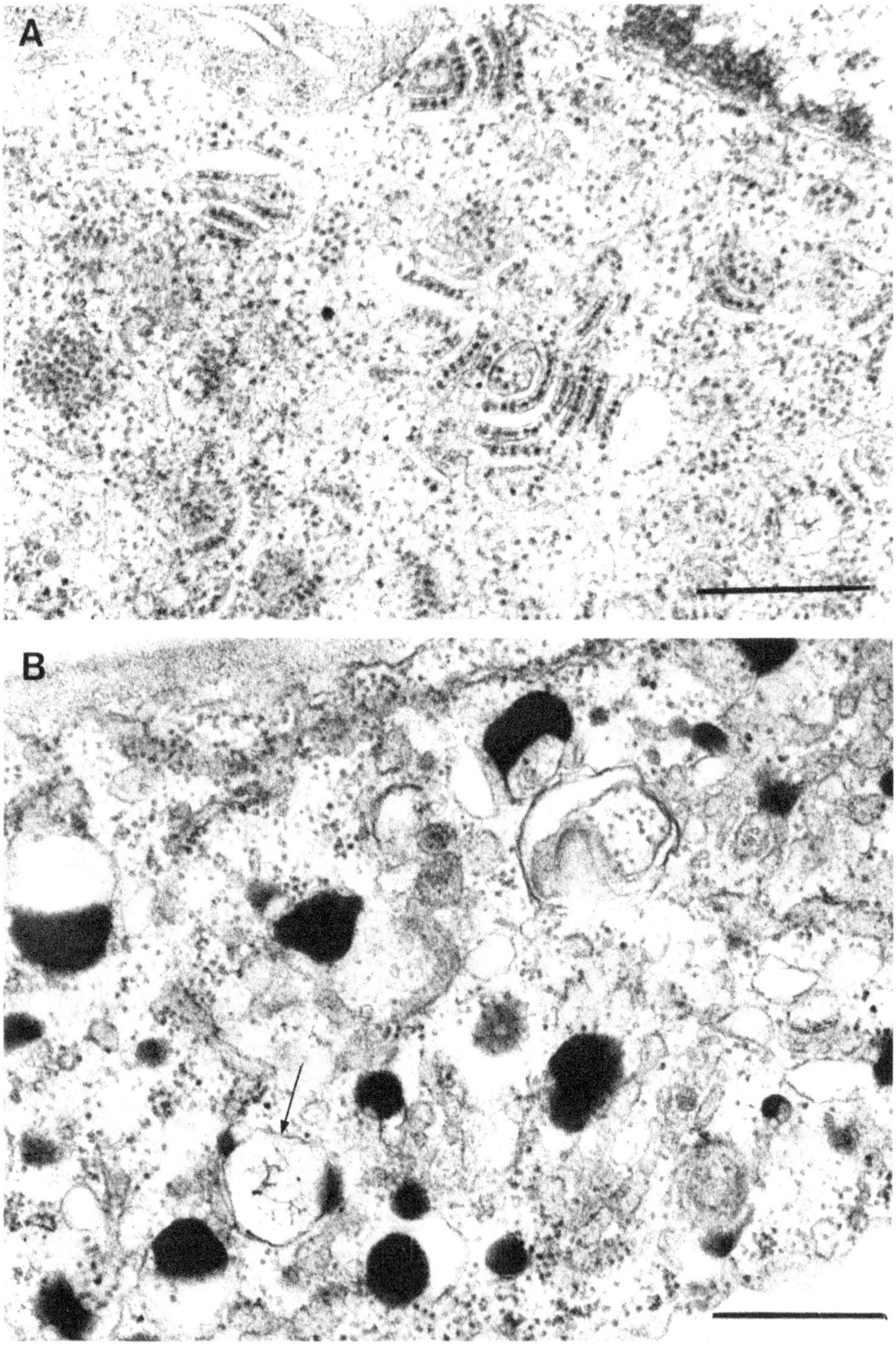

FIGURE 7. A, Thin section as in Figure 6, showing virus particles free in the cytoplasm and some sandwiched between pairs of membranes. B, Thin section of *Cucurbita pepo* (squash) leaves infected with squash mosaic virus (SqMV), showing patches of amorphous electron-dense material and vesicles, one of which contains fibrils (arrow). Bars = 500 nm.

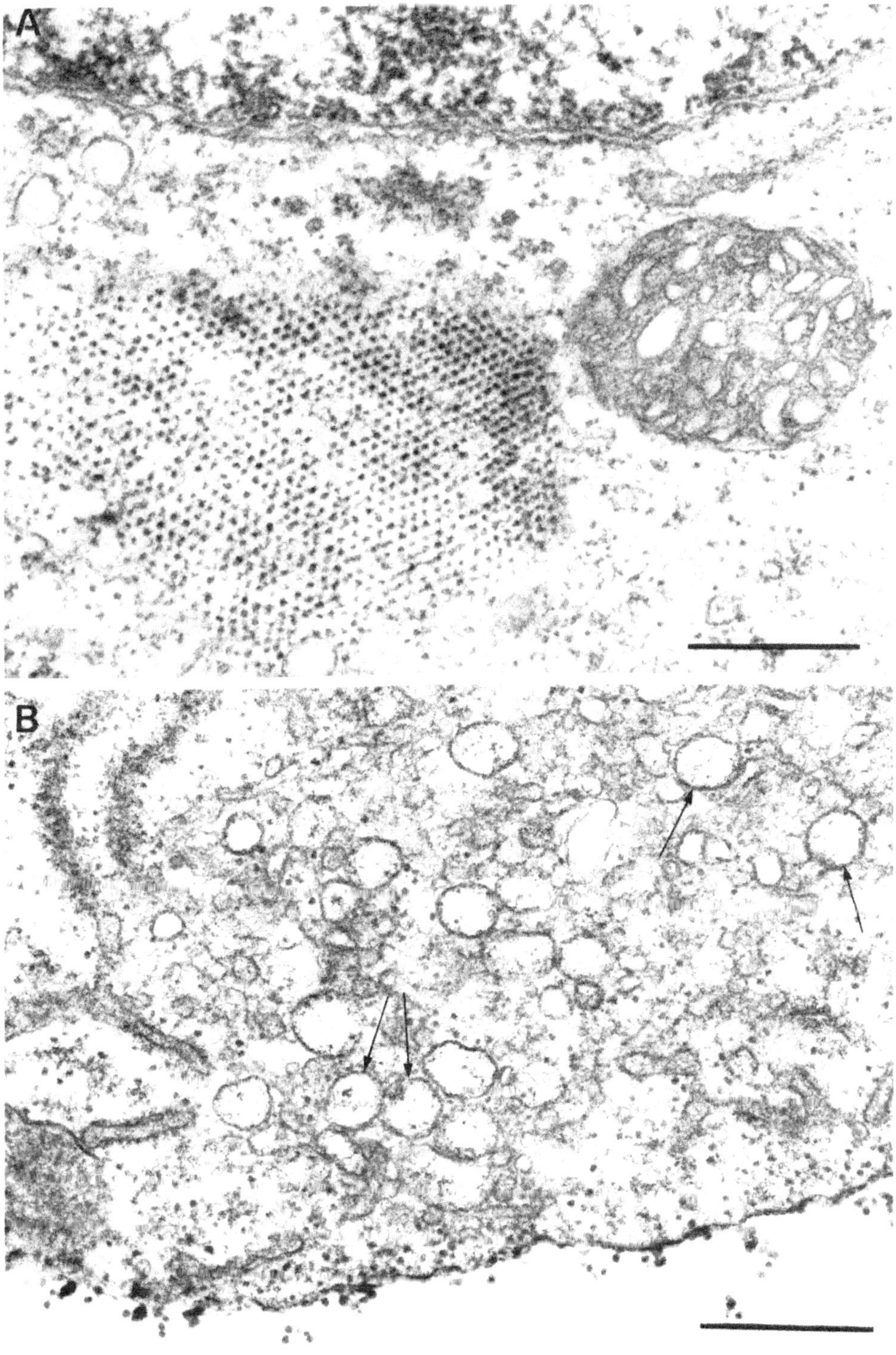

FIGURE 8. Thin sections of a *Vicia faba* (broad bean) leaf cell infected with BBSV. A, Virus particles crystallized in the cytoplasm; B, typical membrane proliferation and fibril-containing vesicles (arrows). (Tissues cleared of ribosomes by RNase treatment.) Bars = 500 nm.

6). In many instances the inclusions contain patches of amorphous electron-dense material (Figure 7B). Many of the inclusion vesicles contain fibrils (Figures 6 through 8). Enzyme cytochemical evidence supports the view that the fibrils are double-stranded RNA,[86] probably the viral replicative form RNA detected in infected plants[94] and shown to co-sediment with the virus-specific inclusions.[79,95] The inclusions can develop in cells infected with only the B component of CPMV, and therefore in the absence of coat protein synthesis.[71] The vesiculated structures are thus almost certainly involved in viral RNA replication but not in the synthesis of coat protein. Cycloheximide but not chloramphenicol inhibits CPMV coat protein synthesis, indicating that this involves the cytoplasmic 80S and not the chloroplast 70S ribosomes.[96]

Comovirus particles have been detected in the cytoplasm and vacuoles of infected cells, either randomly scattered or in crystalline inclusions (Figures 7 through 9). Those scattered in the cytoplasm occur among the vesicles of the vesiculated inclusions or in what appears to be normal cytoplasm (Figures 8B and 10). The particles may be enclosed in long tubules, some of which pass through plasmodesmata and are sometimes associated with unusual outgrowths of the cell walls (Figure 10C). Unless occurring in crystalline arrays (Figures 8A and 9) or in tubules (Figure 10), the virus particles are difficult to distinguish from cytoplasmic ribosomes.[77] However, their identity can be confirmed in sections from tissues in which the ribosomal RNA has been differentially digested with RNase after fixation but prior to embedding[87] (Figures 6 through 8).

The occurrence of crystalline virus aggregates has been considered by Edwardson and Christie[97] to be a useful diagnostic feature of Comovirus-infected plants. However, Russo and colleagues[85] failed to detect any such aggregates in pea and broad bean leaf cells infected with either BBSV or BBTMV although we have occasionally observed such structures in similar material.[98] The reason for this discrepancy is not known but it casts serious doubt on the value of the presence of crystalline virus particle aggregates as diagnostic of Comovirus infection.

The virus particle-containing tubules, cell wall outgrowths, paramural bodies, and membranous vesicles, while being common features of Comovirus-infected cells, have also been observed in cells infected by viruses in several other taxonomic groups, particularly the Nepoviruses,[99] and are therefore of limited diagnostic value. Their function and significance are not clear although a number of speculative suggestions have been made.[99] Paramural bodies, which also occur in cells of healthy plants, are thought to be involved in the synthesis of cell walls;[100] thus they are probably responsible for the development of virus-induced cell wall thickenings usually associated with plasmodesmata in Comovirus-infected cells.

The literature on the cytopathology of BBWV is large.[31,33,35,39,101-110] We can summarize by saying that, as with Comoviruses, the virus particles are found in the cytoplasm, where infection also provokes characteristic proliferation of membranes and vesicles (Figure 11). Organelles are not specifically altered, nor have wall outgrowths, paramural bodies, or the enclosure of virus particles in cytoplasmic tubules been reported. The virus particles (both empty shells and those containing nucleic acid) frequently crystallize in close-packed arrays. In some cases, hexagonally close-packed arrays form monolayers that may be curled in the shape of scrolls or may close upon themselves to form hollow tubes about 80 nm in diameter, with nine particles in cross-section (Figure 12). Many isolates of BBWV serotype I produce such scrolls or tubes, but isolates of type BBWV and of the nasturtium ringspot strain apparently do not.[28] Tubes formed of virus particles have not been demonstrated in plants infected with serotype II isolates (Table 2) or with LMMV,[30] though the cytopathology of serotype II isolates has not been carefully examined. Where tubular arrays of virus particles do occur, they appear to be diagnostic for BBWV.

Crystal forms other than solid close-packing or tubes may occur with BBWV[109] (Figure 13), and a nuclear inclusion has been described in one case.[39]

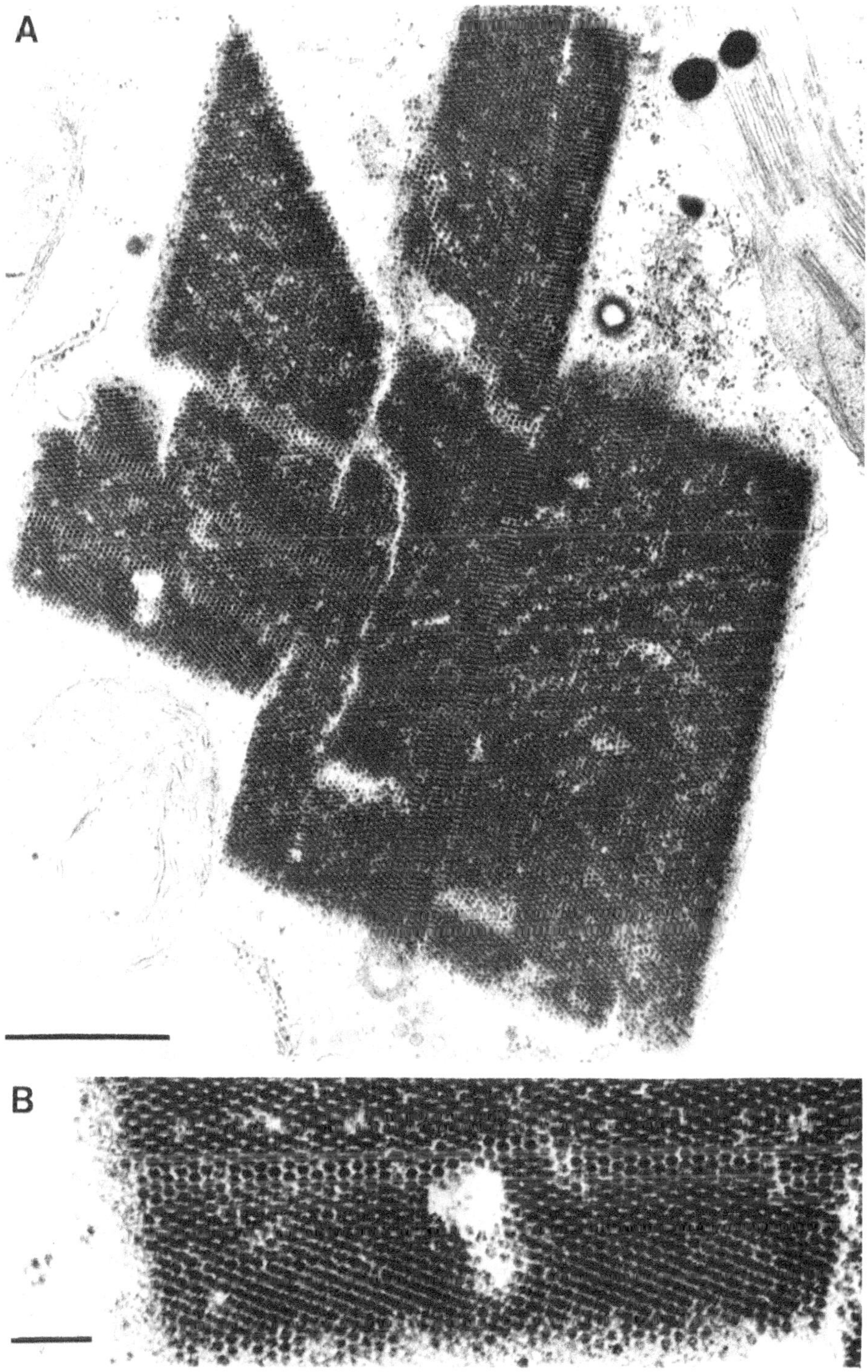

FIGURE 9. A, Thin section of a *Phaseolus vulgaris* (French bean) leaf cell infected with bean rugose mosaic virus (BRMV). Note the large mass of crystallized virus particles; bar = 500 nm. B, Part of A enlarged; bar = 100 nm. (Courtesy of Dr. E. W. Kitajima, Department of Cell Biology, University of Brasilia, Brazil.)

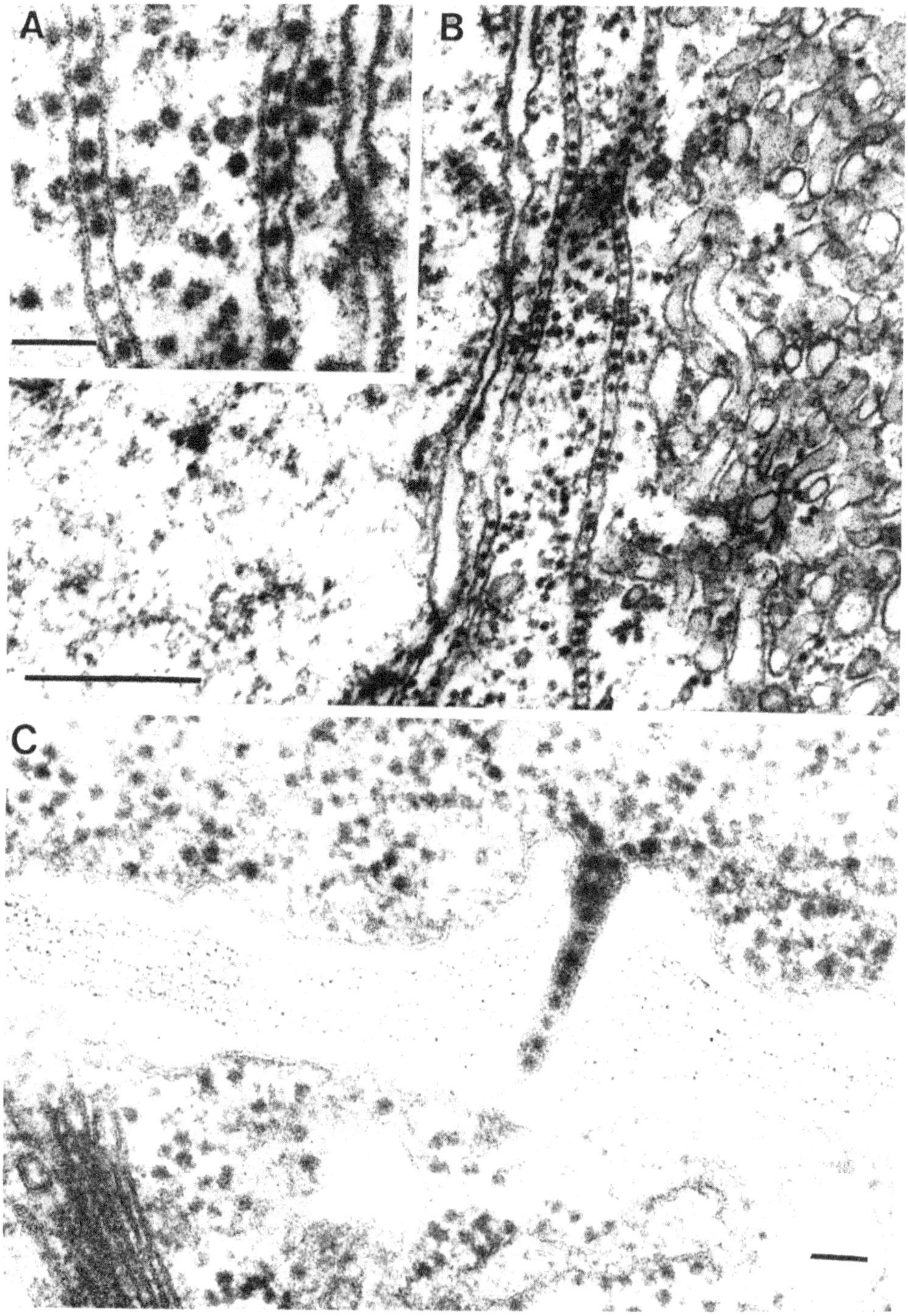

FIGURE 10. Thin sections as in Figure 5. A. Virus particles free in the cytoplasm and between parallel cytoplasmic membranes; bar = 100 nm. B. A mass of vesicles (to the right) and the files of virus particles between parallel membranes; the nucleus is on the left; bar = 500 nm. C. A plasmodesma containing a file of virus particles; bar = 100 nm.

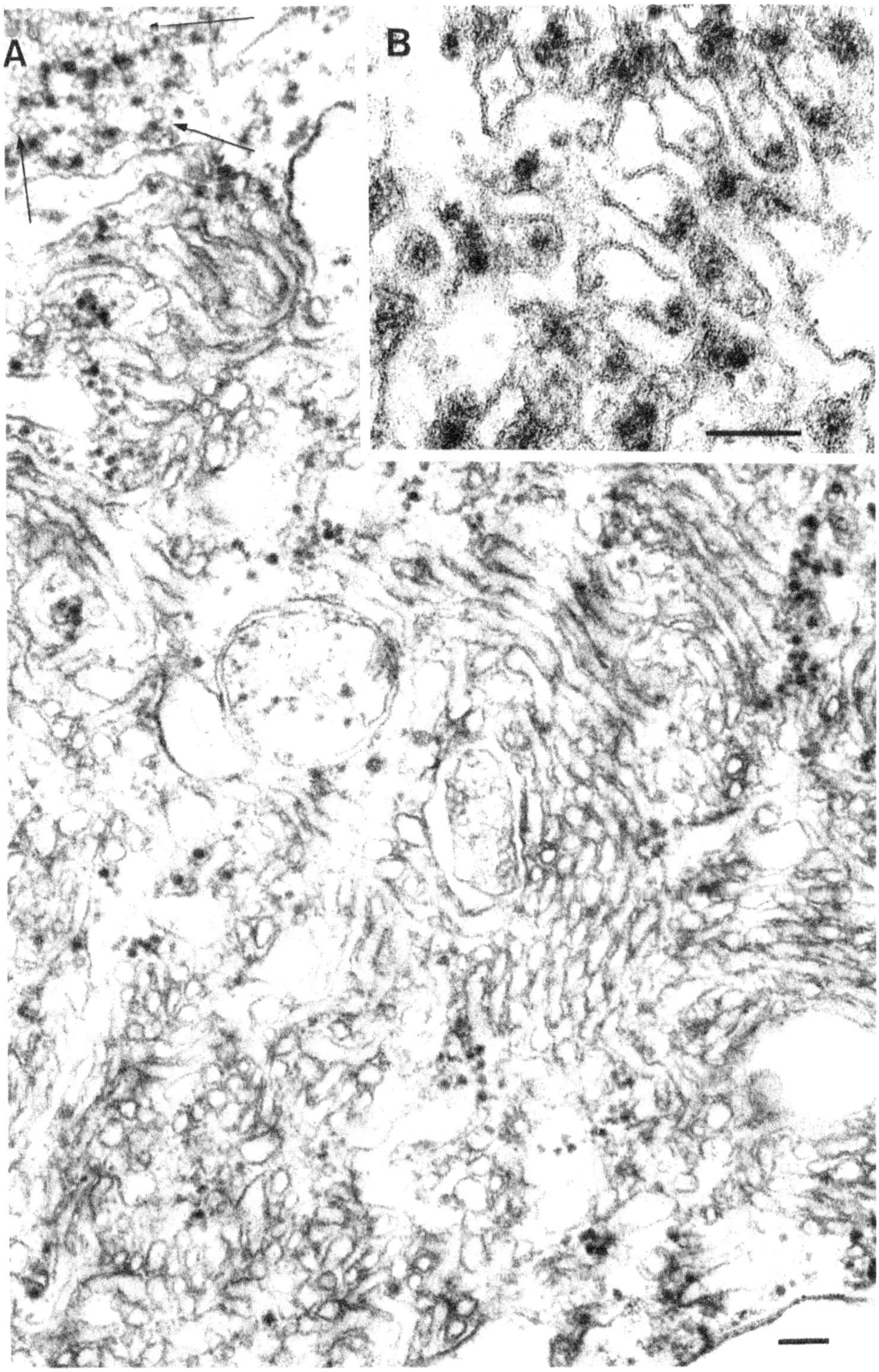

FIGURE 11. Thin sections of a *Nicotiana clevelandii* leaf cell infected with broad bean wilt virus (BBWV). A, Proliferation of membranes in the cytoplasm; particles (mostly empty capsids) are seen at top left (arrows). B, Details of convoluted membranes associated with virus-like particles. Bars = 100 nm.

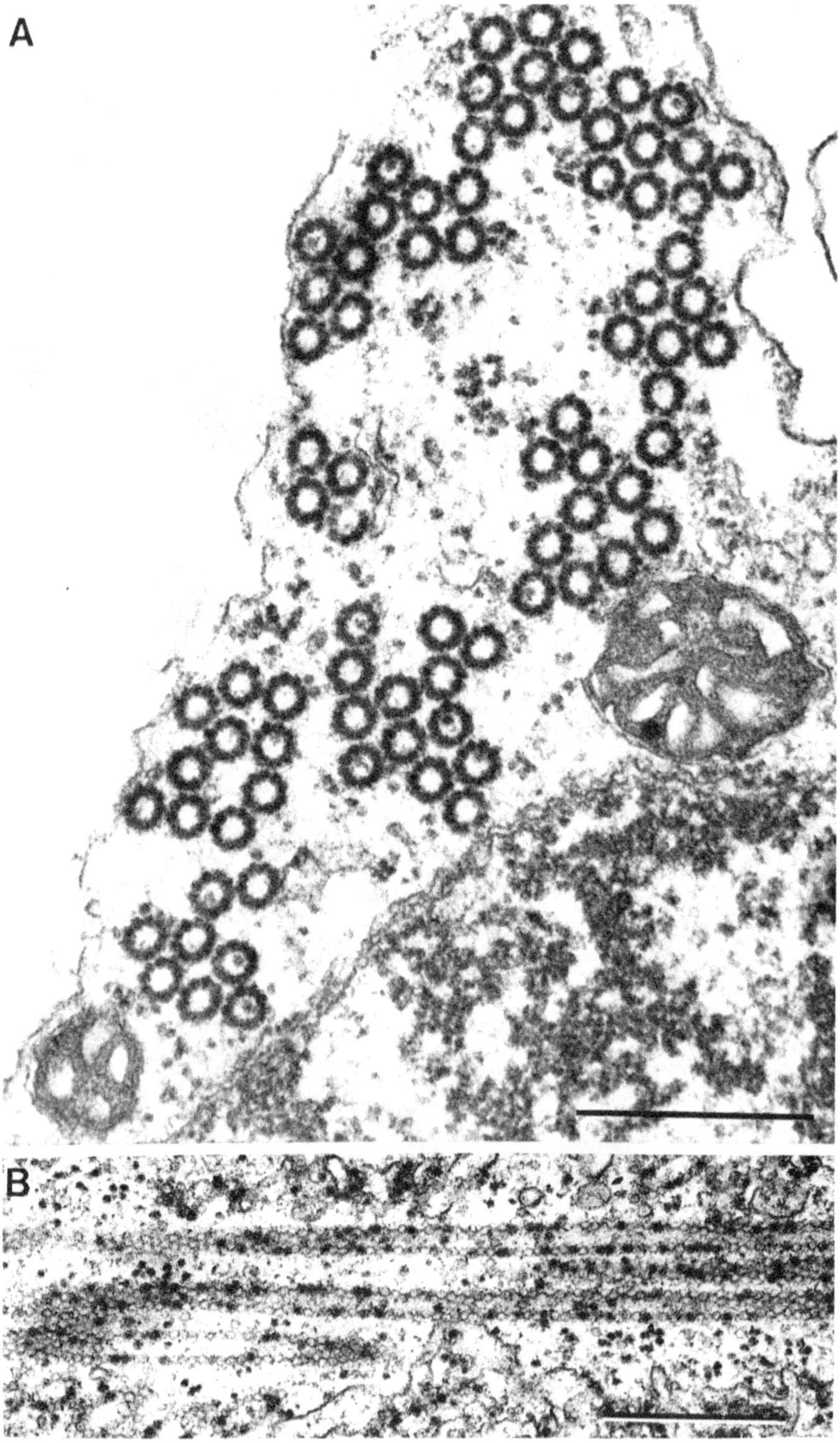

FIGURE 12. A, Thin section of a *Chenopodium quinoa* leaf cell infected with BBWV; the rings in the cytoplasm are cross-sections of tubes constructed of virus particles. (Courtesy of Drs. M. Russo and G. P. Martelli, Istituto di Patologia Vegetale, Università degli Studi, Bari, Italy.) B, Thin section of a BBWV-infected *N. clevelandii* cell showing similar tubes viewed longitudinally. Bars = 500 nm.

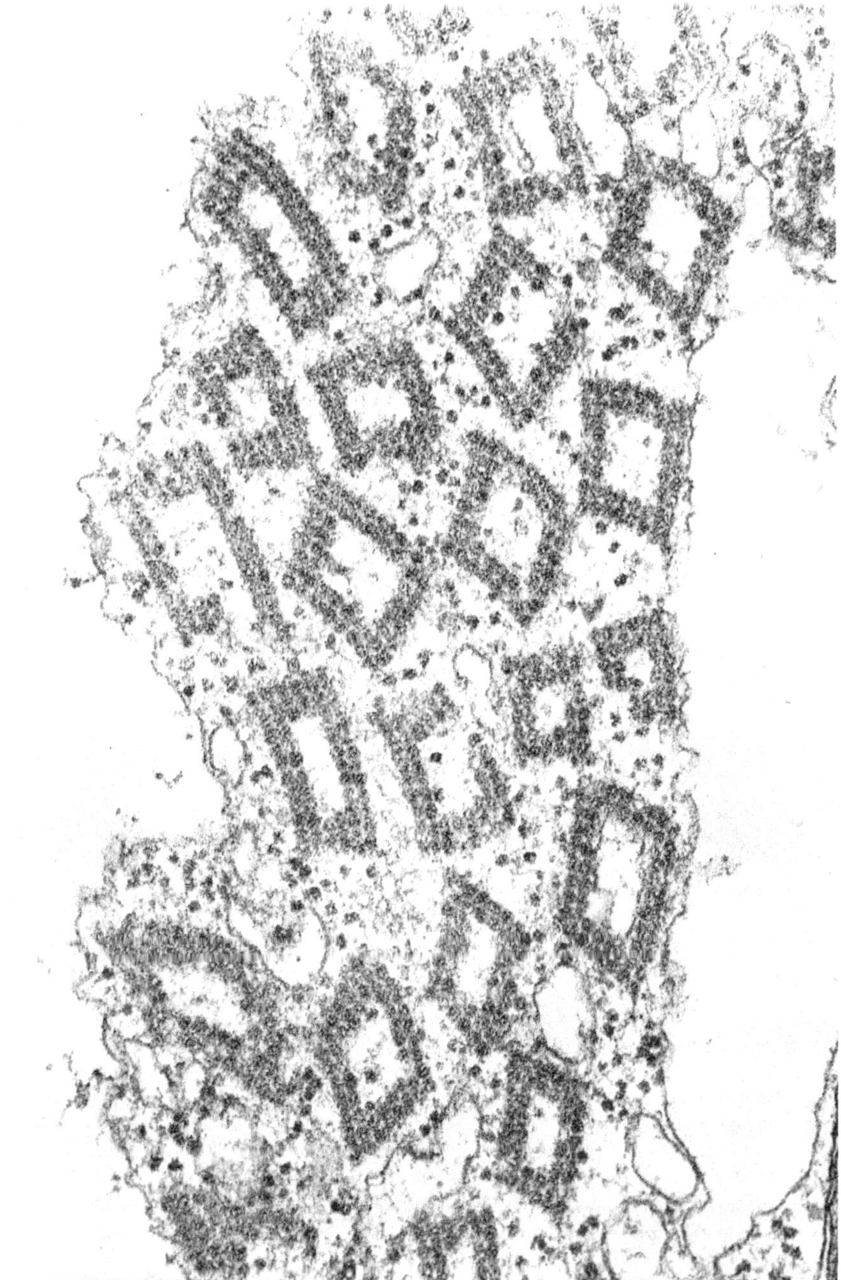

FIGURE 13 Section of a *Vicia faba* (broad bean) leaf cell infected with an Egyptian isolate of BBWV; the virus particles have assembled into tubes of square cross-section. Bar = 500 nm. (Courtesy of Drs. M. Russo and G. P. Martelli, Istituto di Patologia Vegetale, Università degli Studi, Bari, Italy.)

REFERENCES

1. **Matthews, R. E. F.,** Classification and nomenclature of viruses. Fourth report of the International Committee on Taxonomy of Viruses, *Intervirology*, 17, 1, 1982.
2. **Bruening, G.,** Comovirus Group, *CMI/AAB Descriptions of Plant Viruses*, No. 199, 1978.
3. **Stace-Smith, R.,** Comoviruses, in *Handbook of Plant Virus Infections and Comparative Diagnosis*, Kurstak, E., Ed., Elsevier, Amsterdam, 1981, 171.
4. **Van Kammen, A. and Mellema, J. E.,** Comoviruses, in *The Atlas of Insect and Plant Viruses*, Maramorosch, K., Ed., Academic Press, New York, 1977, 167.
5. **Fulton, J. P. and Scott, H. A.,** A serogrouping concept for legume comoviruses, *Phytopathology*, 69, 305, 1979.
6. **Swaans, H. and van Kammen, A.,** Reconsideration of the distinction between the severe and yellow strains of cowpea mosaic virus, *Neth. J. Plant Pathol.*, 79, 257, 1973.
7. **de Jager, C. P.,** Cowpea severe mosaic virus, *CMI/AAB Descriptions of Plant Viruses*, No. 209, 1979.
8. **Pio-Ribeiro, G. and Paguio, O. R.,** Informacões bibliográficas sobre as denominacões "cowpea mosaic virus" e "cowpea severe mosaic virus", *Fitopatol. Brasileira*, 5, 373, 1980.
9. **Lin, M. T., Anjos, J. R. N., and Rios, G. P.,** Serological grouping of cowpea severe mosaic virus isolates from Central Brazil, *Phytopathology*, 71, 435, 1981.
10. **Meiners, J. P., Waterworth, H. E., Lawson R. H., and Smith, F. F.,** Curly dwarf mosaic disease of beans from El Salvador, *Phytopathology*, 67, 163, 1977.
11. **Brunt, A. A., Barton, R. J., and Phillips, S.,** Ullucus virus C, a newly recognized comovirus infecting *Ullucus tuberosus* (Basellaceae), *Ann. Appl. Biol.*, 101, 73, 1982.
12. **Francki, R. I. B.,** Current problems in plant virus taxonomy, in *A Critical Appraisal of Virus Taxonomy*, Matthews, R. E. F., Ed., CRC Press, Boca Raton, Fla., 1983, 63.
13. **Wu, G. J. and Bruening, G.,** Two proteins from cowpea mosaic virus, *Virology*, 46, 596, 1971.
14. **van Kammen, A. and de Jager, C. P.,** Cowpea mosaic virus, *CMI/AAB Descriptions of Plant Viruses*, No. 197, 1978.
15. **Fribourg, C. E., Jones, R. A. C., and Koenig, R.,** Andean potato mottle virus, *CMI/AAB Descriptions of Plant Viruses*, No. 203, 1979.
16. **Semancik, J. S.,** Bean pod mottle virus, *CMI/AAB Descriptions of Plant Viruses*, No. 108, 1972.
17. **Gámez, R.,** Bean rugose mosaic virus, *CMI/AAB Descriptions of Plant Viruses*, No. 246, 1982.
18. **Jones, A. T. and Barker, H.,** Properties and relationships of broad bean stain virus and echtes ackerbohnenmosaik virus, *Ann. Appl. Biol.*, 83, 231, 1976.
19. **Cockbain, A. J. and Bowen, R.,** Diseases of field beans *(Vicia faba L.), Rep. Rothamsted Exp. Station 1974*, Part 1, 234, 1975.
20. **Bowyer, J. W., Dale, J. L., and Behncken, G. M.,** Glycine mosaic virus: a comovirus from Australian native *Glycine* species, *Ann. Appl. Biol.*, 95, 385, 1980.
21. **Valenta, V. and Gressnerová, M.,** Serological relationships among members of the cowpea mosaic virus group, *Acta Virol.*, 10, 182, 1966.
22. **Valenta, V., Gressnerová, M., Marcinka, K., and Nermut, M. V.,** Some properties of pea green mottle virus, a member of the cowpea mosaic group, isolated in Czechoslovakia, *Acta Virol.*, 13, 422, 1969.
23. **Clark, M. F.,** Purification and some properties of a *Comovirus* group virus isolated in New Zealand, *N.Z. J. Agric. Res.*, 15, 846, 1972.
24. **Moore, B. J. and Scott, H. A.,** Quail pea mosaic virus, *CMI/AAB Descriptions of Plant Viruses*, No. 238, 1981.
25. **Campbell, R. N.,** Radish mosaic virus, *CMI/AAB Descriptions of Plant Viruses*, No. 121, 1973.
26. **Valenta, V. and Marcinka, K.,** Red clover mottle virus *CMI/AAB Descriptions of Plant Viruses*, No. 74, 1971.
27. **Campbell, R. N.,** Squash mosaic virus, *CMI/AAB Descriptions of Plant Viruses*, No. 43, 1971.
28. **Taylor, R. H. and Stubbs, L. L.,** Broad bean wilt virus, *CMI/AAB Descriptions of Plant Viruses*, No. 81, 1972.
29. **Doel, T. R.,** Comparative properties of type, nasturtium ringspot and petunia ringspot strains of broad bean wilt virus, *J. Gen. Virol.*, 26, 95, 1975.
30. **Lisa, V., Luisoni, E., Boccardo, G., Milne, R. G., and Lovisolo, O.,** Lamium mild mosaic virus: a virus distantly related to broad bean wilt, *Ann. Appl. Biol.*, 100, 467, 1982.
31. **Rubio-Huertos, M.,** Further studies on ultrastructure of plants infected with petunia ringspot virus, *Protoplasma*, 65, 465, 1968.
32. **Taylor, R. H., Smith, P. R., Reinganum, C., and Gibbs, A. J.,** Purification and properties of broad bean wilt virus, *Aust. J. Biol. Sci.*, 21, 929, 1968.
33. **Frowd, J. A. and Tomlinson, J. A.,** Relationships between a parsley virus, nasturtium ringspot virus, and broad bean wilt virus, *Ann. Appl. Biol.*, 72, 189, 1972.

34. **Boccardo, G. and Conti, M.,** Purification and properties of nasturtium ringspot virus, *Phytopathol. Z.*, 78, 14, 1973.
35. **Sahambi, H. S., Milne, R. G., Cook, S. M., Gibbs, A. J., and Woods, R. D.,** Broad bean wilt and nasturtium ringspot viruses are related, *Phytopathol. Z.*, 76, 158, 1973.
36. **Uyemoto, J. K. and Provvidenti, R.,** Isolation and identification of two serotypes of broad bean wilt virus, *Phytopathology*, 64, 1547, 1974.
37. **Marrou, J., Duteil, M., Leclant, F., Cadilhac, B., and Esvan, C.,** Mise en évidence du virus du flétrissement de la fève broad bean wilt (BBWV) dans les laitues présentant des symptômes de marbrure, *Ann. Phytopathol.*, 8, 461, 1976.
38. **Murant, A. F. and Roberts, I. M.,** An isolate of broad bean wilt virus serotype II from *Helleborus vesicarius*, *Ann. Rep. Scott. Hortic. Res. Inst. 1976*, 1977, p. 83.
39. **Gracia, O. and Feldman, J. M.,** Studies of weed plants as sources of viruses. IV. Broad bean wilt virus in *Sisymbrium irio* and *Plantago lanceolata* in Argentina, *Phytopathol. Z.*, 85, 227, 1976.
40. **Lovisolo, O.,** Virus e piante spontanee. I. "Mosaico lieve del lamium" nuovo virus del tipo maculatura anulare, *Boll. Stn., Patol. Veg. Roma*, 15, 89, 1958.
41. **Crowther, R. A., Geelen, J. L. M. C., and Mellema, J. E.,** A three-dimensional image reconstruction of cowpea mosaic virus, *Virology*, 57, 20, 1974.
42. **Hatta, T. and Francki, R. I. B.,** unpublished results, 1980.
43. **Niblett, C. L. and Semancik, J. S.,** Conversion of the electrophoretic forms of cowpea mosaic virus *in vivo* and *in vitro*, *Virology*, 38, 685, 1969.
44. **Geelen, J. L. M. C., van Kammen, A., and Verduin, B. J. M.,** Structure of the capsid of cowpea mosaic virus. The chemical subunit: molecular weight and number of subunits per particle, *Virology*, 49, 205, 1972.
45. **Bruening, G.,** Plant covirus systems: two-component systems, *Comp. Virol.* 11, 55, 1977.
46. **van Wezenbeek, P., Verver, J., Harmsen, J., Vos, P., and van Kammen, A.,** Primary structure and gene organization of the middle-component RNA of cowpea mosaic virus, *EMBO J.*, 2, 946, 1983.
47. **Lomonossoff, G. P.,** personal communication, 1983.
48. **El Manna, M. M. and Bruening, G.,** Polyadenylate sequences in the ribonucleic acids of cowpea mosaic virus, *Virology*, 56, 198, 1973.
49. **Davies, J. W., Stanley, J., and van Kammen, A.,** Sequence homology adjacent to the 3′-terminal poly(A) of cowpea mosaic virus RNAs, *Nucleic Acids Res.*, 7, 493, 1979.
50. **Daubert, S. D., Bruening, G., and Najarian, R. C.,** Protein bound to the genome RNAs of cowpea mosaic virus, *Eur. J. Biochem.*, 92, 45, 1978.
51. **Stanley, J., Rottier, P., Davies, J. W., Zabel, P., and van Kammen, A.,** A protein linked to the 5′-terminal of both RNA components of the cowpea mosaic virus genome, *Nucleic Acids Res.*, 5, 4505, 1978.
52. **Stanley, J. and van Kammen, A.,** Nucleotide sequences adjacent to the proteins covalently linked to the cowpea mosaic virus genome. Sequence determination after labeling *in vitro* using RNA ligase, *Eur. J. Biochem.*, 101, 45, 1979.
53. **Stanley, J., Goldbach, R. and van Kammen, A.,** The genome-linked protein of cowpea mosaic virus is coded by RNA from the bottom component, *Virology*, 106, 180, 1980.
54. **Najarian, R. C. and Bruening, G.,** Similar sequences from the 5′ end of cowpea mosaic virus RNAs, *Virology*, 106, 301, 1980.
55. **Daubert, S. D. and Bruening, G.,** Genome-associated proteins of Comoviruses, *Virology*, 98, 246, 1979.
56. **Bruening, G.,** The inheritance of top component formation in cowpea mosaic virus, *Virology*, 37, 577, 1969.
57. **de Jager, C. P. and van Kammen, A.,** The relationship between the components of cowpea mosaic virus. III. Location of genetic information for two biological functions in the middle component of CPMV, *Virology*, 41, 281, 1970.
58. **Moore, B. J. and Scott, H. A.,** Properties of a strain of bean pod mottle virus, *Phytopathology*, 61, 831, 1971.
59. **Wood, H. A.,** Genetic complementation between the two nucleoprotein components of cowpea mosaic virus, *Virology*, 49, 592, 1972.
60. **Kassanis, B., White, R. F., and Woods, R. D.,** Genetic complementation between middle and bottom components of two strains of radish mosaic virus, *J. Gen. Virol.* 20, 277, 1973.
61. **Siler, D. J., Babcock, J., and Bruening, G.,** Electrophoretic mobility and enhanced infectivity of a mutant of cowpea mosaic virus, *Virology*, 71, 560, 1976.
62. **de Jager, C. P.,** Genetic analysis of cowpea mosaic virus mutants by supplementation and reassortment tests, *Virology*, 70, 151, 1976.
63. **de Jager, C. P., Zabel, P., van der Beek, C. P., and van Kammen, A.,** Genetic and physiological characterization of a temperature sensitive mutant of cowpea mosaic virus, *Virology*, 76, 164, 1977.
64. **Gopo, J. M. and Frist, R. H.,** Location of the gene specifying the smaller protein of the cowpea mosaic virus capsid, *Virology*, 79, 259, 1977.

65. **Thongmeearkom, P. and Goodman, R. M.,** Complementation and pseudorecombination between ribonucleic acids from two natural isolates of cowpea mosaic virus, *Virology*, 85, 75, 1978.
66. **Oxelfelt, P. and Abdelmoeti, M.,** Genetic complementation between natural strains of red clover mottle virus, *Intervirology*, 10, 78, 1978.
67. **Hiebert, E. and Purcifull, D. E.,** Mapping of the two coat protein genes on the middle RNA component of squash mosaic virus (Comovirus Group), *Virology*, 113, 630, 1981.
68. **Franssen, H., Goldbach, R., Broekhuijsen, M., Moerman, M., and van Kammen, A.,** Expression of middle-component RNA of cowpea mosaic virus: *in vitro* generation of a precursor to both capsid proteins by a bottom-component RNA-encoded protease from infected cells, *J. Virol.*, 41, 8, 1982.
69. **Goldbach, R., Rezelman, G., and van Kammen, A.,** Independent replication and expression of B-component RNA of cowpea mosaic virus, *Nature (London)*, 286, 297, 1980.
70. **Rezelman, G., Goldbach, R., and van Kammen, A.,** Expression of bottom component RNA of cowpea mosaic virus in cowpea protoplasts, *J. Virol.*, 36, 366, 1980.
71. **Rezelman, G., Franssen, H. J., Goldbach, R. W., Ie, T. S., and van Kammen, A.,** Limits to independence of bottom component RNA of cowpea mosaic virus, *J. Gen. Virol.*, 60, 335, 1982.
72. **Pelham, H. R. B.,** Synthesis and proteolytic processing of cowpea mosaic virus proteins in reticulocyte lysates, *Virology*, 96, 463, 1979.
73. **Goldbach, R. W., Schilthuis, T. G., and Rezelman, G.,** Comparison of *in vivo* and *in vitro* translation of cowpea mosaic virus RNAs, *Biochem. Biophys. Res. Commun.*, 99, 89, 1981.
74. **Goldbach, R., Rezelman, G., Zabel, P., and van Kammen, A.,** Expression of the bottom-component RNA of cowpea mosaic virus: evidence that the 60-kilodalton VPg precursor is cleaved into single VPg and a 58 kilodalton polypeptide, *J. Virol.*, 42, 630, 1982.
75. **Zabel, P., Moerman, M., van Straaten, F., Goldbach, R., and van Kammen, A.,** Antibodies against the genome-linked protein VPg of cowpea mosaic virus recognise a 60,000-dalton precursor polypeptide, *J. Virol.*, 41, 1083, 1982.
76. **Goldbach, R. and Rezelman, G.,** Orientation of the cleavage map of the 200-kilodalton polypeptide encoded by the bottom-component RNA of cowpea mosaic virus, *J. Virol.*, 46, 614, 1983.
77. **Milne, R. G.,** Electron microscopy of leaves infected with sowbane mosaic virus and other small polyhedral viruses, *Virology*, 32, 589, 1967.
78. **van der Scheer, C. and Groenewegen, J.,** Structure in cells of *Vigna unguiculata* infected with cowpea mosaic virus, *Virology*, 46, 493, 1971.
79. **de Zoeten, G. A., Assink, A. M., and van Kammen, A.,** Association of cowpea mosaic virus-induced double-stranded RNA with a cytopathological structure in infected cells, *Virology*, 59, 341, 1974.
80. **Kim, K. S., Fulton, J. P., and Scott, H. A.,** Osmiophilic globules and myelinic bodies in cells infected with two comoviruses, *J. Gen. Virol.*, 25, 445, 1974.
81. **Langenberg, W. G. and Schroeder, H. F.,** The ultrastructural appearance of cowpea mosaic virus in cowpea, *J. Ultrastruct. Res.*, 51, 166, 1975.
82. **Kim, K. S. and Fulton, J. P.,** Tubules with viruslike particles in leaf cells infected with bean pod mottle virus, *Virology*, 43, 329, 1971.
83. **Kim, K. S. and Fulton, J. P.,** Fine structure of plant cells infected with bean pod mottle virus, *Virology*, 49, 112, 1972.
84. **Kim, K. S. and Fulton, J. P.,** Plant virus-induced cell wall overgrowth and associated membrane elaboration, *J. Ultrastruct. Res.*, 45, 328, 1973.
85. **Russo, M., Castellano, M. A., and Martelli, G. P.,** The ultrastructure of broad bean stain and broad bean true mosaic virus infections, *J. Submicros. Cytol.*, 14, 149, 1982.
86. **Hatta, T. and Francki, R. I. B.,** Enzyme cytochemical identification of single-stranded and double-stranded RNAs in virus-infected plant and insect cells, *Virology*, 88, 105, 1978.
87. **Hatta, T. and Francki, R. I. B.,** Identification of small polyhedral virus particles in thin sections of plant cells by an enzyme cytochemical technique, *J. Ultrastruct. Res.*, 74, 116, 1981.
88. **Štefanac, Z. and Ljubešić, N.,** Inclusion bodies in cells infected with radish mosaic virus, *J. Gen. Virol.*, 13, 51, 1971.
89. **Honda, Y. and Matsui, C.,** Electron microscopy of intracellular radish mosaic virus, *Phytopathology*, 62, 448, 1972.
90. **Hooper, G. R., Spink, G. C., and Myers, R. L.,** Electron microscopy of leaf enations on chinese white winter radish infected with radish mosaic virus, *Virology*, 47, 833, 1972.
91. **Milne, R. G., Masenga, V., Lenzi, R., and Lovisolo, O.,** Radish mosaic virus in *Eruca sativa* Miller, *Phytopathol. Mediterr.*, 19, 145, 1980.
92. **Tomenius, K. and Oxelfelt, P.,** Ultrastructure of pea leaf cells infected with three strains of red clover mottle virus, *J. Gen. Virol.*, 61, 143, 1982.
93. **Kim, K. S.,** Ultrastructure of plant cells infected with the beetle-transmitted comoviruses, *Fitopatol. Brasileira*, 4, 227, 1979.

94. **van Griensven, L. J. L. D., van Kammen, A., and Rezelman, G.,** Characterization of the double stranded RNA isolated from cowpea mosaic virus-infected *Vigna* leaves, *J. Gen. Virol.*, 18, 359, 1973.
95. **Assink, A-M., Swaans, H., and van Kammen, A.,** The localization of virus-specific double-stranded RNA of cowpea mosaic virus in subcellular fractions of infected *Vigna* leaves, *Virology*, 53, 384, 1973.
96. **Owens, R. A. and Bruening, G.,** The pattern of amino acid incorporation into two cowpea mosaic virus proteins in the presence of ribosome-specific protein synthesis inhibitors, *Virology*, 64, 520, 1975.
97. **Edwardson, J. R. and Christie, R. G.,** Use of virus-induced inclusions in classification and diagnosis, *Ann. Rev. Phytopathol.*, 16, 31, 1978.
98. **Hatta, T. and Francki, R. I. B.,** unpublished results, 1980.
99. **Martelli, G. P.,** Ultrastructural aspects of possible defence reactions in virus-infected plant cells, *Microbiologica*, 3, 369, 1980.
100. **Marchant, R. and Robards, A. W.,** Membrane system associated with the plasmalemma of plant cells, *Ann. Bot.*, 32, 457, 1968.
101. **Juretic, N., Miličic, D., and Schmelzer, K.,** Zur Kenntnis des ringmosaik-Virus der Kapuzinerkresse (nasturtium ringspot virus) und seiner Zell-einschlusskörper, *Acta Bot. Croat.*, 29, 17, 1970.
102. **Hull, R. and Plaskitt, A.,** The *in vivo* behavior of broad bean wilt virus and three of its strains, *Intervirology*, 2, 352, 1973/74.
103. **Miličic, D., Wrischer, M., and Juretić N.,** Intracellular inclusion bodies of broad bean wilt virus, *Phytopathol. Z.*, 80, 127, 1974.
104. **Russo, M. and Martelli, G. P.,** Light and electron microscopy of broad bean wilt virus intracellular inclusions, *J. Submicros. Cytol.*, 7, 335, 1975.
105. **Weidemann, H. L., Lesemann, D., Paul, H. L., and Koenig, R.,** Das broad bean wilt-Virus als Ursache für eine neue Vergilbungskrankheit des Spinats in Deutschland, *Phytopathol. Z.*, 84, 215, 1975.
106. **Cadilhac, B. and Marrou, J.,** Étude ultrastructurale du virus du flétrissement de la fève sur laitue (broad bean wilt - BBWV), *Ann. Phytopathol.*, 8, 471, 1976.
107. **Hein, A., Lesemann, D. E., and Querfurth, G.,** Das broad bean wilt-Virus in Limonium-Kulturen in Süddeutschland, *Phytopathol. Z.*, 89, 340, 1977.
108. **Kishtah, A. A., Russo, M., Tolba, M. A., and Martelli, G. P.,** A strain of broad bean wilt virus isolated from pea in Egypt, *Phytopathol. Mediterr.*, 17, 157, 1978.
109. **Russo, M., Kishtah, A. A., and Martelli, G. P.,** Unusual intracellular aggregates of broad bean wilt virus particles, *J. Gen. Virol.*, 43, 453, 1979.
110. **Vega, J., Gracia, O., and Feldman, J. M.,** Characteristics of Argentine broad bean wilt virus: cytopathological aspects, *Phytopathol. Z.*, 99, 242, 1980.

Chapter 2

NEPOVIRUS GROUP

It was suggested as early as 1963 by Cadman[1] that several viruses formed a natural group based on particle morphology and the ability to be transmitted by nematodes. The name Nepovirus, an acronym for nematode transmitted polyhedral virus, was suggested for the group. The name was one of the first approved by the ICTV for a group of plant viruses, with tobacco ringspot virus (TRSV) as the type member.[2]

Nepoviruses have polyhedral particles 28 to 30 nm in diameter, usually of three types which can be separated into top (T), middle (M), and bottom (B) sedimenting components. The protein shells of all particle types are identical but each contains a different amount of RNA. T particles have no RNA, whereas M and B contain 28 to 30 and 39 to 43% RNA, respectively. All Nepoviruses have bipartite genomes of single-stranded RNA molecules of M_r about 2.1 to 2.8 $\times$ 10^6 (RNA 1) and 1.4 to 2.4 $\times$ 10^6 (RNA 2), which are encapsidated separately.[3,4]

Most Nepoviruses have wide host ranges, and many infect perennial plants and trees. They are easily sap transmitted, at least from herbaceous hosts. A common though not unique characteristic of Nepovirus-infected plants is their ability to stage a partial recovery, during which symptoms become less intense or disappear altogether.[4] The majority of Nepoviruses have been shown to be transmissible to progeny plants by seed and pollen and by nematodes of the genera *Longidorus* or *Xiphinema*.[4,5]

Thin sectioning and electron microscopy of vector nematodes fed upon Nepovirus-infected plants have shown that the virus particles attach to specific sites on the wall of the odontostyle, stylet guide sheath, or esophagus.[4-7]

I. MEMBERS OF THE GROUP AND THEIR RELATIONSHIPS

The group consists of 22 members and a further 5 possible members. The members are recognized primarily by their physical properties, that is to say, the possession of polyhedral particles about 30 nm in diameter sedimenting as two RNA-containing components (M and B) and usually a third component devoid of RNA (T). All Nepoviruses have bipartite genomes and coat proteins with apparent M_r values between 52 and 60 $\times$ 10^3 when determined by electrophoresis in polyacrylamide gels.[3,4]

Most Nepoviruses are not serologically related, and hence subdividing them according to their antigenic properties is not helpful. Martelli and his colleagues[8] suggested that Nepoviruses could be subdivided into four subgroups depending on their physical characteristics; they used the homogeneity or heterogeneity of B component particles, the sedimentation coefficient of M component particles, and the size of viral RNA 2 molecules as the criteria. These divisions are not altogether satisfactory because of our incomplete knowledge of the properties of some Nepoviruses. Furthermore, there are almost certainly many errors in the published particle sedimentation values and RNA molecular sizes. For the present it is convenient to subdivide the nepoviruses into two subgroups, based on whether RNAs 1 and 2 clearly differ in size (difference 0.6 M_r or greater) or are of a similar size (difference 0.4 M_r or less). In Table 1 we have listed the Nepoviruses divided into two subgroups. Some of the viruses have been isolated from very different hosts but are closely related serologically and hence are considered as strains of the same virus. Some of these strains have been referred to as distinct viruses in the past; they are listed in Table 1 under the names of the appropriate type viruses.

Only a very few Nepoviruses, considered to be sufficiently different to warrant distinct

Table 1
MEMBERS OF THE NEPOVIRUS GROUP AND SOME OF THEIR PROPERTIES[a]

Virus and its strains or isolates	S_{20w} of particles T	M	B	M_r of RNAs (× 10^6) RNA 1	RNA 2	M_r of coat protein (× 10^3)	Nematode vector	Ref.
Subgroup 1								
Tobacco ringspot virus TRSV	53	91	126	2.4	1.4	57	*Xiphinema americanum*	3,13
Eucharis mottle								
Anemone necrosis								
Potato black ringspot virus (PBRV)	49	84	117	2.5	1.5	59	Unknown	14
Potato calico								
Arabis mosaic virus (ArMV)	53	93	126	2.4 (2.8)[b]	1.4 (1.3)	54	*X. diversicaudatum*	15
Rasberry yellow dwarf								
Rhubarb mosaic								
Hop line pattern								
Grapevine fan leaf virus (GFLV)	50	86	120	2.4	1.4	54	*X. index* and *X. italiae*	16
Grapevine yellow mosaic								
Grapevine vein banding								
Raspberry ringspot virus (RRV)	50	92	130	2.4 (2.8)	1.4 (1.4)	54	*Longidorus elongatus* *L. macrosoma*	17
English								
Grapevine								
Raspberry Scottish leafcurl								
Red currant ringspot								
Arracacha virus A (AVA)	50	92	125	2.5	1.4	53	Unknown	18
Olive latent ringspot virus (OLRV)	52	100	132	2.3	1.5	60	Unknown	19
Tomato black ring virus (TBRV)	55	97	121	2.5 (2.8)	1.5 (1.7)	57	*L. attenuatus* *L. elongatus*	20
Potato bouquet								
Beet ringspot								
Potato pseudo aucuba								
Lettuce ringspot								
Celery yellow vein								

Cacao necrosis virus (CNV)	54	101	129	2.4	1.5	60	Unknown	21
Grapevine chrome mosaic virus (GCMV)	—[c]	92	117	2.4 (2.1)	1.5 (1.6)	ND[d]	Unknown	22
Artichoke Italian latent virus (AILV)	55	96	121	2.4	1.5	54	*L. apulus*	23
Mulberry ringspot virus (MRV)	50	96	26	(2.8)	(1.5)	ND	*L. martini*	24
Crimson clover latent virus (CCLV)	52	101	122	2.2	1.6	52	Unknown	25
Subgroup 2								
Tomato ringspot virus (ToRSV)	53	119	127	2.6 (2.8)	2.5 (2.4)	58	*X. americanum*	26,27
Grapevine yellow vein								
Peach yellow bud mosaic								
Grapevine Bulgarian latent virus (GBLV)	52	120	127	2.2	2.1	54	Unknown	28
Chicory yellow mottle virus (ChYMV)	51	116	128	2.4	2.0	54	Unknown	29
Parsley strain								
Lucerne Australian latent virus (LALV)	56	128	133	2.4	2.1	55	Unknown	30
Cherry leafroll virus (CLRV)	52	114	132	2.3(2.8)	2.1 (2.3)	54	*X. diversicaudatum* *X. coxi*	31—33
Elm mosaic								
Golden elderberry								
Blackberry								
Dogwood								
Birch								
Raspberry								
Rhubarb								
Walnut								
Myrobalan latent ringspot virus (MLRV)	50	109	125	2.4 (2.8)	1.9 (2.0)	53	Unknown	12,34
Peach rosette mosaic virus (PRMV)	52	115	134	2.5	2.2	57	*X. americanum*	35,36
Hibiscus latent ringspot virus (HLRV)	51	114	132	2.5	2.2	54	Unknown	37

Table 1 (continued)
MEMBERS OF THE NEPOVIRUS GROUP AND SOME OF THEIR PROPERTIES[a]

	S_{20w} of particles			M_r of RNAs (× 10^6)				
Virus and its strains or isolates	T	M	B	RNA 1	RNA 2	M_r of coat protein (× 10^3)	Nematode vector	Ref.
Artichoke yellow ringspot virus (AYRV)	ND	ND	ND	2.2	1.9	53	Unknown	38

[a] Viruses bracketed are distantly related serologically.
[b] Values in parentheses from determinations by polyacrylamide-gel electrophoresis under denaturing conditions.
[c] No top component detected.
[d] ND, not determined.

Table 2
POSSIBLE MEMBERS OF THE NEPOVIRUS GROUP AND SOME OF THEIR PROPERTIES

Virus	S_{20w} of particles T	M	B	M_r of RNAs (× 10^6) RNA 1	RNA 2	M_r of coat protein (× 10^3)	Nematode vector	Ref.
Strawberry latent ringspot virus (SLRV)	58	94	126	2.6 (2.9)[a]	1.6 (1.4)	44, 29	*Xiphinema diversicaudatum* and *X. coxi*	41
Cherry rasp leaf virus (CRLV)	56	96	128	2.0	1.5	24, 22.5	*X. americanum*	42
Artichoke vein banding virus (AVBV)	56	92	124	2.1	1.5	27.5, 24.5, 22	Unknown	43
Tomato top necrosis virus (TTNV)	52	102	126	ND[b]	ND	ND	Unknown	44
Satsuma dwarf virus (SDV)	—[c]	119	129	1.9	1.7	42, 21	Unknown	45

[a] Values from determinations by polyacrylamide-gel electrophoresis under denaturing conditions.
[b] ND, not determined.
[c] No top component detected.

names, have been shown to be distantly related serologically. These distant relationships are indicated by brackets in Table 1. Myrobalan latent ringspot virus (MLRV) has been reported to be serologically related to tomato black ring virus (TBRV);[9,10] however, the relationship could not be confirmed in either of two other laboratories.[11,12]

There are five viruses with many properties characteristic of Nepoviruses (Table 2), the inclusion of which in the group is in doubt because of anomalous data on their coat proteins. The dissociated proteins migrate as two or more bands with mobilities corresponding to polypeptides much smaller than the M_r of 52 to 60 × 10^3 characteristic of accepted Nepoviruses (Table 1). However, this may not be a basic difference since with TRSV,[39] olive latent ringspot virus (OLRV)[19], and tomato ringspot virus (ToRSV)[40], preparations of dissociated virus protein contain polypeptides other than the typical major component. With TRSV there is strong evidence that the major polypeptide is a tetramer of a much smaller polypeptide which is thought to be the basic coat protein subunit[39] (see Section II.A). It seems possible that at least some of the tentative members of the Nepovirus group are viruses whose coat proteins form aggregates of the subunit other than tetramers. With strawberry latent ringspot virus (SLRV), for example, it is conceivable that the polypeptides of M_r 29 and 44 × 10^3 detected in polyacrylamide gels (Table 2) may be dimers and trimers of the true coat protein subunit. If this is so, then the M_r of the subunit monomer would be about 14.5 × 10^3, which is close to that suggested for TRSV.[39] A similar situation could exist with the coat protein of satsuma dwarf virus (SDV). However, this explanation does not appear likely with cherry rasp leaf virus (CRLV) or artichoke vein banding virus (AVBV).

II. VIRUS STRUCTURE AND COMPOSITION

A. Particle Structure

Electron micrographs of negatively stained Nepovirus particles such as those in Figure 1 show polyhedra about 30 nm in diameter. The B, M, and T particles are indistinguishable in structure, protein composition, and antigenic properties. The T particles are always devoid of RNA whereas M particles always contain a single molecule of RNA 2. On the other hand, B particles of some Nepoviruses such as TRSV[46] can contain a single molecule of

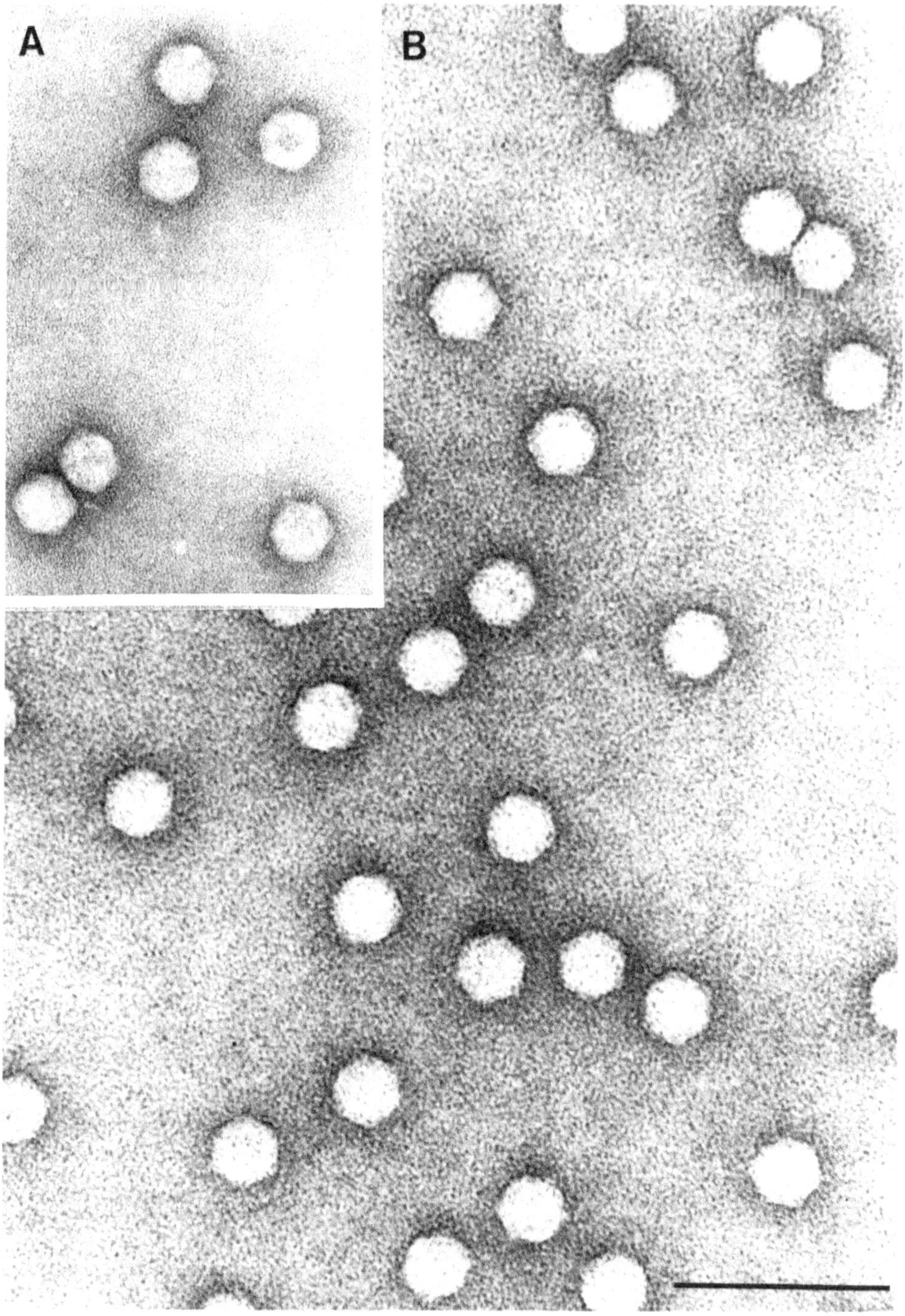

FIGURE 1. Purified particles of A. tobacco ringspot virus (TRSV) and B. tomato ringspot virus (ToRSV), both stained in uranyl acetate and photographed at the same magnification. Bar = 100 nm.

RNA 1 or two molecules of RNA 2, but others such as TBRV[17] always encapsidate a single molecule of RNA 1.

In many electron micrographs of Nepoviruses, some particles are seen to be penetrated by negative stain; this has sometimes been taken as evidence for the presence of T particles. However, it has been shown with TRSV that both penetrated and unpenetrated particles can occur in preparations of pure T, M, and B components, and that the conditions of staining affect the proportion of particles penetrated.[48]

There has been some controversy regarding the structure of Nepovirus particles. For TRSV, an icosahedral model with T = 4 symmetry was derived[49] from electron micrographs subjected to the rotation technique for enhancement of image detail.[50] Similar studies on arabis mosaic virus (ArMV) indicated the same symmetry.[51] However, Mayo and his colleagues[52] estimated by polyacrylamide-gel electrophoresis (PAGE) that the size of the TRSV coat protein subunit was 57×10^3 M_r, thus making a T = 1 rather than a T = 4 structure more plausible. The T = 1 structure would be built of 60 subunits and would have an M_r of 3.4×10^6, in agreement with the value obtained from Svedberg's equation.[52] More recently, Chu and Francki[39] have obtained evidence suggesting that the 57×10^3 M_r protein is a stable tetramer of a polypeptide of M_r about 13×10^3. A T = 4 protein shell consisting of 240 monomeric subunits would have the same mass as a T = 1 shell made of 60 tetramers. Thus the existence of the 13×10^3 M_r monomer would satisfy both the chemical and electron microscopic constraints.

X-ray diffraction data from crystallized TRSV particles have recently been obtained extending to at least 3.3 Å resolution.[53] Further X-ray work may decide between the T = 1 and T = 4 Nepovirus architectures.

B. The Viral Genome

The Nepovirus genome is a bipartite in that both RNA 1 and RNA 2 are essential for infectivity.[54] Although the RNA sizes of all the Nepoviruses have been determined, the values obtained (Table 1) and also those for the possible Nepoviruses (Table 2) should be considered with caution. Nearly all the published values have been estimated from relative electrophoretic mobilities in gels, a procedure that can be subject to considerable errors, especially when the RNAs are electrophoresed at different degrees of denaturation. Probably the best estimates have been obtained by Murant and his colleagues[55] using agarose-gel electrophoresis and electron microscopy after denaturing the RNAs with glyoxal.

Nepovirus genome RNAs have covalently linked proteins (VPgs) at their 5′ termini;[56,57] these are essential for infectivity but not for in vitro translation of the viral RNAs.[58,59] Mayo and his colleagues[56] analyzed the VPgs of five Nepoviruses and SLRV and found them all to differ. However, the VPgs from strains of the same viruses appeared to be very similar. It was also shown that the VPgs from RNA 1, RNA 2, and the satellite RNA of TBRV all produced indistinguishable peptide maps.[56] At their 3′ ends, all Nepoviruses investigated have polyadenylate tails.[60]

Numerous pseudorecombinants have been constructed in vitro from closely related Nepoviruses.[47,61-65] It has been shown with several of the viruses that RNA 1 determines seed transmissibility and some host reactions. On the other hand, RNA 2 determines the serological properties of the virus (i.e., codes for the coat protein), ability to be transmitted by nematodes, and also some host reactions.[62-67]

Nepoviruses require both RNA 1 and RNA 2 for normal infection and virus multiplication. However, when protoplasts are inoculated with only TBRV RNA 1, some replication of this RNA takes place.[68] Furthermore the synthesized RNA 1 is fully functional when mixed with RNA 2. This implies that the new RNA 1 must have been synthesized complete with its VPg attached, which in turn indicates that the VPg is coded for by RNA 1. In similar experiments the replication of RNA 2 alone could not be demonstrated.[68]

RNA-to-RNA annealing experiments have established that RNA 1 and RNA 2 of TRSV have distinct nucleotide sequences. However, the two RNAs appear to have some (probably about 900) nucleotides in common.[69] Similar experiments have shown that RNA 2 isolated from M component particles has base sequences indistinguishable from those of RNA 2 from the B particles.[69] It seems safe to assume that the RNAs 1 and 2 of other Nepoviruses also differ in their nucleotide sequences.

RNAs 1 and 2 of TBRV have been translated in vitro into distinct protein products, each of which was separated by PAGE into a large number of polypeptides. The largest of these polypeptides translated from RNA 1 and RNA 2 were of M_r about 220 × 10^3 and 160 × 10^3, respectively, accounting for more than 80% of the coding capacity of the viral genome.[70] The significance of the smaller polypeptides is obscure at present but it seems likely that the Nepovirus genome is translated into polyproteins which are then processed into smaller functional proteins like those of the Comoviruses (see Chapter 1).

C. Satellite RNAs

Several Nepoviruses have been found to contain satellite RNAs dependent for their replication on the genomes of their helper viruses.[71] They are encapsidated in particles indistinguishable from those of their helper viruses.

At least 24 distinct isolates of TRSV have been found to contain satellite RNAs (S-TRSV) in their capsids.[72,73] The most common size of S-TRSV is of M_r about 0.12 × 10^6. However, the S-TRSV has also been detected as a series of apparently covalently linked multimeric forms consisting of up to 12 repeating monomer sequences.[72]

S-TRSV interferes with the replication of its helper in that lesions produced by the combination are smaller than those induced by TRSV alone.[74,75] S-TRSV does not appear to have any nucleotide sequences in common with its helper virus RNA[76,77] and lacks the VPg and polyadenylate at its 5′ and 3′ termini, respectively.[72] The mode of replication of the satellite is interesting as it appears to involve a heterogeneous, high-molecular-weight double-stranded RNA. This apparently replicative form RNA can be as much as 20 times larger than the size expected for the monomeric double-stranded form.[78] It has been suggested from electron microscopic evidence that S-TRSV replicates by a rolling circle mechanism.[79]

Larger satellite RNAs of M_r between 0.41 and 0.5 × 10^6 have been detected in isolates of TBRV and MLRV.[9,11,12,80] Unlike S-TRSV, the satellite RNA (RNA 3) of TBRV can be translated in vitro into a polypeptide of M_r about 48 × 10^3 which has also been detected in infected protoplasts.[81] RNA 3 contains a VPg at its 5′ end[57] and is polyadenylated at its 3′ end.[60] Also, unlike S-TRSV RNA, TBRV RNA 3 appears to have no effect on the replication or symptom production by its helper, except that it decreases the number of local lesions formed on the inoculated leaves of French beans and *Chenopodium amaranticolor*.[71,80]

III. CYTOPATHOLOGY

In thin sections of plant cells, the recognition of Nepovirus particles presents some difficulties because of their resemblance to ribosomes.[82] The method of removing ribosomes from sections by RNase treatment of aldehyde-fixed tissues[83] is also not completely satisfactory, because the RNA in some Nepovirus particles can be digested, leaving empty protein shells. Nevertheless, apparently intact TRSV particles have been observed in sections of RNase-treated tissues cleared of ribosomes (Figure 2A). In such sections, empty capsids can also sometimes be observed[83] and it is not clear if they were present as such originally, or were M or B component particles whose RNA had been digested.

There are two other circumstances in which virus particles can be distinguished from ribosomes. First, Nepovirus-infected cells often contain virus particles in tubules scattered in the cytoplasm or associated with plasmodesmata, through which they often penetrate

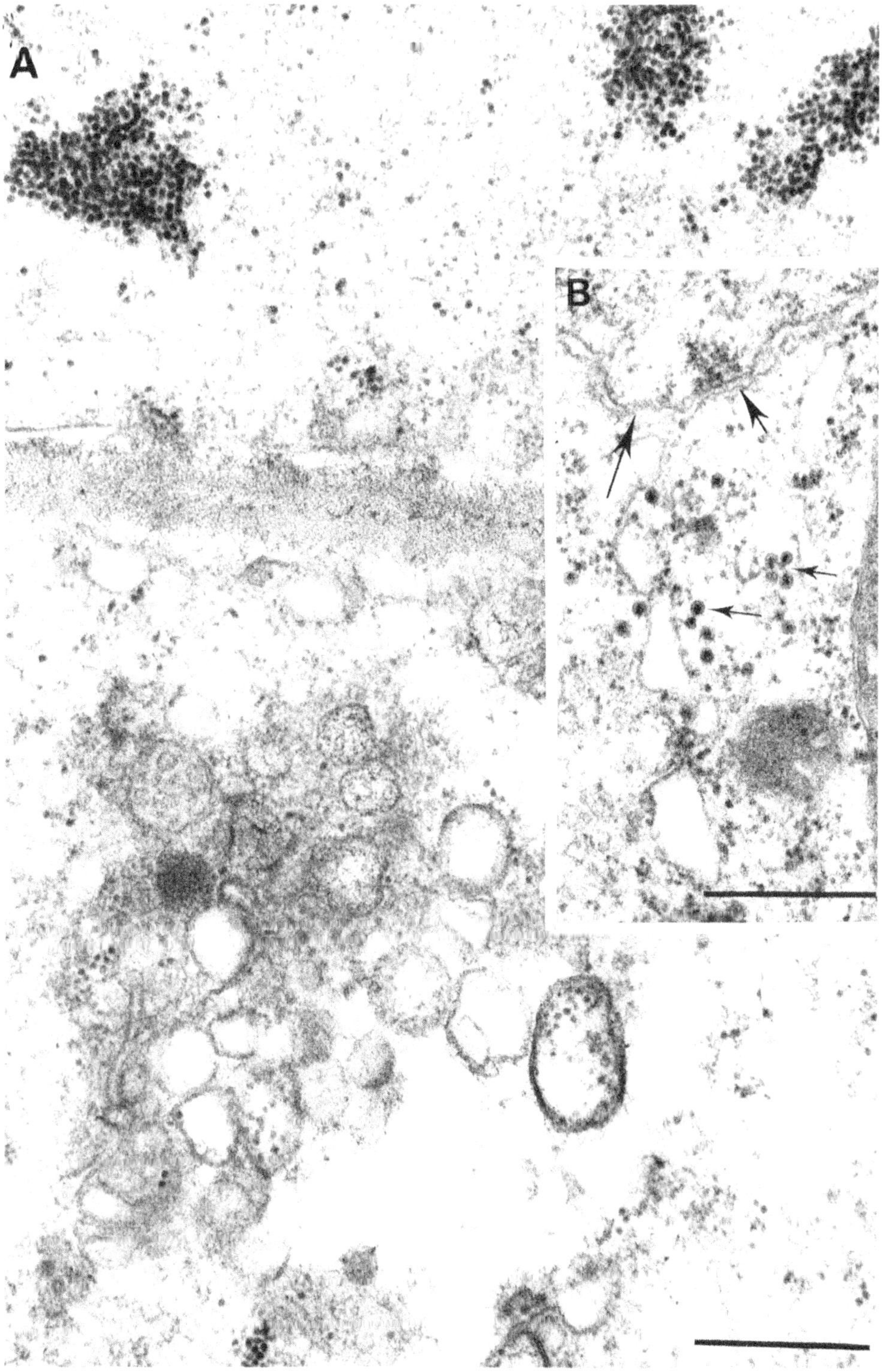

FIGURE 2. A, Thin section of *Cucumis sativus* (cucumber) leaf cells infected with tobacco ringspot virus (TRSV). (Tissue cleared of ribosomes by RNase treatment.) The upper cell shows aggregates of virus particles in the cytoplasm and the one below shows virus-induced vesicles. B, Part of a cucumber leaf cell showing the possible formation of vesicles from the nuclear membrane (large arrows) and some TRSV-containing tubules seen in transverse section (small arrows). Bars = 500 nm.

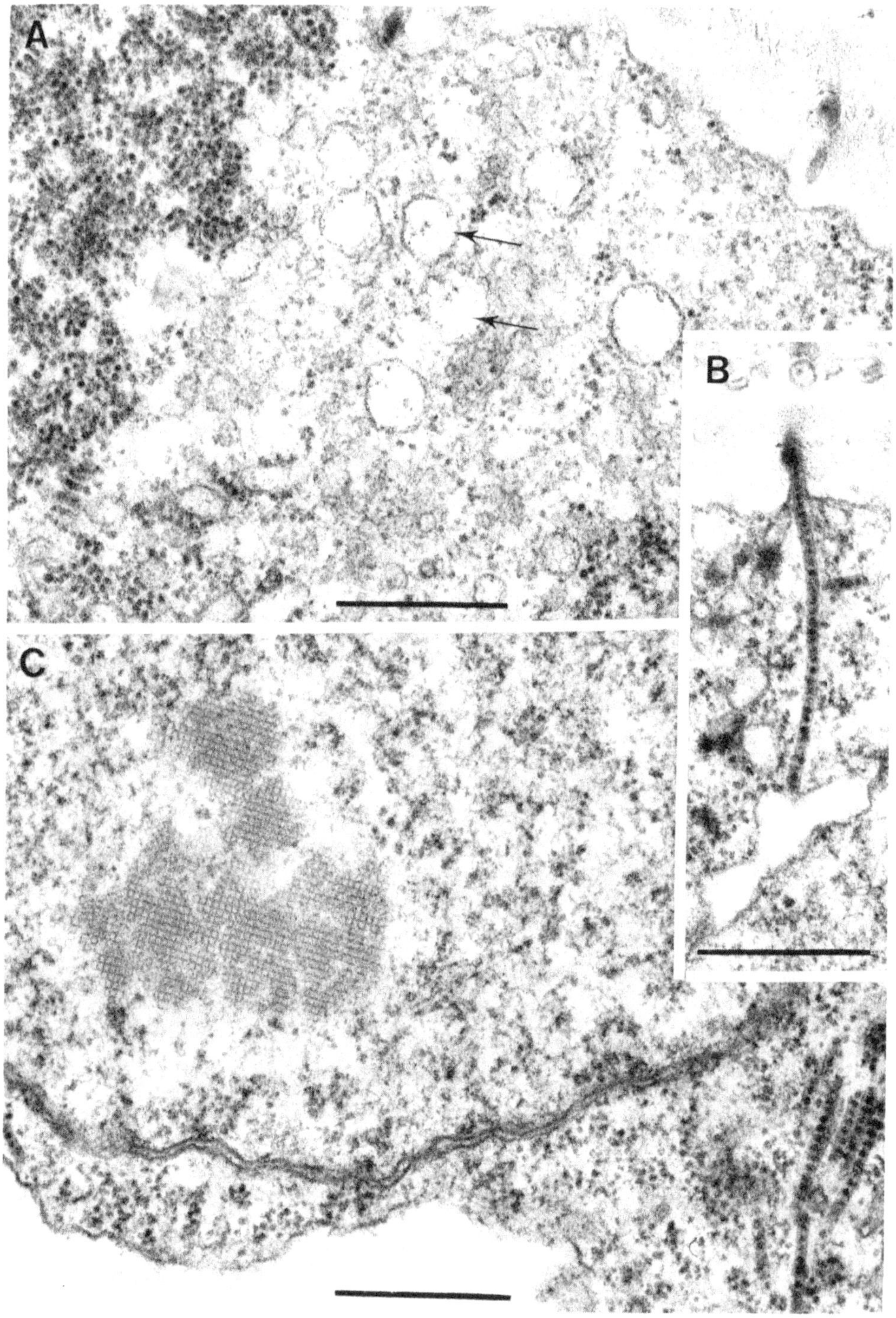

FIGURE 3. Thin sections of *Cucumis sativus* (cucumber) leaf cells infected with TRSV showing A, aggregates of virus particles and virus-induced vesicles containing fibrils (arrows). B, a virus-containing tubule associated with a plasmodesma, and C, a cell containing crystalline aggregates of empty capsids in the nucleus and virus-containing tubules in the cytoplasm. Bars = 500 nm.

(Figures 3B and 4). Second, in some instances, small partially ordered or crystalline aggregates of virus particles can be seen in the cytoplasm (Figures 2A and 3A) or vacuoles (Figure 4A).[62,82,84-92]

Cytoplasmic tubules containing Nepovirus particles were first observed in sections of apical meristem extracts of *Nicotiana rustica* infected with cherry leafroll virus (CLRV).[93] They are very characteristic of cells infected by all the Nepoviruses whose cytopathology has been investigated. Their function, if any, is unknown but it seems unlikely that they could be involved in virus multiplication. Atchison and Francki[94] presented evidence that there is probably no replication of TRSV in bean root tip cells, in which tubules containing virus particles were nevertheless very numerous.[82,86] The frequent penetration of plasmodesmata by virus-induced tubules (Figure 4) in infected meristematic tissues of TRSV-infected plants[82,86,89] and the speed with which the virus reaches the meristems[95] suggest that the tubules may be involved in translocation of Nepoviruses from cell to cell.

The significance of virus particles accumulating in the vacuoles of Nepovirus-infected cells is also not clear. The apparently intact tonoplasts and absence of gross cytoplasmic abnormalities in many cells whose vacuoles contain virus particles[82,89] indicate that the cells are still metabolically functional. Some electron micrographs indicate a possible mechanism by which virus particles may be extruded from the cytoplasm into the vacuole without damage to the tonoplast.[82]

The ability to infect all tissues of a plant, including the meristems, appears to be a general characteristic of Nepoviruses; an example is CLRV.[96,97] Other than the presence of virus particles in cytoplasmic tubules, plasmodesmata and vacuoles, no morphological changes have been reported in the cellular organization of meristematic cells.[82,89] On the other hand, extensive changes have been observed in infected mesophyll cells, including severe modifications of chloroplast structure[84,85] and the development of cell wall protrusions.[91,97] These changes occur at advanced stages of infection; they may be caused by virus-induced senescence of the cells rather than Nepovirus-specific effects.

At early stages of infection, cells of Nepovirus-infected plants develop extensive regions of cytoplasm containing diffuse inclusions, often near the nuclei (Figures 2A and 3A), but the major cell organelles remain unaffected.[3,85,88,92,97,98] The inclusions consist of complex membranous structures some of which form vesicles (Figures 2 and 3A) containing strands similar to those shown to be double-stranded RNA in cells infected with a Comovirus.[99] Prominent Golgi bodies are often seen among the vesicles,[3,98] suggesting that they may be involved in the membrane proliferation, and it has been suggested[92] that virus multiplication and particle assembly may take place in the vesicles. Virus particles are often seen at the periphery of the inclusions. Aggregates of empty protein shells were observed in the nuclei of artichoke yellow ringspot virus (AYRV)-infected cells,[92] a phenomenon also observed in TRSV-infected cells (Figure 3C).

All the electron microscopic observations on Nepovirus-infected cells indicate that virus replication takes place in the cytoplasm, probably in or near the virus-induced inclusions. For TRSV-infected cells this generalization is supported by the finding that TRSV-specific double-stranded RNA and RNA-dependent RNA polymerase remained in the cytoplasmic fraction of extracts from which the large organelles had been removed by centrifugation. Furthermore, it was demonstrated that virus coat protein synthesis was resistant to chloramphenicol and sensitive to cycloheximide, indicating the involvement of cytoplasmic and not chloroplast ribosomes in the synthesis of viral coat protein.[98]

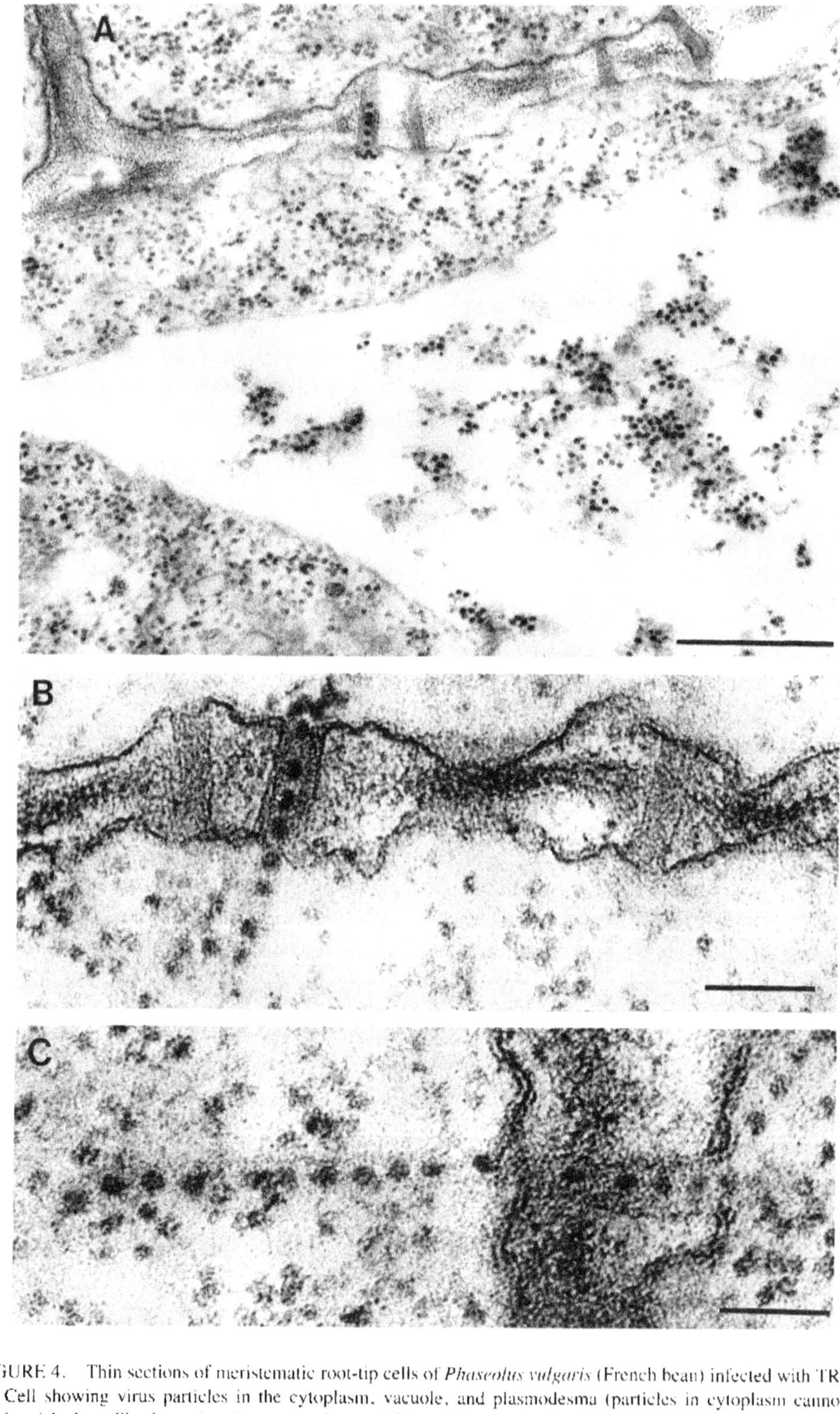

FIGURE 4. Thin sections of meristematic root-tip cells of *Phaseolus vulgaris* (French bean) infected with TRSV. A, Cell showing virus particles in the cytoplasm, vacuole, and plasmodesma (particles in cytoplasm cannot be distinguished readily from the ribosomes); bar = 500 nm. B and C. Details of plasmodesma containing virus particles and their extension into the cytoplasm in tubules. Bars = 100 nm. (Courtesy of Dr. E. M. Davison, Western Australian Department of Agriculture, Perth.)

REFERENCES

1. **Cadman, C. H.,** Biology of soil-borne viruses, *Ann. Rev. Phytopathol.*, 1, 143, 1963.
2. **Wildy, P.,** Classification and nomenclature of viruses. First report of the International Committee on Nomenclature of Viruses, *Monogr. Virol.*, 5, 1, 1971.
3. **Francki, R. I. B. and Hatta, T.,** Nepovirus (tobacco ringspot virus) Group, in *The Atlas of Insect and Plant Viruses*, Maramorosch, K., Ed., Academic Press, New York, 1977, 221.
4. **Murant, A. F.,** Nepoviruses, in *Handbook of Plant Virus Infections and Comparative Diagnosis*, Kurstak, E., Ed., Elsevier, Amsterdam, 1981, 197.
5. **Martelli, G. P.,** Some features of nematode-borne viruses and their relationship with the host plants, *Nematode Vectors of Plant Viruses*, Lamberti, F., Taylor, C. E., and Seinhorst, J. W., Eds., *Life Sciences*, Vol. 2, (NATO Advanced Study Inst., Ser. A), Plenum Press, New York, 1975, 223.
6. **Taylor, C. E. and Robertson, W. M.,** Acquisition, retention and transmission of viruses by nematodes, in *Nematode Vectors of Plant Viruses*, Lamberti, F., Taylor, C. E., and Seinhorst, J. W., Eds., *Life Sciences*, Vol. 2, (NATO Advanced Study Inst., Ser. A), Plenum Press, New York, 1975, 253.
7. **Taylor, C. E., Robertson, W. M., and Roca, F.,** Specific association of artichoke Italian latent virus with the odontostyle of its vector, *Longidorus attenuatus*, *Nematol. Mediterr.*, 4, 23, 1976.
8. **Martelli, G. P., Quacquarelli, A., Gallitelli, D., Savino, V., and Piazzolla, P.,** A tentative grouping of nepoviruses, *Phytopathol. Mediterr.*, 17, 145, 1978.
9. **Dunez, J. and Delbos, R.,** Myrobalan latent ring spot, a bipartite genome virus and strain of tomato black ring virus, *Mitt. Biol. Bundesanst. Land Forstwirtsch. Berlin-Dahlem*, 170, 23, 1976.
10. **Delbos, R., Dunez, J., and Fisac, R.,** The RNAs of three strains of tomato black ring virus, *Ann. Microbiol. Inst. Pasteur*, 172A, 101, 1976.
11. **Murant, A. F. and Taylor, M.,** Estimates of molecular weights of nepovirus RNA species by polyacrylamide gel electrophoresis under denaturing conditions, *J. Gen. Virol.*, 41, 53, 1978.
12. **Gallitelli, D., Piazzolla, P., Savino, V., Quacquarelli, A., and Martelli, G. P.,** A comparison of myrobalan latent ringspot virus with other nepoviruses, *J. Gen. Virol.*, 53, 57, 1981.
13. **Stace-Smith, R.,** Tobacco ringspot virus, *CMI/AAB Descriptions of Plant Viruses*, No. 17, 1970.
14. **Salazar, L. F. and Harrison, B. D.,** Potato black ringspot virus, *CMI/AAB Descriptions of Plant Viruses*, No. 206, 1979.
15. **Murant, A. F.,** Arabis mosaic virus, *CMI/AAB Descriptions of Plant Viruses*, No. 16, 1970.
16. **Quacquarelli, A., Gallitelli, D., Savino, V., and Martelli, G. P.,** Properties of grapevine fanleaf virus, *J. Gen. Virol.*, 32, 349, 1976.
17. **Murant, A. F.,** Raspberry ringspot virus, *CMI/AAB Descriptions of Plant Viruses*, No. 198, 1978.
18. **Jones, R. A. C. and Kenten, R. H.,** Arracacha virus A, *CMI/AAB Descriptions of Plant Viruses*, No. 216, 1980.
19. **Savino, V., Gallitelli, D., and Barba, M.,** Olive latent ringspot virus, a newly recognized virus infecting olive in Italy, *Ann. Appl. Biol.*, 103, 243, 1983.
20. **Murant, A. F.,** Tomato black ring virus, *CMI/AAB Descriptions of Plant Viruses*, No. 38, 1970.
21. **Kenten, R. H.,** Cacao necrosis virus, *CMI/AAB Descriptions of Plant Viruses*, No. 173, 1977.
22. **Martelli, G. P. and Quacquarelli, A.,** Grapevine chrome mosaic virus, *CMI/AAB Descriptions of Plant Viruses*, No. 103, 1972.
23. **Martelli, G. P., Rana, G. L., and Savino, V.,** Artichoke Italian latent virus, *CMI/AAB Descriptions of Plant Viruses*, No. 176, 1977.
24. **Tsuchizaki, T.,** Mulberry ringspot virus, *CMI/AAB Descriptions of Plant Viruses*, No. 142, 1975.
25. **Kenten, R. H., Cockbain, A. J., and Woods, R. D.,** Crimson clover latent virus — a newly recognized seed-borne virus infecting crimson clover *(Trifolium incarnatum)*, *Ann. Appl. Biol.*, 96, 79, 1980.
26. **Stace-Smith, R.,** Tomato ringspot virus, *CMI/AAB Descriptions of Plant Viruses*, No. 18, 1970.
27. **Schneider, I. R., White, R. M., and Civerolo, E. L.,** Two nucleic acid-containing components of tomato ringspot virus, *Virology*, 57, 139, 1974.
28. **Martelli, G. P., Quacquarelli, A., and Gallitelli, D.,** Grapevine Bulgarian latent virus, *CMI/AAB Descriptions of Plant Viruses*, No. 186, 1978.
29. **Quacquarelli, A., Martelli, G. P., and Vovlas, C.,** Chicory yellow mottle virus, *CMI/AAB Descriptions of Plant Viruses*, No. 132, 1974.
30. **Jones, A. T. and Forster, R. L. S.,** Lucerne Australian latent virus, *CMI/AAB Descriptions of Plant Viruses*, No. 225, 1980.
31. **Cropley, R. and Tomlinson, J. A.,** Cherry leaf roll virus, *CMI/AAB Descriptions of Plant Viruses*, No. 80, 1971.
32. **Jones, A. T. and Mayo, M. A.,** The two nucleoprotein particles of cherry leaf roll virus, *J. Gen. Virol.*, 16, 349, 1972.
33. **Walkey, D. G. A., Stace-Smith, R., and Tremaine, J. H.,** Serological, physical, and chemical properties of strains of cherry leaf roll virus, *Phytopathology*, 63, 566, 1973.

34. **Dunez, J., Delbos, R., and Dupont, G.,** Myrobalan latent ringspot virus, *CMI/AAB Descriptions of Plant Viruses*, No. 160, 1976.
35. **Dias, H. F.,** Peach rosette mosaic virus, *CMI/AAB Descriptions of Plant Viruses*, No. 150, 1975.
36. **Dias, H. F. and Allen, W. R.,** Characterization of the single protein and two nucleic acids of peach rosette mosaic virus, *Can. J. Bot.*, 58, 1747, 1980.
37. **Brunt, A. A., Barton, R. J., and Phillips, S.,** Hibiscus latent ringspot virus, *CMI/AAB Descriptions of Plant Viruses*, No. 233, 1981.
38. **Rana, G. L., Gallitelli, D., Kyriakopoulou, P. E., Russo, M., and Martelli, G. P.,** Host range and properties of artichoke yellow ringspot virus, *Ann. Appl. Biol.*, 96, 177, 1980.
39. **Chu, P. W. G. and Francki, R. I. B.,** The chemical subunit of tobacco ringspot virus coat protein, *Virology*, 93, 398, 1979.
40. **Chu, P. W. G., Francki, R. I. B., and Hatta, T.,** Some properties of tomato ringspot virus isolated from *Pentas laceolata* in South Australia, *Plant Pathol.*, 32, 353, 1983.
41. **Murant, A. F.,** Strawberry latent ringspot virus, *CMI/AAB Descriptions of Plant Viruses*, No. 126, 1974.
42. **Stace-Smith, R. and Hansen, A. J.,** Cherry rasp leaf virus, *CMI/AAB Descriptions of Plant Viruses*, No. 159, 1976.
43. **Gallitelli, D., Rana, G. L., and Di Franco, A.,** Il virus della scolorazione perinervale del cariofo, *Phytopathol. Mediterr.*, 17, 1, 1978.
44. **Bancroft, J. B.,** Tomato top necrosis virus, *Phytopathology*, 58, 1360, 1968.
45. **Usugi, T. and Saito, Y.,** Satsuma dwarf virus, *CMI/AAB Descriptions of Plant Viruses*, No. 208, 1979.
46. **Diener, T. O. and Schneider, I. R.,** The two components of tobacco ringspot virus nucleic acid: origin and properties, *Virology*, 29, 100, 1966.
47. **Randles, J. W., Harrison, B. D., Murant, A. F., and Mayo, M. A.,** Packaging and biological activity of the two essential RNA species of tomato black ring virus, *J. Gen. Virol.*, 36, 187, 1977.
48. **Davison, E. M. and Francki, R. I. B.,** Observation on negatively stained tobacco ringspot virus and its two RNA-deficient components, *Virology*, 39, 235, 1969.
49. **Chambers, T. C., Francki, R. I. B., and Randles, J. W.,** The fine structure of gladiolus virus, *Virology*, 25, 15, 1965.
50. **Markham, R., Frey, S., and Hills, G. J.,** Methods for the enhancement of image detail and accentuation of structure in electron microscopy, *Virology*, 20, 88, 1963.
51. **Agrawal, H. O.,** The morphology of arabis mosaic virus, *J. Ultrastruct. Res.*, 17, 84, 1967.
52. **Mayo, M. A., Murant, A. F., and Harrison, B. D.,** New evidence on the structure of nepoviruses, *J. Gen. Virol.*, 12, 175, 1971.
53. **Heuss, K. L., Murthy, M. R. N., and Argos, P.,** Crystallization of tobacco ringspot virus, *J. Mol. Biol.*, 153, 1161, 1981.
54. **Harrison, B. D., Murant, A. F., and Mayo, M. A.,** Evidence for two functional RNA species in raspberry ringspot virus, *J. Gen. Virol.*, 16, 339, 1972.
55. **Murant, A. F., Taylor, M., Duncan, G. H., and Raschké, J. H.,** Improved estimates of molecular weight of plant virus RNA by agarose gel electrophoresis and electron microscopy after denaturation with glyoxal, *J. Gen. Virol.*, 53, 321, 1981.
56. **Mayo, M. A., Barker, H., and Harrison, B. D.,** Specificity and properties of the genome-linked proteins of nepoviruses, *J. Gen. Virol.*, 59, 149, 1982.
57. **Koenig, I. and Fritsch, C.,** A protein linked at the 5′ end of satellite and genomic tomato black ring virus RNAs: a study of *in vitro* translation after protease treatment, *J. Gen. Virol.*, 60, 343, 1982.
58. **Harrison, B. D. and Barker, H.,** Protease-sensitive structure needed for infectivity of nepovirus RNA, *J. Gen. Virol.*, 40, 711, 1978.
59. **Chu, P. W. G., Boccardo, G., and Francki, R. I. B.,** Requirement of a genome-associated protein of tobacco ringspot virus for infectivity but not for *in vitro* translation, *Virology*, 109, 428, 1981.
60. **Mayo, M. A., Barker, H., and Harrison, B. D.,** Polyadenylate in the RNA of five nepoviruses, *J. Gen. Virol.*, 43, 603, 1979.
61. **Harrison, B. D., Murant, A. F., and Mayo, M. A.,** Two properties of raspberry ringspot virus determined by its smaller RNA, *J. Gen. Virol.*, 17, 137, 1972.
62. **Harrison, B. D., Murant, A. F., Mayo, M. A., and Roberts, I. M.,** Distribution of determinants for symptom production, host range and nematode transmissibility between the two RNA components of raspberry ringspot virus, *J. Gen. Virol.*, 22, 233, 1974.
63. **Harrison, B. D. and Murant, A. F.,** Nematode transmissibility of pseudo-recombinant isolates of tomato black ring virus, *Ann. Appl. Biol.*, 86, 209, 1977.
64. **Haber, S. and Hamilton, R. I.,** Distribution of determinants for symptom production, nucleoprotein component distribution and antigenicity of coat protein between the two RNA components of cherry leaf roll virus, *J. Gen. Virol.*, 50, 377, 1980.
65. **Jones, A. T. and Duncan, G. H.,** The distribution of some genetic determinants in the two nucleoprotein particles of cherry leaf roll virus, *J. Gen. Virol.*, 50, 269, 1980.

66. **Hanada, K. and Harrison, B. D.**, Effects of virus genotype and temperature on seed transmission of nepoviruses, *Ann. Appl. Biol.*, 85, 79, 1977.
67. **Harrison, B. D. and Hanada, K.**, Competitiveness between genotypes of raspberry ringspot virus is mainly determined by RNA-1, *J. Gen. Virol.*, 31, 455, 1976.
68. **Robinson, D. J., Barker, H., Harrison, B. D., and Mayo, M. A.**, Replication of RNA-1 of tomato black ring virus independently of RNA-2, *J. Gen. Virol.*, 51, 317, 1980.
69. **Rezaian, M. A. and Francki, R. I. B.**, Replication of tobacco ringspot virus. II. Differences in nucleotide sequences between the viral RNA components, *Virology*, 59, 275, 1974.
70. **Fritsch, C., Mayo, M. A., and Murant, A. F.**, Translation products of genome and satellite RNAs of tomato black ring virus, *J. Gen. Virol.*, 46, 381, 1980.
71. **Murant, A. F. and Mayo, M. A.**, Satellites of plant viruses, *Ann. Rev. Phytopathol.*, 20, 49, 1982.
72. **Kiefer, M. C., Daubert, S. D., Schneider, I. R., and Bruening, G.**, Multimeric forms of satellite of tobacco ringspot virus RNA, *Virology*, 121, 262, 1982.
73. **Rezaian, M. A.**, Three low molecular weight RNA species detected in tobacco ringspot virus, *Virology*, 100, 400, 1980.
74. **Schneider, I. R.**, Satellite-like particle of tobacco ringspot virus that resembles tobacco ringspot virus, *Science*, 166, 1627, 1969.
75. **Schneider, I. R.**, Characteristics of a satellite-like virus of tobacco ringspot virus, *Virology*, 45, 108, 1971.
76. **Bruening, G.**, personal communication, 1983.
77. **Rezaian, M. A. and Jackson, A. O.**, Low-molecular-weight RNAs associated with tobacco ringspot virus are satellites, *Virology*, 114, 534, 1981.
78. **Schneider, I. R. and Thompson, S. M.**, Double-stranded nucleic acids found in tissue infected with the satellite of tobacco ringspot virus, *Virology*, 78, 453, 1977.
79. **Sogo, J. M. and Schneider, I. R.**, Electron microscopy of double-stranded nucleic acids found in tissue infected with the satellite of tobacco ringspot virus, *Virology*, 117, 401, 1982.
80. **Murant, A. F., Mayo, M. A., Harrison, B. D., and Goold, R. A.**, Evidence for two functional RNA species and a "satellite" RNA in tomato black ring virus, *J. Gen. Virol.*, 19, 275, 1973.
81. **Fritsch, C., Mayo, M. A., and Murant, A. F.**, Translation of the satellite RNA of tomato black ring virus *in vitro* and in tobacco protoplasts, *J. Gen. Virol.*, 40, 587, 1978.
82. **Crowley, N. C., Davison, E. M., Francki, R. I. B., and Owusu, G. K.**, Infection of bean root-meristems by tobacco ringspot virus, *Virology*, 39, 322, 1969.
83. **Hatta, T. and Francki, R. I. B.**, Identification of small polyhedral virus particles in thin sections of plant cells by an enzyme cytochemical technique, *J. Ultrastruct. Res.*, 74, 116, 1981.
84. **Gerola, F. M., Bassi, M., and Betto, E.**, Some observations on the shape and localization of different viruses in experimentally infected plants, and on the fine structure of the host cells. I. Arabis mosaic virus in *Chenopodium amaranticolor*, *Caryologia*, 18, 353, 1965.
85. **Gerola, F. M., Bassi, M., and Belli, G. G.**, Some observations on the shape and localization of different viruses in experimentally infected plants, and on the fine structure of the host cells. IV. Arabis mosaic virus in *Petunia hybrida* Hort., *Caryologia*, 19, 481, 1966.
86. **Davison, E. M.**, Cell to cell movement of tobacco ringspot virus, *Virology*, 37, 694, 1969.
87. **De Zoeten, G. A. and Gaard, G.**, Possibilities for inter- and intracellular translocation of some icosahedral plant viruses, *J. Cell Biol.*, 40, 814, 1969.
88. **Roberts, I. M. and Harrison, B. D.**, Inclusion bodies and tubular structures in *Chenopodium amaranticolor* plants infected with strawberry latent ringspot virus, *J. Gen. Virol.*, 7, 47, 1970.
89. **Roberts, D. A., Christie, R. G., and Archer, M. C.**, Infection of apical initials in tobacco shoot meristems by tobacco ringspot virus, *Virology*, 42, 217, 1970.
90. **Hibino, H., Tsuchizaki, T., Usugi, T., and Saito, J.**, Fine structure and developmental process of tubules induced by mulberry ringspot virus and satsuma ringspot virus infection, *Ann. Phytopathol. Soc. Jpn.*, 43, 255, 1977.
91. **Halk, E. L. and McGuire, J. M.**, Translocation of tobacco ringspot virus in soybean, *Phytopathology*, 63, 1291, 1973.
92. **Russo, M., Martelli, G. P., Rana, G. L., and Kyriakopoulou, P. E.**, The ultrastructure of artichoke yellow ringspot virus infections, *Microbiologica*, 1, 81, 1978.
93. **Walkey, D. G. A. and Webb, M. J. W.**, Virus in plant apical meristems, *J. Gen. Virol.*, 3, 311, 1968.
94. **Atchison, B. A. and Francki, R. I. B.**, The source of tobacco ringspot virus in root-tip tissue of bean plants, *Physiol. Plant Pathol.*, 2, 105, 1972.
95. **Francki, R. I. B.**, unpublished results, 1969.
96. **Walkey, D. G. A. and Webb, M. J. W.**, Tubular inclusion bodies in plants infected with viruses of the NEPO type, *J. Gen. Virol.*, 7, 159, 1970.
97. **Jones, A. T., Kinninmonth, A. M., and Roberts, I. M.**, Ultrastructural changes in differentiated leaf cells infected with cherry leaf roll virus, *J. Gen. Virol.*, 18, 61, 1973.

98. **Rezaian, M. A., Francki, R. I. B., Chu, P. W. G., and Hatta, T.**, Replication of tobacco ringspot virus. III. Site of virus synthesis in cucumber cotyledon cells, *Virology*, 74, 481, 1976.
99. **Hatta, T. and Francki, R. I. B.**, Enzyme cytochemical identification of single-stranded and double stranded RNAs in virus-infected plant and insect cells, *Virology*, 88, 105, 1978.

Chapter 3

PEA ENATION MOSAIC VIRUS GROUP

Pea enation mosaic virus (PEMV) is the sole member of the group.[1,2] It has polyhedral particles about 28 nm in diameter sedimenting as two components of 112S (B component) and 99S (M component). The particles contain single-stranded, positive-sense RNA molecules of M_r about 1.7 and 1.3 × 10^6, respectively; both are essential for infectivity. The particles are built from one polypeptide of M_r about 22 × 10^3 and both components contain 28 to 30% RNA.

PEMV is transmitted by aphids in a persistent manner, but the virus probably does not multiply in the vector. The virus can also be experimentally transmitted by mechanical inoculation.[2-4] A low frequency of seed transmission has been reported.[2]

I. VIRUS STRUCTURE AND COMPOSITION

A. Particle Structure

The reported diameter of PEMV particles varies from 22 to 30 nm.[2-11] The M and B particles differ in their appearance in electron micrographs in that the B particles are often hexagonal or pentagonal in outline, whereas M particles are less regular in shape (Figure 1). German and de Zoeten[12] observed that M and B particles were 25 and 28 nm in diameter, respectively, whereas Hull and Lane[11] were unable to detect any significant size differences. B and M components of PEMV preparations can be separated by electrophoresis in polyacrylamide gels[11,13] and yet they migrate as a single boundary in free electrophoresis.[14] This has been interpreted to indicate that the particles differ in their hydrated diameters.[4] The particles also differ in stability; for example, M but not B particles are degraded in CsCl or phosphotungstate.[7,12]

Both M and B components of PEMV appear to contain about 28 to 30% of RNA[11,14] and to be of M_r about 4.38 and 5.57 × 10^6, respectively.[11] Only one major polypeptide of M_r 21.8 × 10^3 was detected in PEMV preparations both by polyacrylamide gel electrophoresis[15] and amino acid analyses accompanied by peptide mapping.[15] From these data, Hull and Lane[11] calculated that each B particle contains about 180 polypeptides and each M particle only 150. This indicates that B particles are icosahedra with T = 3 surface lattice symmetry, a structure which would be consistent with electron microscopic observations.[9] If M particles of PEMV do indeed contain only 150 polypeptides, their structure becomes difficult to explain in terms of the principles laid down for icosahedral symmetry.[16] However, Hull[3] has suggested an interesting model in which the polypeptides would not be quasi-equivalently arranged, but adjacent hexamer morphological subunits would be shared. A particle so constructed would be expected to be less stable than a regular T = 3 structure, and this could account for the observed differences in stability between B and M particles already mentioned.

An interesting difference has been reported between the particles of PEMV strains that are aphid-transmissible and those that are not.[13,17] In addition to the coat protein subunit of M_r about 22 × 10^3 present in all strains, particles of the aphid-transmissible strains also contain small amounts of a larger polypeptide, of M_r about 56 × 10^3. Furthermore, electrophoresis of nontransmissible PEMV preparations in polyacrylamide gels resolved only two components corresponding to M and B particles whereas particles in transmissible virus preparations migrated as numerous boundaries. It was concluded that the M_r 56 × 10^3 protein was responsible for aphid transmissibility and that the numerous electrophoretic components were produced because increasing amounts of this protein associated with a

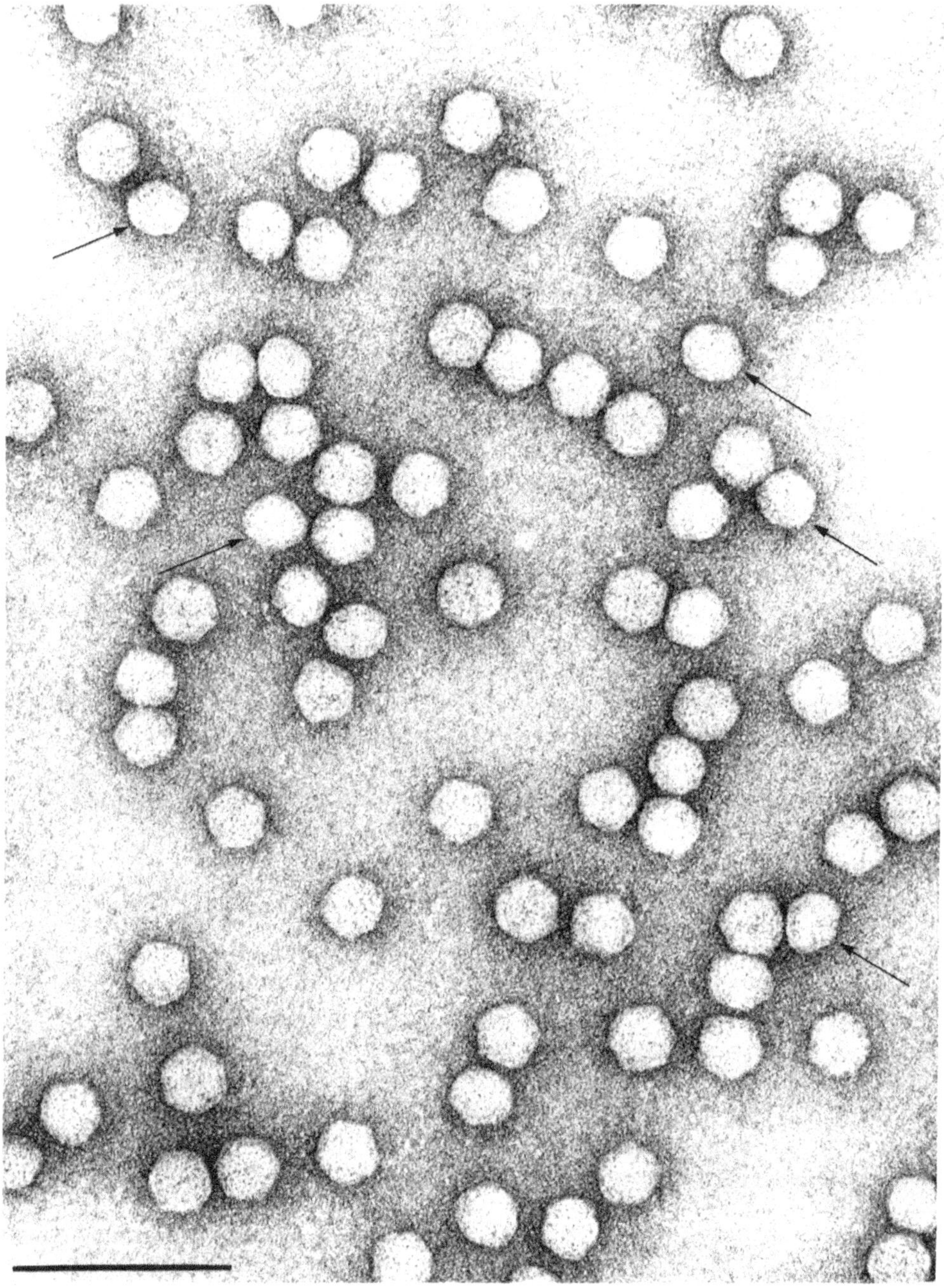

FIGURE 1. Purified particles of pea enation mosaic virus (PEMV) stained in uranyl acetate. Bottom component (B) particles are regular in outline whereas middle component (M) particles (possible candidates arrowed) are less regular and possibly smaller. Bar = 100 nm.

virus particle would proportionally decrease its electrophoretic mobility. The barrier to transmission of PEMV by its aphid vector is probably associated with inability to enter the salivary glands.[18] It may be that this entry is controlled by the M_r 56 × 10^3 protein.

B. The Viral Genome

Early studies on the infectivity of PEMV generated some controversy. Whereas Bozarth and Chow[5] believed that only B particles were infective, other workers maintained that each type of particle was potentially infective.[6,7,19,20] Hull and Lane[11] finally reported convincing evidence that both particles were essential for infectivity and hence that PEMV has a bipartite genome. In the earlier studies, separation of the two components was almost certainly inadequate. Sucrose density gradients were used which Hull[3] reported to be so inefficient that even after three cycles of centrifugation, the B and M components were still cross-contaminated. Separation by polyacrylamide gels had far better resolution.[11]

It has been demonstrated that B particles of PEMV contain a larger RNA (RNA 1) than M particles (RNA 2). The size estimates of RNA 1 vary from M_r 1.4 to 2.3 × 10^6 and those of RNA 2 from 1.1 to 1.7 × 10^6, depending on the method of analysis.[11,12,19] However, there also appear to be differences between strains.[13] By construction of pseudorecombinants between RNA 1 and RNA 2 of different PEMV strains, Hull and Lane[11] determined that the coat protein gene is on RNA 1. PEMV and Dianthoviruses seem to be the only positive-sense, single-stranded RNA viruses with segmented genomes that carry their coat protein genes on the largest genome segment, all others appear to carry them on the smallest segment.[21]

Some strains of PEMV have been shown to contain a third RNA species (RNA 3) that does not seem necessary for infectivity.[11,19] Size estimates of this RNA vary from M_r 0.11 to 0.36 × 10^6.[11,12,19] German and de Zoeten[12] suggested that RNA 3 arises as a degradation product of RNAs 1 and 2, but the possibility that RNA 3 may be a satellite RNA does not seem to have been explored.

PEMV RNAs do not have poly A tracts at their 3′ ends.[22] However, like at least some Comoviruses, Nepoviruses, Potyviruses, and Sobemoviruses, they do have a covalently linked protein (VPg) at their 5′ ends.[23]

II. CYTOPATHOLOGY

Cells of PEMV-infected plants contain virus particles scattered in the nuclei and to a lesser extent in the cytoplasm and vacuoles (Figures 2 and 3); sometimes the particles also form crystalline arrays in the cytoplasm.[24,25] From these observations, it was suggested that PEMV multiplies in the nucleus, from which some of the particles invade the cytoplasm. Subsequently, vesicles were found in the perinuclear spaces of PEMV-infected plant cells, apparently developed from the inner nuclear membrane.[26] Similar vesicles, also observed in the cytoplasm, may have originated in the perinuclear spaces[20] (Figures 2 and 3).

The proposal that PEMV replicates in the nucleus also finds support from the results of biochemical studies. It has been shown that both PEMV-specific double-stranded RNA and virus-induced RNA-dependent RNA polymerase are located in nuclei of infected cells.[27,28] Furthermore, it was demonstrated that PEMV RNA could undergo initial stages of replication in isolated pea nuclei.[29] RNA-dependent RNA polymerase was detected in the nuclei after priming with PEMV RNA, and at least some of the product of the enzyme activity hybridized to viral RNA.[29]

PEMV, which is circulative in its aphid vector, has been studied extensively in thin sections of infective insects. Virus particles have been seen to be concentrated in the lumen of the gut.[24,26,30] Shikata and colleagues[24] also observed virus particles in the fat bodies of insects fed on infected plants, and Harris and Bath[30] detected particles in viroplasm-like

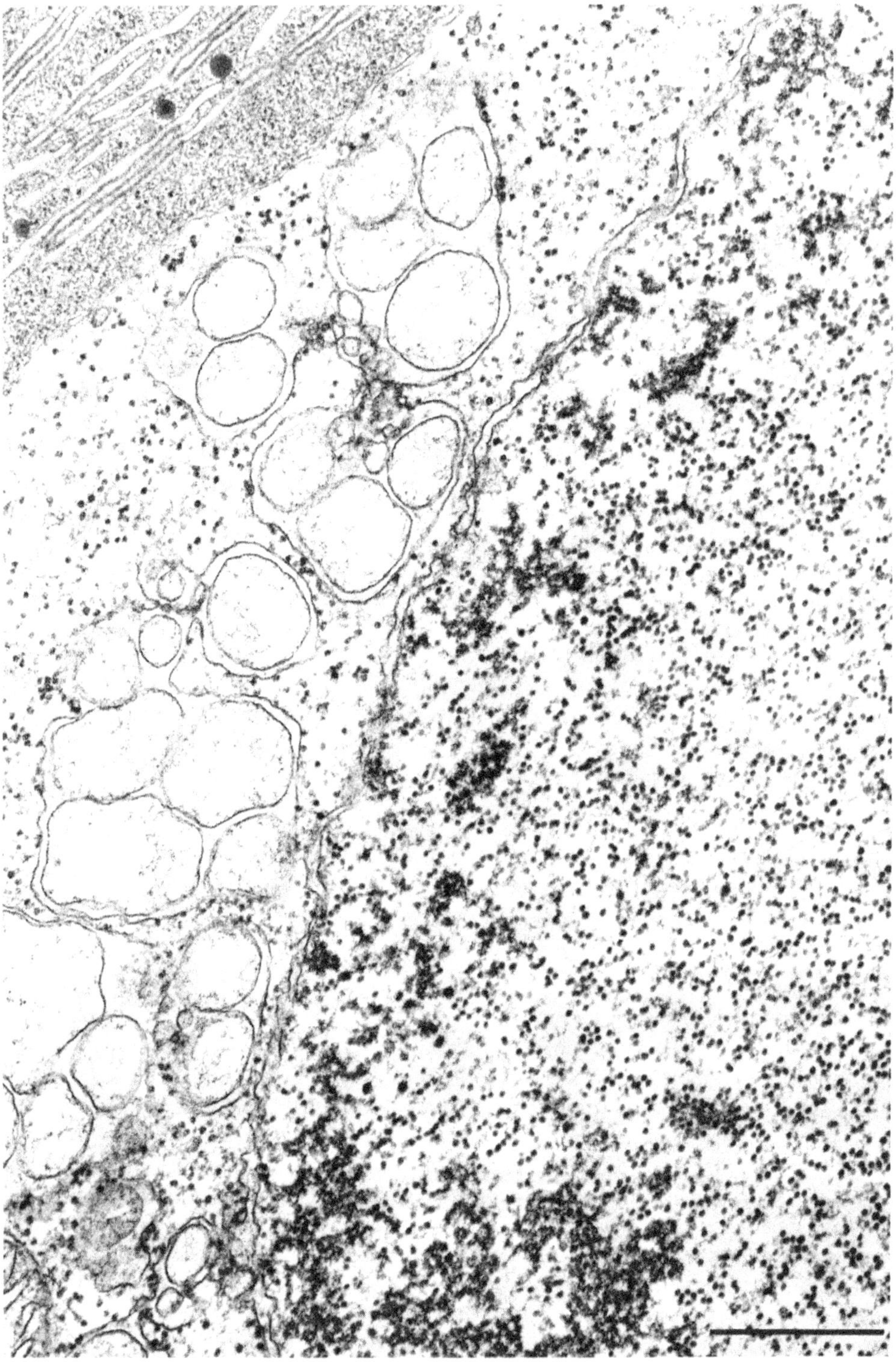

FIGURE 2. Thin section of a leaf of *Pisum sativum* (pea) infected with PEMV. Note the large vesicles near the nucleus and virus particles in both nucleus and cytoplasm. Bar = 500 nm. (Courtesy of Drs. M. Russo and G. P. Martelli, Istituto di Patologia Vegetale, Università degli studi, Bari, Italy.)

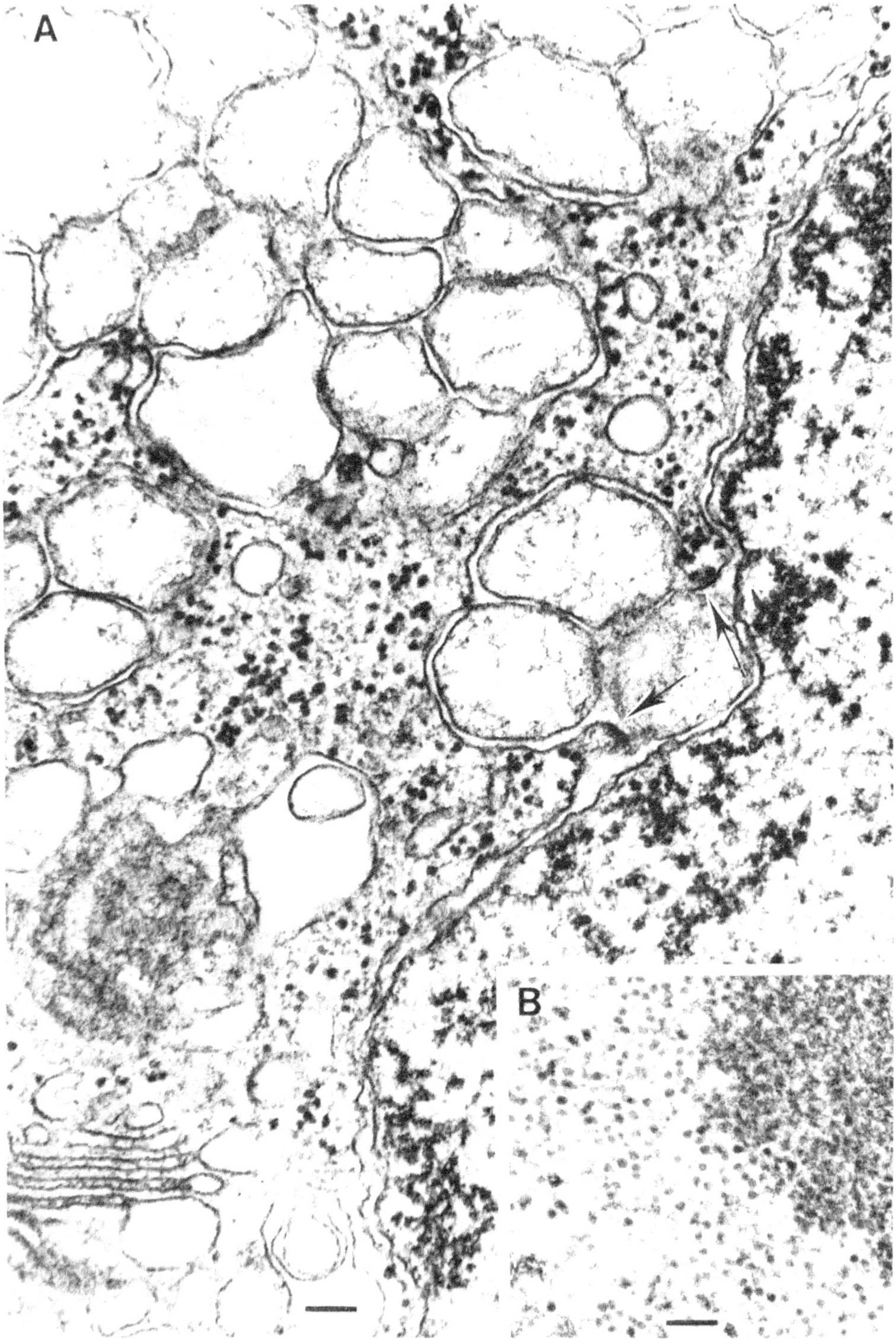

FIGURE 3. Material as in Figure 2. A, Fibril-containing vesicles apparently derived from the nuclear membrane (arrows point to the outer nuclear membrane continuous with a vesicle); B, virus particles in the nucleus. Bars = 500 nm. (Courtesy of Drs. M. Russo and G. P. Martelli, Instituto di Patologia Vegetale, Università degli Studi, Bari, Italy.)

areas, cytoplasm, and nuclei of midgut epithelial cells as well as in the cytoplasm of midgut muscle cells and hemocytes. In subsequent studies, Harris and colleagues[18] examined aphids injected with purified PEMV preparations directly into the hemocoele. Virus particles were later observed in the salivary system including the basal laminae of both primary and accessory glands and the plasma membrane cisternae of accessory gland cells. These observations suggest that membrane flow is responsible for transport of virus particles from the hemocoele to the salivary duct lumina. Weighing up all the evidence from biological and electron microscopic studies, Harris and colleagues[18] concluded that PEMV, although circulative in its aphid vector, probably did not multiply in its tissues.

REFERENCES

1. **Matthews, R. E. F.,** Classification and nomenclature of viruses. Fourth report of the International Committee on Taxonomy of Viruses, *Intervirology*, 17, 1, 1982.
2. **Peters, D.,** Pea enation mosaic virus, *CMI/AAB Descriptions of Plant Viruses*, No. 257, 1982.
3. **Hull, R.,** Properties of an aphid-borne virus: pea enation mosaic virus, in *Aphids as Virus Vectors*, Harris, K. F. and Maramorosch, K., Eds., Academic Press, New York, 1977, 137.
4. **Hull, R.,** Pea enation mosaic virus, in *Handbook of Plant Virus Infections and Comparative Diagnosis*, E., Kurstak, Ed., Elsevier, Amsterdam, 1981, 239.
5. **Bozarth, R. F. and Chow, C. C.,** Pea enation mosaic virus: differential suspension of virus components in buffer and sucrose, *Virology*, 36, 506, 1968.
6. **Gibbs, A. J., Harrison, B. D., and Woods, R. D.,** Purification of pea enation mosaic virus, *Virology*, 29, 348, 1966.
7. **Izadpanah, K. and Shepherd, R. J.,** Purification and properties of the pea enation mosaic virus, *Virology*, 28, 463, 1966.
8. **Musil, M., Marcinka, K., and Ciampar, F.,** Some properties of pea enation mosaic virus, *Acta Virol.*, 14, 285, 1970.
9. **Farro, A. and Rassel, A.,** Caractéristiques en microscopie électronique d'un isolement du virus de la mosaäique "énation" du pois, *Ann. Phytopathol.*,3, 53, 1971.
10. **Vovlas, C. and Rana, G. L.,** Le virosi delle piante ortensi in Puglia. VII. *Lens esculenta* Moench. ospite naturale del virus del mosaico con enazioni del pisello, *Phytopathol. Mediterr.*, 11, 97, 1972.
11. **Hull, R. and Lane, L. C.,** The unusual nature of the components of a strain of pea enation mosaic virus, *Virology*, 55, 1, 1973.
12. **German, T. L. and de Zoeten, G. A.,** Purification and properties of the replicative forms and replicative intermediates of pea enation mosaic virus, *Virology*, 66, 172, 1975.
13. **Adam, G., Sander, E., and Shepherd, R. J.,** Structural differences between pea enation mosaic virus strains affecting transmissibility by *Acyrthosiphon pisum* (Harris), *Virology*, 92, 1, 1979.
14. **Shepherd, R. J., Wakeman, R. J., and Ghabrial, S. A.,** Preparation and properties of the protein and nucleic acid components of pea enation mosaic virus, *Virology*, 35, 255, 1968.
15. **Hill, J. H. and Shepherd, R. J.,** Molecular weights of plant virus coat proteins by polyacrylamide gel electrophoresis, *Virology*, 47, 817, 1972.
16. **Caspar, D. L. D. and Klug, A.,** Physical principles in the construction of regular viruses, *Cold Spring Harbor Symp. Quant. Biol.*, 27, 1, 1962.
17. **Hull, R.,** Particle differences related to aphid-transmissibility of a plant virus, *J. Gen. Virol.*, 34, 183, 1977.
18. **Harris, K. F., Bath, J. E., Thottappilly, G., and Hooper, G. R.,** Fate of pea enation mosaic virus in PEMV-injected pea aphids, *Virology*, 65, 148, 1975.
19. **Gonsalves, D. and Shepherd, R. J.,** Biological and physical properties of the two nucleoprotein components of pea enation mosaic virus and their associated nucleic acids, *Virology*, 48, 709, 1972.
20. **Mahmood, K. and Peters, D.,** Purification of pea enation mosaic virus and the infectivity of its components, *Neth. J. Plant Pathol.*, 79, 138, 1973.
21. **Lane, L. C.,** The nucleic acids of multipartite, defective and satellite plant viruses, in *Nucleic Acids in Plants*, Vol. II, Hall, T. C. and Davies, J. W., Eds., CRC Press, Boca Raton, Fla., 1979, 65.

22. **German, T. L., de Zoeten, G. A., and Hall, T. C.,** Pea enation mosaic virus genome RNA contains no polyadenylate sequences and cannot be aminoacylated, *Intervirology*, 9, 226, 1978.
23. **Reisman, D. and de Zoeten, G. A.,** A covalently linked protein at the 5'-ends of the genomic RNAs of pea enation mosaic virus, *J. Gen. Virol.*, 62, 187, 1982.
24. **Shikata, E., Maramorosch, K., and Granados, R. R.,** Electron microscopy of pea enation mosaic virus in plants and aphid vectors, *Virology*, 29, 426, 1966.
25. **Shikata, E. and Maramorosch, K.,** Electron microscopy of pea enation mosaic virus in plant cell nuclei, *Virology*, 30, 439, 1966.
26. **De Zoeten, G. A., Gaard, G., and Diez, F. B.,** Nuclear vesiculation associated with pea enation mosaic virus-infected plant tissue, *Virology*, 48, 638, 1972.
27. **De Zoeten, G. A., Powell, C. A., Gaard, G., and German, T. L.,** *In situ* localization of pea enation mosaic virus double-stranded ribonucleic acid, *Virology*, 70, 459, 1976.
28. **Powell, C. A., de Zoeten, G. A., and Gaard, G.,** The localization of pea enation mosaic virus-induced RNA-dependent RNA polymerase in infected peas, *Virology*, 78, 135, 1977.
29. **Powell, C. A. and de Zoeten, G. A.,** Replication of pea enation mosaic virus RNA in isolated pea nuclei, *Proc. Natl. Acad. Sci. U.S.A.*, 74, 2919, 1977.
30. **Harris, K. F. and Bath, J. E.,** The fate of pea enation mosaic virus in its pea aphid vector, *Acyrthosiphon pisum* (Harris), *Virology*, 50, 778, 1972.

Chapter 4

DIANTHOVIRUS GROUP

The Dianthoviruses form the most recently established group of plant viruses.[1] Carnation ringspot virus (CRSV) is the type member from which the group name is derived, *Dianthus* being the generic name of carnation. Particles of viruses in this group are stable, isometric, and about 34 nm in diameter. They are constructed from identical coat protein subunits of M_r about 40 × 10^3. The viral genome consists of two single-stranded, positive-sense RNAs of M_r about 1.5 and 0.5 × 10^6, packaged so that all nucleoprotein particles have the same sedimentation rate of about 135S and buoyant density in CsCl of 1.37 g/cm^3. The viruses are readily transmitted by mechanical inoculation and they are soil-borne although no vector has been identified. There is no evidence of seed transmission.[2,3]

I. MEMBERS OF THE GROUP

At present only three members have been recognized besides CRSV, red clover necrotic mosaic virus (RCNMV), and sweet clover necrotic mosaic virus (SCNMV).[1] Clover primary leaf necrosis virus (CPLNV)[4] has now been shown to be serologically related to SCNMV but the relationship appears sufficiently remote for the two to be considered as distinct viruses.[5]

CRSV has been found naturally only in carnation and in *Stellaria media*, and RCNMV, CPLNV, and SCNMV only in species within the Leguminosae. However, all the viruses have rather wide experimental host ranges.[2,3,6,7] CRSV and RCNMV appear to spread naturally by leaf contact and through infected soil without specific vectors. On the other hand, there is a report of transmission of CRSV by the nematodes *Longidorus* and *Xiphinema*,[2] and a suggestion that RCNMV may be transmitted by the fungus *Olpidium*.[8] These observations need to be confirmed.

II. VIRUS STRUCTURE AND COMPOSITION

A. Particle Structure

Particles of CRSV and RCNMV appear very similar when negatively stained and examined in the electron microscope. More extensive studies have been done with RCNMV, and it was observed that particles stained in uranyl acetate are slightly larger (about 34 nm) than those stained in neutral phosphotungstate (about 31 nm).[9] Particles stained in uranyl acetate have rounded outlines and granular surfaces (Figure 1). They are indistinguishable from the particles of Tombusviruses and a number of viruses which have not yet been classified, such as galinsoga mosaic virus.[10] There is no information from electron microscopy or other studies about the arrangement of the protein subunits.

The capsids of both CRSV and RCNMV are composed of a single coat polypeptide of M_r about 38 to 40 × 10^3 as determined by polyacrylamide-gel electrophoresis.[2,9] Amino acid analysis of CRSV coat protein indicates that there are 347 residues in each polypeptide of M_r 38 × 10^3.[11]

Dianthovirus particles sediment at about 135S.[2,3] RCNMV has a buoyant density in CsCl of about 1.36 g/cm^3.[9] CRSV particles are of M_r about 7 × 10^6 as determined by physical methods.[11]

B. The Viral Genome

Infectivity of Dianthoviruses is dependent on both the RNA components.[5,9,12,13] With

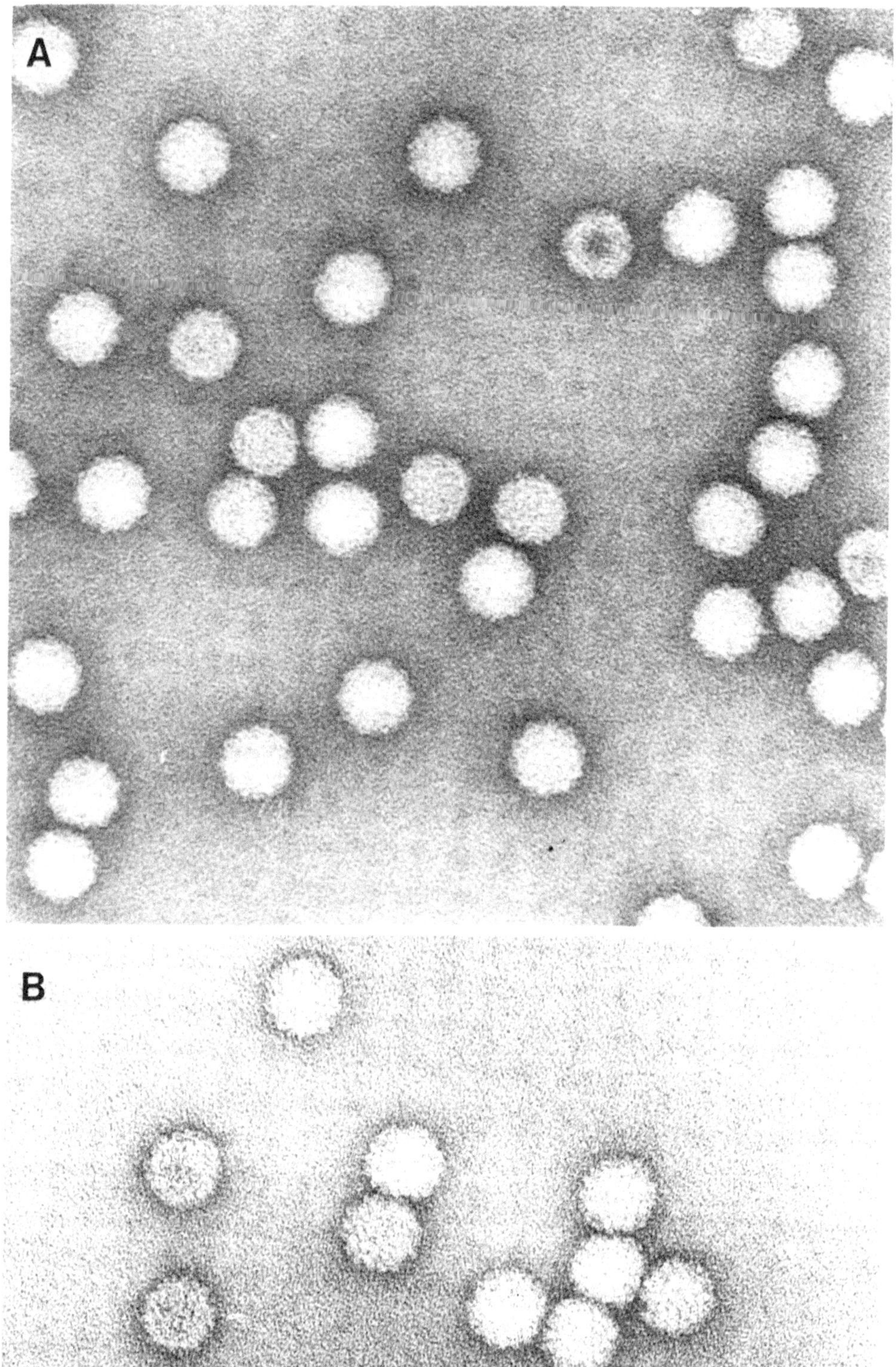

FIGURE 1. A and B. Two different preparations of purified red clover necrotic mosaic virus (RCNMV) stained in uranyl acetate and photographed at the same magnification. Bar = 100 nm.

RCNMV, no base sequence homology was detected by cDNA hybridization tests between the larger RNA of M_r about 1.5×10^6 (RNA 1) and the smaller one of 0.5×10^6 (RNA 2).[9] Similar studies also failed to detect significant homology between the two RNAs of CRSV.[14] Hybridization experiments also revealed that there is no more than 12% base sequence homology between the RNAs of CRSV and RCNMV.[14]

It appears that RNA 1 carries the coat protein gene since pseudorecombinants constructed from RCNMV, CPLNV, and SCNMV RNAs were shown to have the serological specificity of the RNA 1 donor.[13] Furthermore, in vitro translation studies also indicate that, in RCNMV, the gene for coat protein is on its RNA 1, and it appears that a subgenomic mRNA is involved in coat protein synthesis.[15] The chemical properties of Dianthovirus RNAs do not appear to have been studied in detail.

There are no reliable data regarding the packaging of the RNA in virus particles, but it seems most likely that Dianthoviruses have two types of particles, one containing a single molecule of RNA 1 and the other with three molecules of RNA 2. This would give both particles the same sedimentation properties and buoyant densities as has been observed with preparations of both CRSV and RCNMV.[2,3,9,11]

III. CYTOPATHOLOGY

Plant cells infected with RCNMV contain numerous virus particles scattered in the cytoplasm and vacuole (Figure 2). Occasionally, nuclei of infected cells also contain a few particles (Figure 3). Particles have never been observed to aggregate to form crystals in infected cells, though in heavily infected cells they can form irregular aggregates.[16]

Membranous vesicles and patches of densely stained amorphous material (Figures 2 and 3) often develop in the cytoplasm of infected cells. As far as we are aware, the cytopathological effects of CRSV and SCNMV have not been investigated, but a micrograph of a red clover leaf cell infected with CPLNV has been published which shows accumulation of virus particles in the cytoplasm and some vesiculation in the chloroplasts.[4]

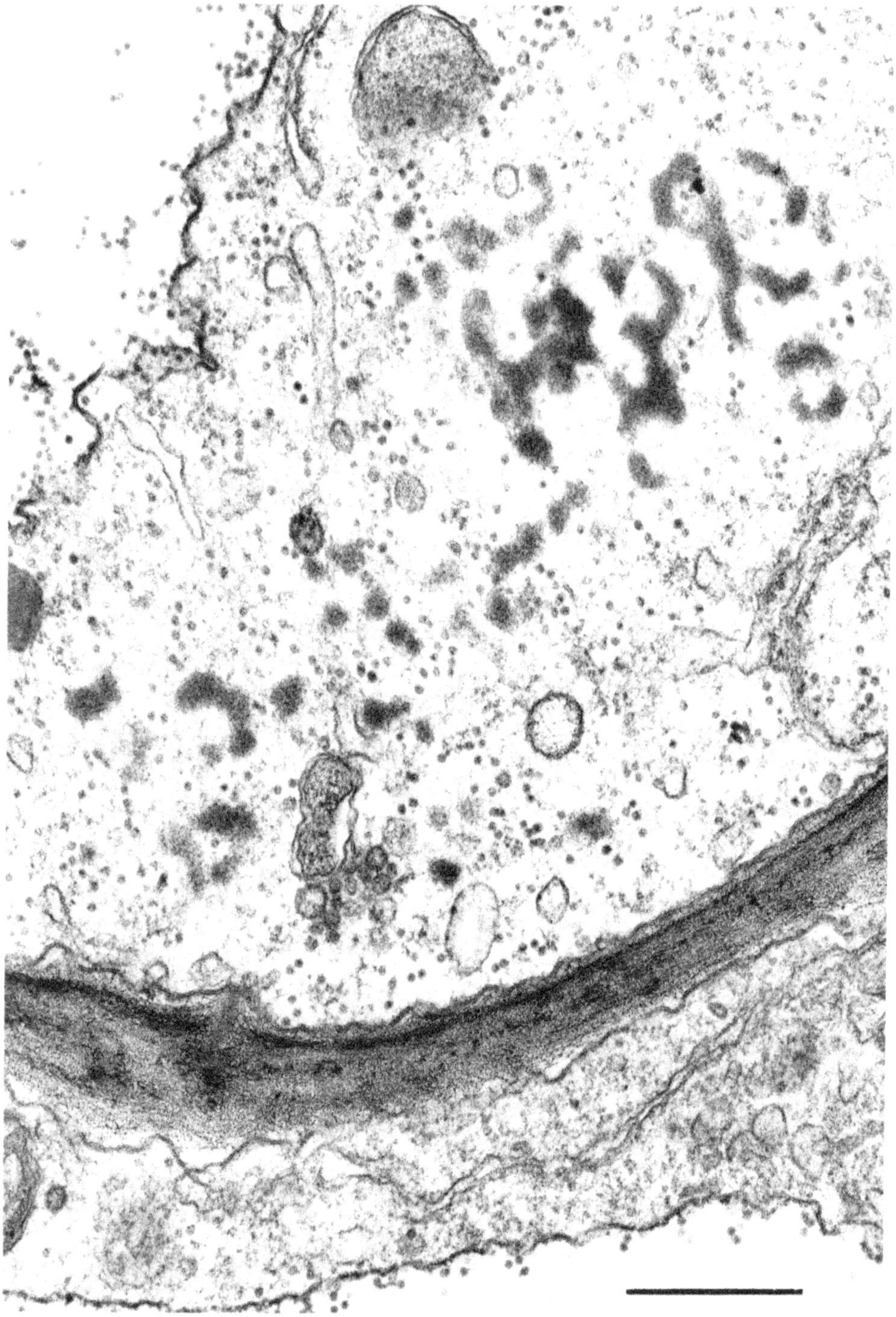

FIGURE 2. Thin section of RCNMV-infected *Nicotiana clevelandii* leaf cells showing numerous virus particles in the cytoplasm and vacuoles. (Tissue cleared of ribosomes by RNase treatment.) In the cytoplasm, patches of virus-specific electron-dense material can also be seen. Bar = 500 nm.

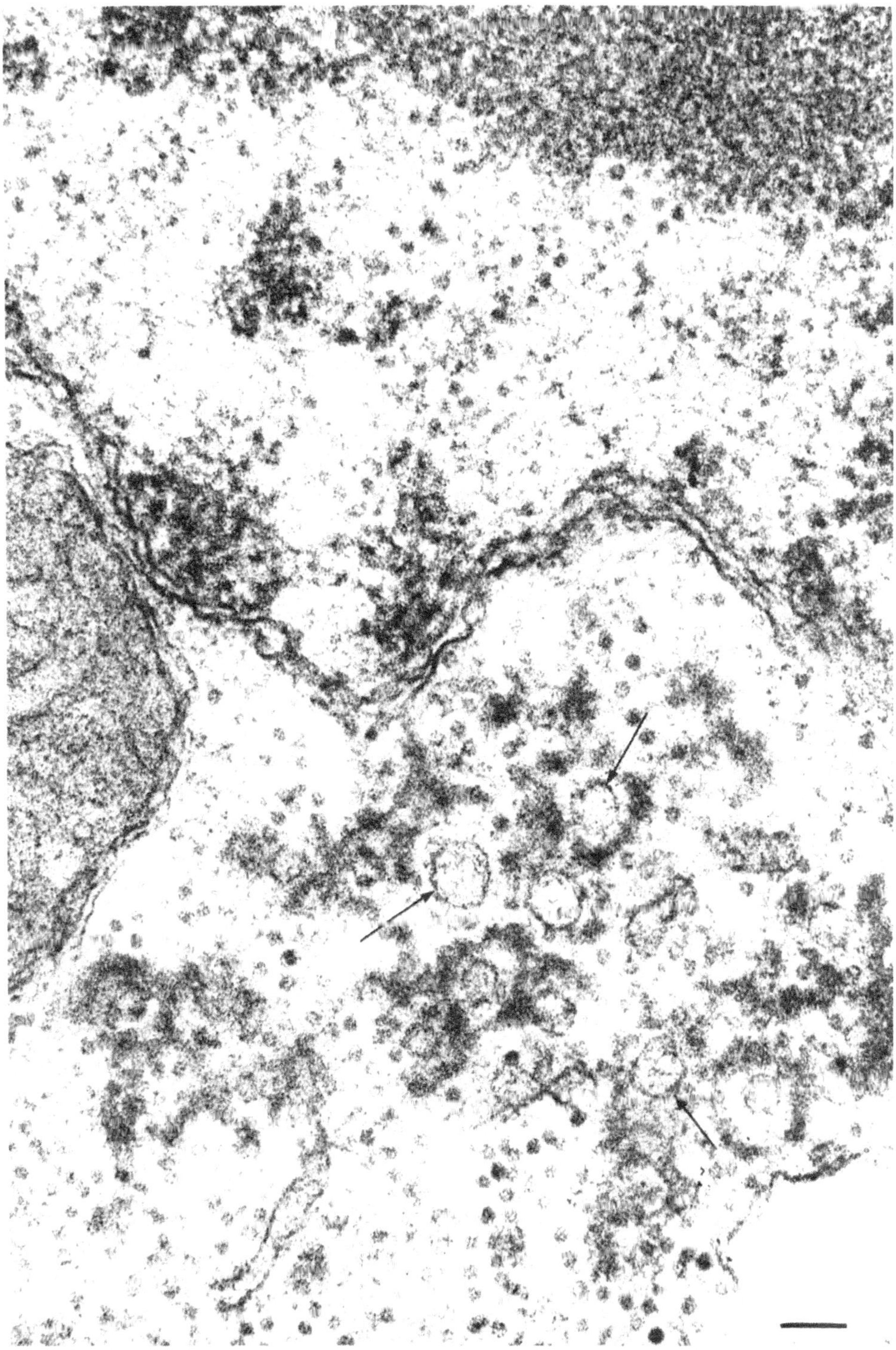

FIGURE 3. Thin section of a *Nicotiana clevelandii* leaf cell infected with RCNMV showing virus particles in the nucleus and cytoplasm. Small vesicles specific to infection (arrows) and patches of electron-dense material can also be seen in the cytoplasm. (Tissue cleared of ribosomes by RNase treatment.) Bar = 100 nm.

REFERENCES

1. **Matthews, R. E. F.,** Classification and nomenclature of viruses. Fourth report of the International Committee on Taxonomy of Viruses, *Intervirology*, 17, 1, 1982.
2. **Hollings, M., and Stone, O. M.,** Carnation ringspot virus, *CMI/AAB Descriptions of Plant Viruses*, No. 21, 1970.
3. **Hollings, M., and Stone, O. M.,** Red clover necrotic mosaic virus, *CMI/AAB Descriptions of Plant Viruses*, No. 181, 1977.
4. **Ragetli, H. W. J., and Elder, M.,** Characteristics of clover primary leaf necrosis virus, a new spherical isolate from *Trifolium pratense*, *Can. J. Bot.*, 55, 2122, 1977.
5. **Hiruki, C.,** personal communication, 1983.
6. **Rüdel, M., Querfurth, G., and Paul, H. L.,** Natural occurrence of carnation ringspot virus in *Stellaria media* (L.) Cyrill. from vineyards detected for the first time, *Nachr. Dtsch. Pflanzenschutzdienstes (Braunschweig)*, 29, 59, 1977.
7. **Kleinhempel, H., Gruber, G., and Kegler, H.,** Investigations of carnation ringspot virus in fruit trees, *Acta Phytopathol. Acad. Sci. Hung.*, 15, 107, 1980.
8. **Bowen, R. and Plumb, R. T.,** The occurrence and effects of red clover necrotic mosaic virus in red clover *(Trifolium pratense)*, *Ann. Appl. Biol.*, 91, 227, 1979.
9. **Gould, A. R., Francki, R. I. B., Hatta, T., and Hollings, M.,** The bipartite genome of red clover necrotic mosaic virus, *Virology*, 108, 499, 1981.
10. **Hatta, T., Francki, R. I. B., and Grivell, C. J.,** Particle morphology and cytopathology of galinsoga mosaic virus, *J. Gen. Virol.*, 64, 687, 1983.
11. **Kalmakoff, J. and Tremaine, J. H.,** Some physical and chemical properties of carnation ringspot virus, *Virology*, 33, 10, 1967.
12. **Dodds, J. A., Tremaine, J. H., and Ronald, W. P.,** Some properties of carnation ringspot virus single- and double-stranded ribonucleic acid, *Virology*, 83, 322, 1977.
13. **Okuno, T., Hiruki, C., Rao, D. V., and Figueiredo, G. C.,** Genetic determinants distributed in two genomic RNAs of sweet clover necrotic mosaic, red clover necrotic mosaic and clover primary leaf necrosis viruses, *J. Gen. Virol.*, 64, 1907, 1983.
14. **Mossop, D. W.,** personal communication, 1983.
15. **Morris-Krsinich, B. A. M., Forster, R. L. S., and Mossop, D. W.,** Translation of red clover necrotic mosaic virus RNA in rabbit reticulocyte lysate: identification of the virus coat protein cistron on the larger RNA strand of the bipartite genome, *Virology*, 124, 349, 1983.
16. **Hatta, T. and Francki, R. I. B.,** Identification of small polyhedral virus particles in thin sections of plant cells by an enzyme cytochemical technique, *J. Ultrastruct. Res.*, 74, 116, 1981.

Chapter 5

CUCUMOVIRUS GROUP

Cucumoviruses derive their name from the type member cucumber mosaic virus (CMV). They constitute a small but coherent group of serologically distantly related viruses.[1,2] The viruses have isometric particles about 29 nm in diameter, each built from 180 identical protein subunits of M_r about 24 × 10^3. They encapsidate four single-stranded RNA molecules; the three largest, of M_r about 1.3, 1.1, and 0.8 × 10^6, constitute the tripartite genome; the smallest one, of M_r about 0.35 × 10^6, is the subgenomic coat protein messenger.

The viruses are transmitted by aphids in a nonpersistent manner to a wide range of plants. They can also be transmitted by mechanical inoculation and are seed borne in a number of hosts.[2] CMV has been isolated from no less than 775 species of 174 genera in 52 families of flowering plants[3] and can be transmitted by more than 60 species of aphids.[4]

I. MEMBERS OF THE GROUP AND THEIR RELATIONSHIPS

A number of viruses other than CMV have been described as members or likely members of the group. However, only two have been characterized sufficiently to be considered as true members of the group: tomato aspermy virus (TAV)[5] and peanut stunt virus (PSV).[6] Robinia mosaic virus (RobMV)[7] and clover blotch virus (CBV)[8] were originally described as distinct viruses, but recent studies have shown that they are closely related to each other and to PSV, indicating that they can be considered as strains of PSV.[9]

Cowpea ringspot virus (CPRSV) was described as a possible member of the Cucumovirus group on the basis of some of its physical properties.[10] Antiserum to this virus failed to react with either CMV or TAV preparations but the tests did not preclude a distant relationship since the antiserum was relatively low titered. Similarly, antisera to CMV and TAV failed to react with purified preparations of CPRSV but the significance of these tests is not clear since homologous titers were not specified. Additional data on the physical and chemical properties of the virus are also needed before its taxonomic position is clarified.

Although it is generally accepted that CMV, TAV, and PSV are distinct viruses, their relationships have been a matter of some controversy. The most common approach to tracing relationships among these viruses has been serological comparison, and CMV and TAV have been studied more thoroughly than PSV. Early workers failed to agree on whether CMV and TAV are serologically related.[11-16] Probably, the lack of agreement was due to the use of different strains of the viruses and to the lack of high-titered antisera because of the rather poor immunogenicity of the viruses.[16,17] More recent work with relatively high-titered antisera produced to aldehyde-fixed virus preparations indicates that there is a distant relationship between many strains of CMV and TAV.[18-20] Both these viruses also appear to be distantly related to PSV.[18] However, interrelationships among the three viruses are complicated by considerable serological variation among strains. For example, strains of CMV studied by Devergne and Cardin[18,21] were shown to fall into two distinct groups which differed from each other by about as much as they did from PSV.

More recently, base sequence homology studies have been used in attempts to trace relationships among Cucumoviruses. Gonda and Symons[22] compared the sequences of each of the four RNAs of three strains of CMV and one strain of TAV by hybridizing them to complementary DNAs (cDNAs) synthesized in vitro with avian myeloblastosis virus reverse transcriptase.[23] They demonstrated that the RNA sequences of two serologically closely related CMV strains were very similar, whereas the similarity was less between strains showing greater serological separation. Furthermore, RNA sequence homology between

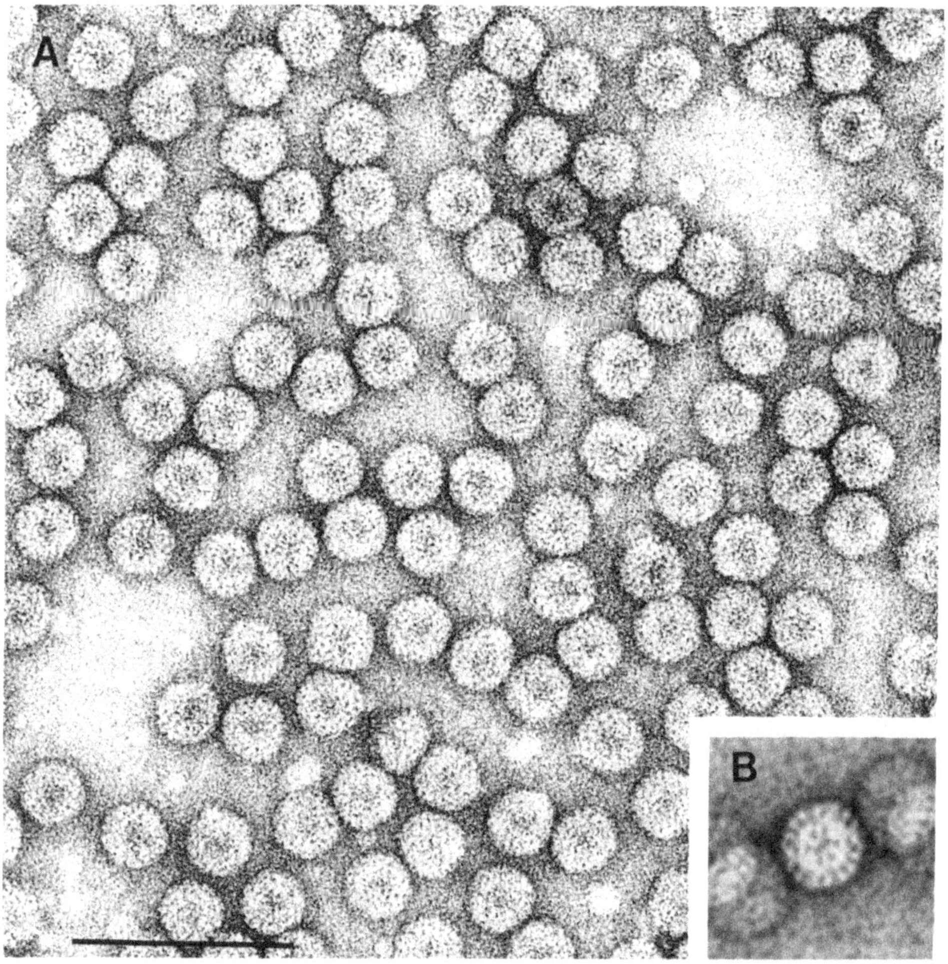

FIGURE 1. A, Purified particles of cucumber mosaic virus (CMV) in uranyl acetate stain; B, a CMV particle viewed down a twofold symmetry axis and subjected to n = 2 photographic rotation. Bar = 100 nm.

CMV and TAV, which are serologically very distantly related, was only barely detectable.[22] Piazzolla and his colleagues[24] used RNA-RNA hybridization methods to compare the RNA base sequences of eighteen CMV and three PSV isolates. They detected no homologies between CMV and PSV and found that with the exception of one strain, the CMVs fell into two distinct groups. Each group of strains had RNAs whose base sequences were virtually indistinguishable whereas no significant homology was detected between RNAs of strains belonging to the two groups. Some of the CMV strains used were those compared serologically by Devergne and Cardin[21] and their assignment into two groups appears consistent using the two criteria.

II. VIRUS STRUCTURE AND COMPOSITION

A. Particle Structure

Particles of CMV, TAV, and PSV appear indistinguishable in the electron microscope;[25] they are isometric with a diameter of about 29 nm when stained in uranyl acetate (Figure 1). Electron microscopy of CMV particles in this stain (Figure 1) together with physical and chemical data indicate that they are composed of 180 protein subunits arranged in pentamer-hexamer clusters with T = 3 surface lattice symmetry.[26] Neutral phosphotungstate is unsuitable for staining Cucumovirus particles since it degrades them.[27]

The Cucumovirus capsid is built from a single species of polypeptide. The electrophoretic mobility of CMV and TAV coat proteins in SDS-polyacrylamide gels are very similar.[15] Their reported sizes range from M_r 24 × 10^3 to M_r 26.3 × 10^3, as estimated from electrophoretic mobilities and amino acid compositions.[15,28-31] The most reliable value appears to be 26.2 × 10^3; this has been calculated from the amino acid composition predicted from the nucleotide sequence of the coat protein gene of CMV RNA.[32] The derived amino acid composition agrees well with that determined by chemical analysis.[32]

Neutron small-angle scattering experiments of CMV in solution indicate that the particle has a protein shell with external and internal diameters of 29 and 16.5 nm.[33] The viral RNA also appears to be arranged as an inner layer which partially penetrates the protein shell; it has external and internal diameters of about 20 and 12 nm, respectively. In this model, the particle has a central hole about 12 nm in diameter which corresponds approximately to the electron-dense centers (about 13 nm) of many particles seen in electron micrographs of uranyl acetate stained virus preparations.[25-27]

It was also concluded from neutron small-angle studies that the packing of the protein subunits of the CMV capsid is not very dense, leaving 32 relatively large holes in the centers of each hexamer and pentamer cluster.[33] The presence of these holes may explain the observed sensitivity of encapsidated CMV RNA to ribonuclease.[34] The stability of CMV particles appears to depend largely on protein-RNA interactions which can readily be disrupted in low concentrations of ionic detergents such as SDS or high concentrations of neutral chloride salts.[27,35,36] Both CMV and TAV have been reversibly dissociated by varying the concentration of lithium chloride in solution;[35] moreover, the RNA of one virus can be reassociated with the protein of the other.[37] The importance of protein-RNA interactions for the integrity of the Cucumovirus particle can be seen from its disintegration as the encapsidated RNA is digested with ribonuclease.[34,35,38,39]

CMV particles sediment as a single boundary with $S_{20w} = 98.6 - 1.04C$, where C is the virus concentration in mg/mℓ. M_r estimates have varied between 5.0 and 6.7 × 10^6.[27,28,40,41] Recently, a value of between 5.8 to 6.1 × 10^6 has been obtained from diffusion coefficient measurements by laser light scattering.[4]

B. The Viral Genome

Although the physical heterogeneity of CMV RNA was observed soon after the virus was purified and characterized,[42,43] its significance was not appreciated till some years later.[44,45] Since CMV particles sediment as a single homogeneous component, it was logical to assume that the heterogeneity of the viral RNA was the result of scission of the largest component. However, it is now well established that the three largest encapsidated RNAs (RNAs 1, 2, and 3) are all necessary for infection[44,45] and consist of molecules with unique nucleotide sequences.[23] The smallest encapsidated RNA (RNA 4) is not required for infectivity; its nucleotide sequence appears to be identical to part of the sequence of RNA 3 at the 3′ terminus and it can function in vitro as mRNA for virus coat protein synthesis.[46,47] All the four RNAs can be charged at the 3′ end with tyrosine by plant-derived amino-acyl tRNA synthetases.[48] Furthermore, all have the same 138 3′ terminal nucleotide residues for which a "clover leaf" t-RNA-like structure has been proposed.[49] Similar 3′ terminal sequences are present on TAV RNA.[50] These RNAs also closely resemble those of three Bromoviruses, brome mosaic (BMV), cowpea chlorotic mottle (CCMV), and broad bean mottle (BBMV), including in their proposed 3′ terminal secondary structure.[51]

There has been controversy regarding the exact size of the CMV RNAs; values ranging between 1.01 to 1.35, 0.89 to 1.21, 0.68 to 0.93, and 0.33 to 0.38 × 10^6 for RNAs 1 to 4, respectively, have been reported from determination by PAGE.[44,52,53] Kaper and Re[54]

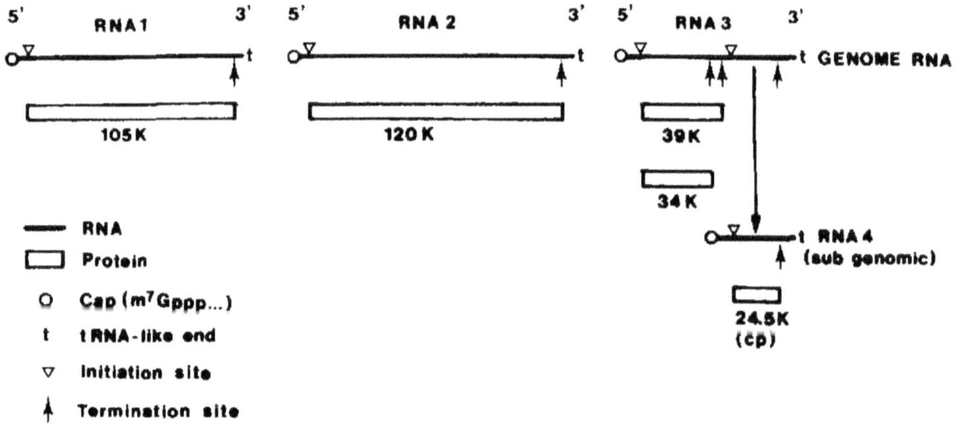

FIGURE 2. Physical map of the organization of the Cucumovirus genome.

argued for the lower values using data on the molecular weight of the virus particle and its percentage of RNA. On the other hand, the higher values are supported by polyacrylamide-gel determinations using phage MS2 RNA and the two *E. coli* r-RNAs as markers,[4] all of which have been sequenced.[55-57] The complete sequence of CMV RNA 3 revealed that it consists of 2216 nucleotides and hence has an M_r of 0.75×10^6.[32] From these data it can be deduced that RNA 4 consists of 1049 nucleotides and hence has an M_r of 0.36×10^6.

The four RNAs of PSV and TAV are essentially similar in size to those of CMV although some small differences have been observed.[15,45,58] The lack of significant heterogeneity in the sedimentation of Cucumovirus particles can be accounted for by the existence of three classes of particles: one encapsidating a single molecule of RNA 1, a second containing one molecule of RNA 2, and a third enclosing one molecule each of RNA 3 and RNA 4.[59]

Pseudorecombinants have been constructed in vitro from different Cucumovirus isolates to determine the distribution of genes on the three segments of their genomes. Only limited data were obtained from pseudorecombinants of CMV and TAV because, whereas RNA 3 was interchangeable between the two viruses, RNAs 1 and 2 were not.[60] Nevertheless, experiments with pseudorecombinants involving exchanges between the RNAs 3 demonstrated that the coat protein gene was on this RNA.[61] This has been confirmed by data from pseudorecombinants from other Cucumovirus isolates.[62-64] It was also shown that aphid transmissibility of CMV is determined by RNA 3.[63] The genetic determinants for some host plant reactions are determined by RNA 2, some by RNA 3, and yet others by the interactions of RNAs 2 and 3.[62,64-66]

The results of in vitro translation studies on CMV RNA revealed that RNAs 1 and 2 are monocistronic, specifying proteins of M_r about 105 and 120×10^3, respectively.[47] RNA 3, on the other hand, is translated into one protein of M_r about 35×10^3, although it also codes for the coat protein of M_r 26.2×10^3,[32] which is translated in vitro only by RNA 4.[47,67] Based on in vitro translation studies, the genetic organization of CMV RNA is summarized in Figure 2. Only the coat protein has been detected in virus-infected protoplasts or leaf tissues.[68-70] The possible functions of the other three proteins remain obscure.

C. Satellite RNA

Small, encapsidated, single-stranded RNA molecules of M_r about 1×10^5 have been detected in a number of CMV and PSV isolates.[71-75] Their total dependence on a Cucumovirus for replication and their unique base sequences indicate that they are satellite RNA molecules.[76-81] The multiplication of two of these, one isolated from CMV (CARNA 5) and another from PSV (PARNA 5), was not supported by the other helper virus.[72] However, the replication of another satellite RNA isolated from CMV could be supported by either CMV or TAV, although it multiplied more efficiently in the presence of CMV.[82] Although satellite RNAs are dependent on their helper viruses for both replication and encapsidation, they can remain dormant in plants for some days when inoculated alone, but begin to replicate after a second inoculation with the helper.[79,83]

Some satellite RNAs have been shown to reduce the amount of helper virus produced in most host species and to induce milder symptoms;[73,82,84] however, CARNA 5 has been observed to increase the severity of symptoms in tomato, producing a lethal necrotic disease.[84,85] Some satellite RNA isolates have also been shown to increase the severity of CMV-induced symptoms in other hosts.[73-75] There seems to be a wide variety of Cucumovirus satellite RNAs with differing biological properties and primary structure. The complete nucleotide sequences of two RNA isolates have been determined, those of CARNA 5[80] and an Australian satellite RNA[81] with 335 and 336 residues, respectively. These satellite RNAs exert quite distinct effects on symptom development induced by CMV,[82,84,85] though they have 88% sequence homology.[81]

III. CYTOPATHOLOGY

Although all three Cucumoviruses infect a wide range of plants inducing prominent disease symptoms, most cytopathological studies have been done on plants infected by CMV. Early investigators studying CMV-infected plant cells by electron microscopy were frustrated by the difficulty of distinguishing the virus particles from cytoplasmic ribosomes.[86] However, a number of authors observed the presence of virus-like particles in the cytoplasm and/or the vacuoles of infected cells.[87-89] There are also some reports, although fewer, of the occurrence of similar particles in the nuclei[90-92] and chloroplasts.[93] The particles were identified as CMV particles by their doughnut appearance and irregular or crystalline aggregations not seen in healthy plants. The aggregated particles found in CMV-infected cells and identified as virus particles were indistinguishable from those in thin sections of pellets of purified CMV.[94] Virus-like particle aggregates have also been observed in the cytoplasm of cells infected with TAV and PSV.[95-97] Lawson and Hearon[95] observed virus-like particles in the plasmodesmata of TAV-infected tissues.

In trying to distinguish CMV particles from ribosomes in the cytoplasm of infected cells, Honda and Matsui[91] floated discs of leaf tissue on phosphate buffer, after which the ribosomes but not the virus particles were found to have lost their electron density. Although useful for detecting virus particles in cells, this method subjects tissues to abnormal conditions and offers no guarantee that other artifacts have not been introduced. More recently, Hatta and Francki[98] reported that CMV and TAV can be distinguished from ribosomes by their greater resistance to digestion with pancreatic RNase in aldehyde-fixed tissues under appropriate conditions. Using this technique it was shown that numerous CMV and TAV particles are scattered throughout the cytoplasm of infected cells (Figure 3).

Studies of *Nicotiana tabacum* and *N. clevelandii* leaves infected with six strains of CMV and three of TAV, using the RNase method,[98,99] showed that the distribution of virus particles in infected cells was quite variable.[100] Cells of all plants infected by any of the virus strains contained virus particles scattered throughout the cytoplasm (Figure 3), although with all strains their concentrations were higher in *N. clevelandii* than *N. tabacum*. The large amount of virus distributed in this way may account for previous reports that in infected cells the cytoplasm contains higher concentrations of ribosomes than in cells of healthy plants.[87] Crystalline aggregates of virus particles were only detected frequently in the cytoplasm of *N. clevelandii* cells infected with the T strain of CMV (Figure 4A), and less often with the C strain of TAV. Vacuoles of cells infected with either the T strain of CMV or the V strain of TAV also contained crystals of virus particles (Figure 5A); ordered arrays of virus particles in cells infected by several other TAV strains could be induced by plasmolysis (Figure 5B).

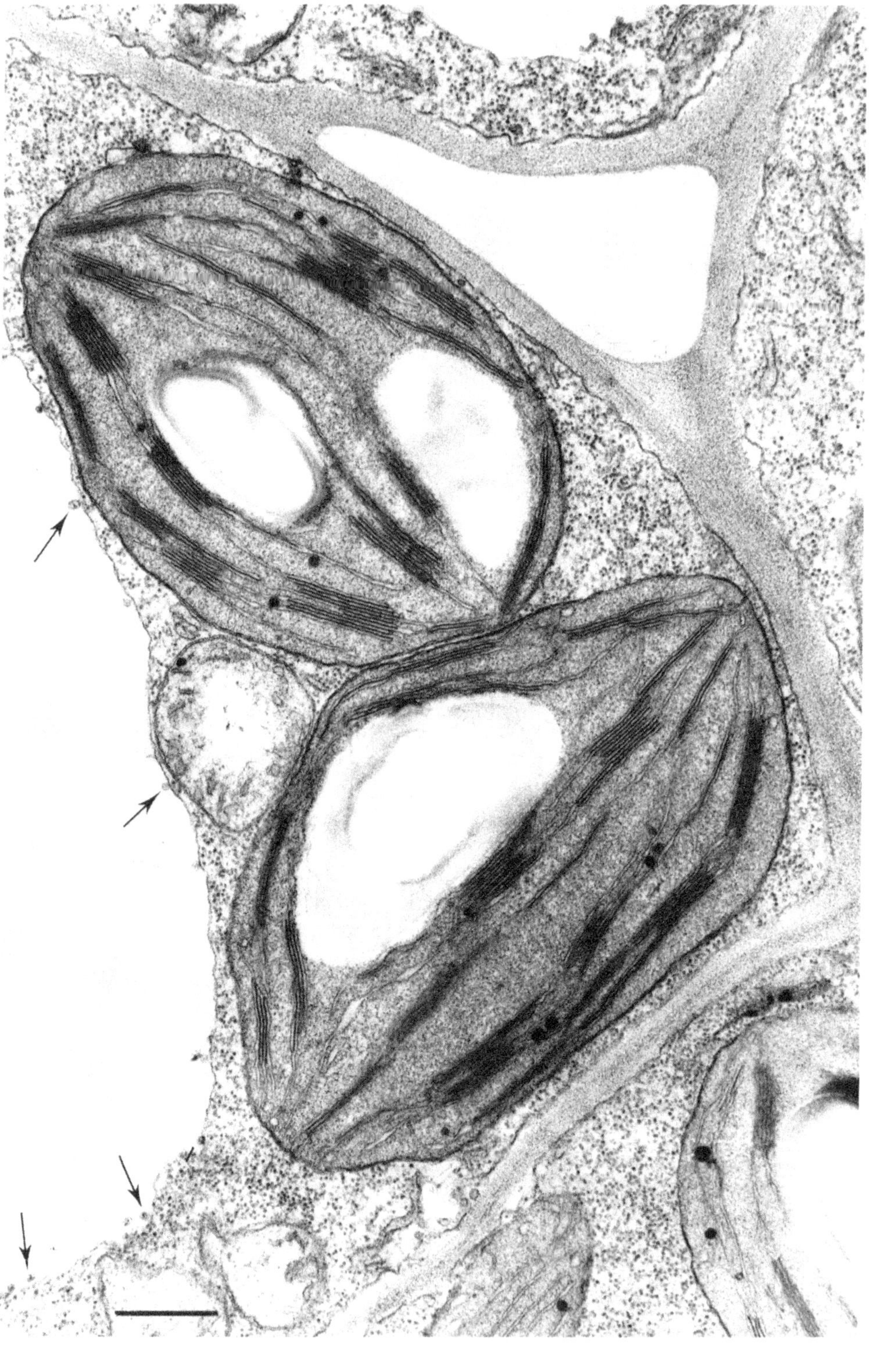

FIGURE 3. Thin section of tobacco (*Nicotiana tabacum*) leaf cells infected with tomato aspermy virus (TAV) showing numerous virus particles throughout the cytoplasm and virus-specific vesicles attached to the tonoplast (arrows). (Tissue cleared of ribosomes by RNase treatment.) Bar = 500 nm.

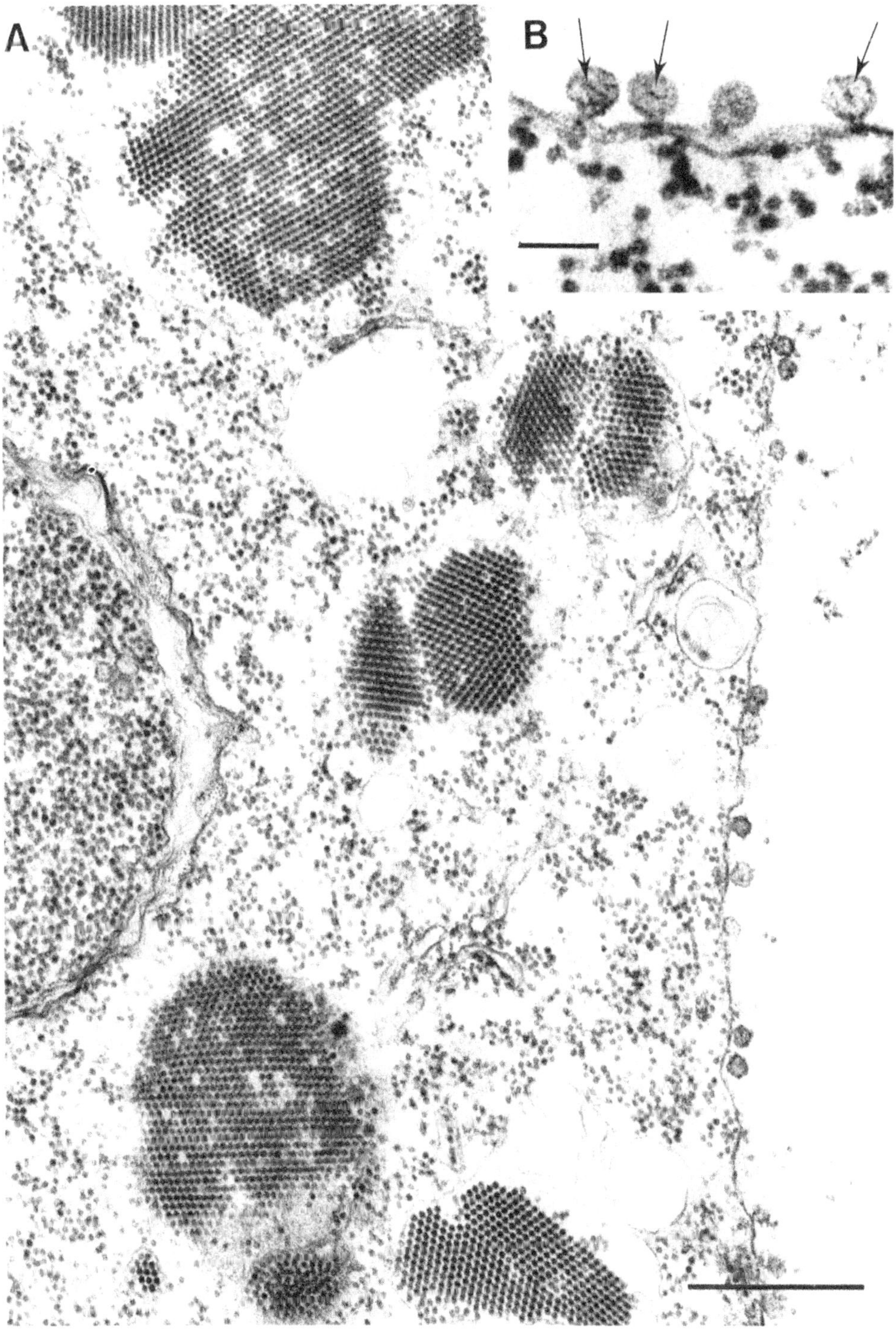

FIGURE 4. Thin section of a *Nicotiana clevelandii* leaf cell infected with the T strain of CMV. In A, note the virus-specific vesicles attached to the tonoplast and the crystalline aggregates of virus particles in the cytoplasm. Such crystals are unusual with most Cucumovirus strains. Bar = 500 nm. B, Note the characteristic vesicles associated with the tonoplast; densely staining fibrillar material reminiscent of nucleic acid (arrows) can be seen in some of them. Bar = 100 nm.

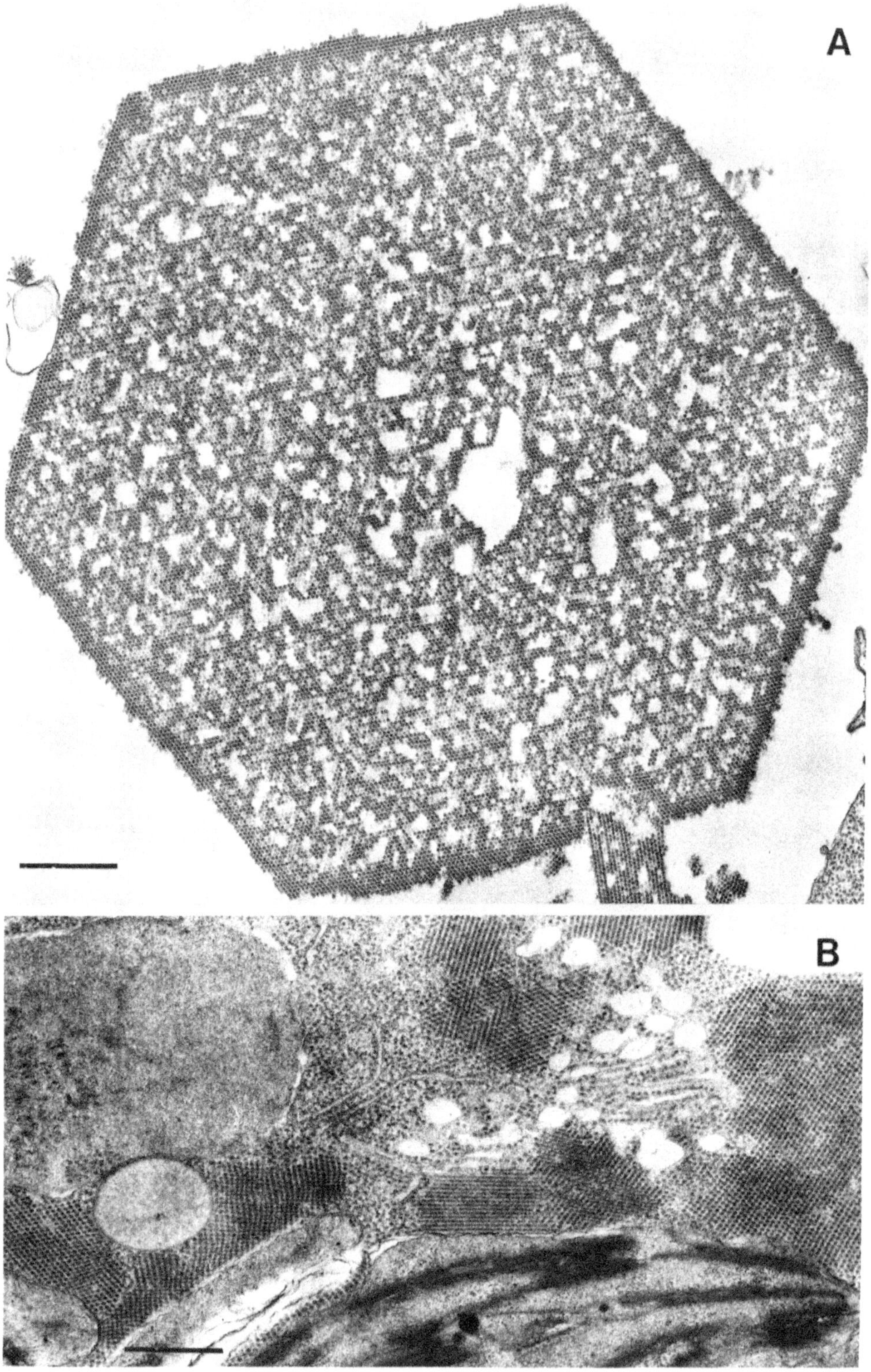

FIGURE 5. A, Material as in Figure 4, showing a large crystal of virus particles in the cell vacuole; B, thin section of a *N. clevelandii* leaf cell infected with the N strain of TAV, showing virus particles which have crystallized after plasmolysis of the leaf prior to fixation. Bars = 500 nm.

Virus-like particles in nuclei were detected with confidence in *N. clevelandii* but not in *N. tabacum* infected with the T strain of CMV (Figure 6). The particle distribution in *N. clevelandii* cells infected with the C strain of TAV were unusual in that some of the particles formed long, rod-like aggregates (Figure 7A). In *N. glutinosa* infected with the N strain of TAV, tubules containing virus particles were seen passing through the plasmodesmata (Figure 7B), rather like those associated with Comoviruses and Nepoviruses (see Chapters 1 and 2).

A number of cytopathic effects, other than the accumulation of virus particles, have been observed in plant cells infected by Cucumoviruses. These include various abnormalities in the chloroplasts and golgi bodies, and the development of myelin-like or multivesicular structures and membrane-limited aggregates of tubules.[87,89,91,93-96] Ehara and Misawa[93] reported detailed studies on chloroplast structure in CMV-infected tobacco leaf cells. They observed that the abnormalities were dependent on the developmental stage of the chloroplasts at the time of infection. The most deranged were those which had been at the proplastidial stage at the time of inoculation. They had poorly developed lamellae and few grana, and some contained tubular networks of myelin-like structures. Both normal and abnormal chloroplasts were often seen in the same leaf cell. It was concluded that the structural changes seen were not the direct effects of virus multiplication.

Studies by Hatta and Francki[100,101] of plant cells infected by six strains of CMV and three of TAV indicate that the only consistent structural change in infected cells is the development of characteristic small vesicles associated with the tonoplast (Figures 3 and 4). Enzyme cytochemical evidence indicated that the vesicles contain double-stranded RNA; they may therefore be the sites of viral RNA replication.[101] Although a number of other structural changes were observed in infected cells, they were not consistent in cells infected by the various virus strains and hence cannot be considered to be characteristic of Cucumovirus infection. Darkly staining cytoplasmic inclusions (Figure 7D) reminiscent of the viroplasms of Caulimoviruses (see Volume I, Chapter 2) were seen in some cells infected by the three strains of TAV and the U strain of CMV. Outgrowths of the cell walls of plant cells infected with the Q strain of CMV and the N strain of TAV (Figure 7C) were also seen; they appear similar to those reported in cells infected by viruses belonging to a number of groups.

Cells in chlorotic tissues of young leaves infected by the majority of the strains which induced clear mosaic symptoms did not exhibit any consistent and severe chloroplast abnormalities (Figures 8A through 8C). However, chloroplasts of cells infected with the M strain of CMV, which induces a bright yellow mosaic,[102] were smaller than normal with poorly developed grana and small starch granules (Figure 8D).

Little is known about the events leading to Cucumovirus morphogenesis. From the limited data available, it seems that viral RNA may replicate in the virus-induced tonoplast vesicles[101] and that virus particles are assembled and accumulate in the cytoplasm. In some cases mature particles are also found in vacuoles and nuclei. Biochemical studies have revealed the presence of double-stranded RNAs corresponding to the four viral RNAs, and also to satellite RNA, in CMV-infected protoplasts[103] and leaf tissue.[104] Virus infection-specific RNA replicase has also been detected in extracts from infected plants, both in the particulate and soluble fractions.[105,106] However, highly purified preparations of the enzyme did not appear to contain full-length translation products of any of the CMV genomic RNAs, and the cellular location of the enzyme is not known.[107]

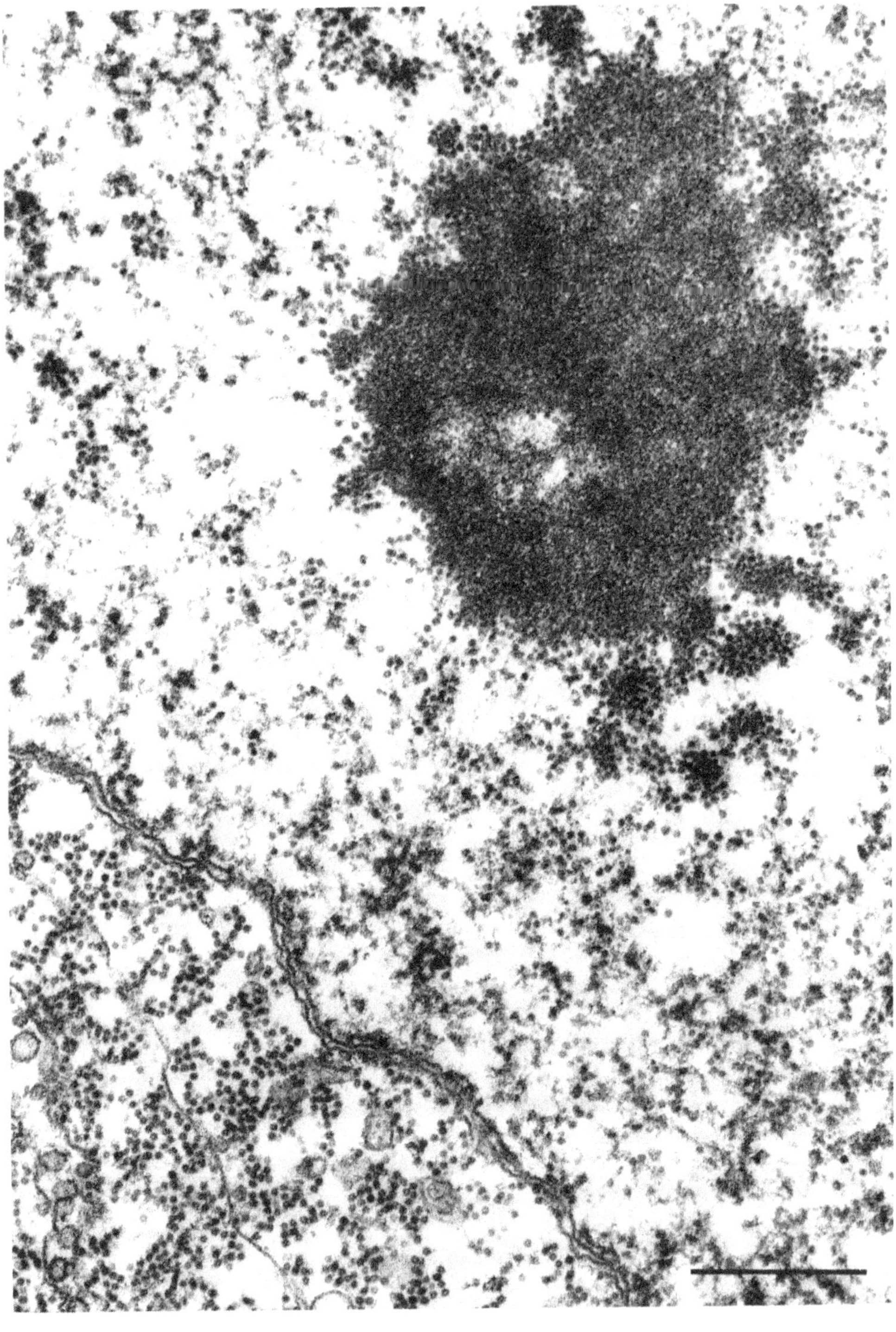

FIGURE 6. Thin section of a *N. clevelandii* leaf cell infected with the T strain of CMV showing virus particles in both nucleus and cytoplasm. (Tissue cleared of ribosomes by RNase treatment.) Bar = 500 nm.

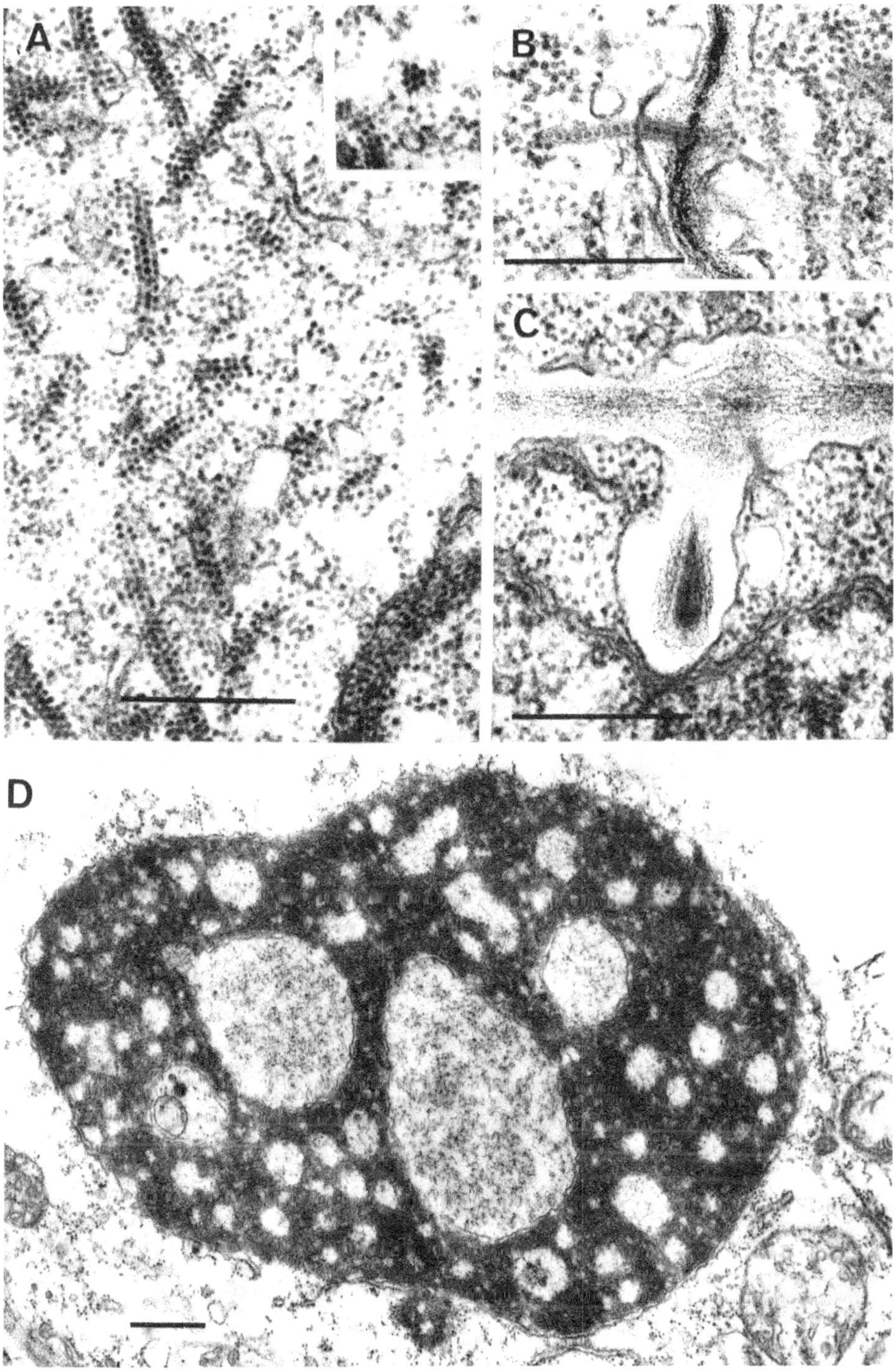

FIGURE 7. A. Thin section of a *N. clevelandii* cell infected with the C strain of TAV, showing chain-like aggregates of virus particles in the cytoplasm; inset shows such an aggregate in transverse section, consisting of seven closely-packed particles. (Tissue cleared of ribosomes by RNase treatment.) B, Thin section of a *N. glutinosa* cell infected with the N strain of TAV showing a file of virus particles passing through the plasmodesma. C, Material as in B, showing a cell wall outgrowth associated with a plasmodesma. D, Thin section of a *N. clevelandii* cell infected with the V strain of TAV, showing a viroplasm-like cytoplasmic inclusion. (Tissue cleared of ribosomes by RNase treatment.) Bars = 500 nm.

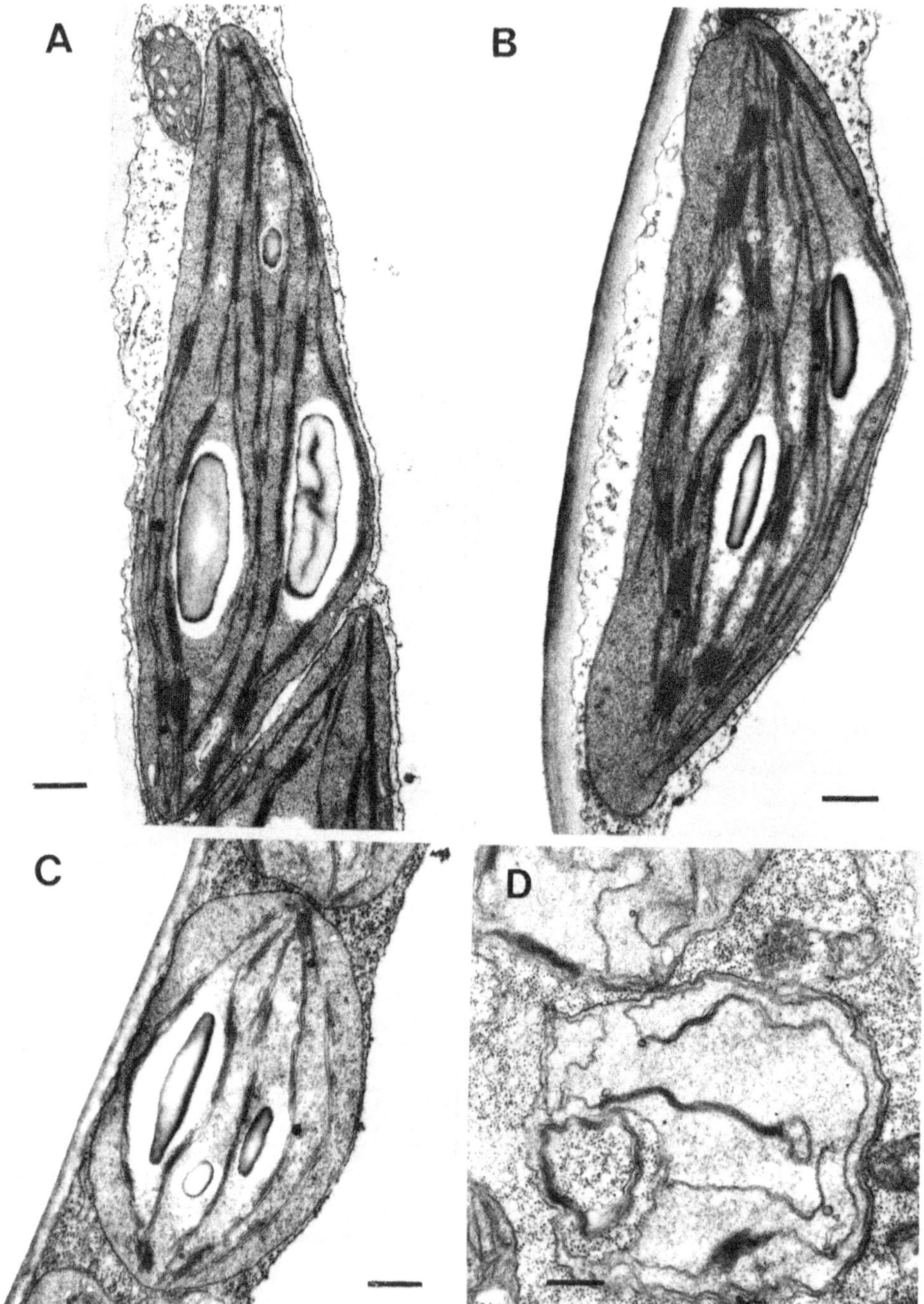

FIGURE 8. Thin sections of *N. clevelandii* chloroplasts illustrating some effects of Cucumovirus infection. A, Healthy; B, infected with the U strain of CMV; C, infected with the T strain of CMV; D, infected with the M strain of CMV. Bars = 500 nm.

REFERENCES

1. **Matthews, R. E. F.,** Classification and nomenclature of viruses. Fourth report of the International Committee on Taxonomy of Viruses, *Intervirology*, 17, 1, 1982.
2. **Kaper, J. M. and Waterworth, H. E.,** Cucumoviruses, in *Handbook of Plant Virus Infections and Comparative Diagnosis*, Kurstak, E. Ed., Elsevier, Amsterdam, 1981, 257.
3. **Douine, L., Quiot, J. B., Marchoux, G., and Archange, P.,** Recensement des espèces végétales sensibles au virus de la mosaique du concombre (CMV). Etude bibliographique, *Ann. Phytopathol.*, 11, 439, 1979.
4. **Francki, R. I. B., Mossop, D. W., and Hatta, T.,** Cucumber mosaic virus, *CMI/AAB Descriptions of Plant Viruses*, No. 213, 1979.
5. **Hollings, M. and Stone, O. M.,** Tomato aspermy virus, *CMI/AAB Descriptions of Plant Viruses*, No. 79, 1971.
6. **Mink, G. I.,** Peanut stunt virus, *CMI/AAB Descriptions of Plant Viruses*, No. 92, 1972.
7. **Schmelzer, K.,** Robinia mosaic virus, *CMI/AAB Descriptions of Plant Viruses*, No. 65, 1971.
8. **Musil, M., Marcinka, K., and Mucha, V.,** Some properties of clover blotch virus, *Acta. Virol.*, 19, 437, 1975.
9. **Richter, J., Proll, E., and Musil, M.,** Serological relationships between robinia mosaic, clover blotch and peanut stunt viruses, *Acta Virol.*, 23, 489, 1979.
10. **Phatak, H. C., Diaz-Ruiz, J. R., and Hull, R.,** Cowpea ringspot virus: a seed transmitted cucumovirus, *Phytopathol. Z.*, 87, 132, 1976.
11. **van Slogteren, D. H. M.,** Multiple antigenicity in tobacco plants infected with viruses, notably strains of cucumber mosaic virus, *Proc. 2nd Int. Sci. Tobacco Cong.*, 144, 1958.
12. **Grogan, R. G., Uyemoto, J. K., and Kimble, K. A.,** Evidence that tomato aspermy and cucumber mosaic viruses are serologically unrelated, *Virology*, 21, 36, 1963.
13. **Lawson, R. H.,** Relationships among tomato aspermy, aspermy-related viruses from Chrysanthemum, and two strains of cucumber mosaic virus, *Virology*, 32, 357, 1967.
14. **Mink, G. I.,** Serological relationships among cucumber mosaic virus, tomato aspermy type viruses, and peanut stunt virus, *Phytopathology*, 59, 1889, 1969.
15. **Habili, N. and Francki, R. I. B.,** Comparative studies on tomato aspermy and cucumber mosaic viruses. I. Physical and chemical properties, *Virology*, 57, 392, 1974.
16. **Habili, N. and Francki, R. I. B.,** Comparative studies on tomato aspermy and cucumber mosaic viruses. IV. Immunogenic and serological properties, *Virology*, 64, 421, 1975.
17. **Francki, R. I. B. and Habili, N.,** Stabilization of capsid structure and enhancement of immunogenicity of cucumber mosaic virus (Q strain) by formaldehyde, *Virology*, 48, 309, 1972.
18. **Devergne, J. C. and Cardin, L.,** Relations sérologiques entre cucumovirus (CMV, TAV, PSV), *Ann. Phytopathol.*, 7, 255, 1975.
19. **Devergne, J. C., Cardin, L., Burckard, J., and van Regenmortel, M. H. V.,** Comparison of direct and indirect ELISA for detecting antigenically related cucumoviruses, *J. Virol. Meth.*, 3, 193, 1981.
20. **Rao, A. L. N., Hatta, T., and Francki, R. I. B.,** Comparative studies on tomato aspermy and cucumber mosaic viruses. VII. Serological relationships reinvestigated, *Virology*, 116, 318, 1982.
21. **Devergne, J. C. and Cardin, L.,** Contribution a l'étude du virus de la mosaique du concombre (CMV), *Ann. Phytopathol.*, 5, 409, 1973.
22. **Gonda, T. J. and Symons, R. H.,** The use of hybridization analysis with complementary DNA to determine the RNA sequence homology between strains of plant viruses: its application to several strains of cucumoviruses, *Virology*, 88, 361, 1978.
23. **Gould, A. R. and Symons, R. H.,** Determination of the sequence homology between the four RNA species of cucumber mosaic virus by hybridization analysis with complementary DNA, *Nucl. Acids Res.*, 4, 3787, 1977.
24. **Piazzolla, P., Diaz-Ruiz, J. R., and Kaper, J. M.,** Nucleic acid homologies of eighteen cucumber mosaic virus isolates determined by competition hybridization, *J. Gen. Virol.*, 45, 361, 1979.
25. **Tolin, S. A.,** Cucumovirus (cucumber mosaic virus) group, in *The Atlas of Insect and Plant Viruses*, Maramorosch, K., Ed., Academic Press, New York, 1977, 303.
26. **Finch, J. T., Klug, A., and van Regenmortel, M. H. V.,** The structure of cucumber mosaic virus, *J. Mol. Biol.*, 24, 303, 1967.
27. **Francki, R. I. B., Randles, J. W., Chambers, T. C., and Wilson, S. B.,** Some properties of purified cucumber mosaic virus (Q strain), *Virology*, 28, 729, 1966.
28. **van Regenmortel, M. H. V., Hendry, D. A., and Baltz, T.,** A reexamination of the molecular size of cucumber mosaic virus and its coat protein, *Virology*, 49, 647, 1972.
29. **Hill, J. H. and Shepherd, R. J.,** Molecular weights of plant virus coat proteins by polyacrylamide gel electrophoresis, *Virology*, 47, 817, 1972.
30. **Stace-Smith, R. and Tremaine, J. H.,** Biophysical and biochemical properties of tomato aspermy virus, *Virology*, 51, 401, 1973.

31. **Linnasalmi, A. and Toiviainen, A.,** The chemical composition of Finnish cucumber mosaic virus (CMV) and cucumber green mottle mosaic virus (CGMMV), *Ann. Agr. Fenn.*, 13, 79, 1974.
32. **Gould, A. R. and Symons, R. H.,** Cucumber mosaic virus RNA 3. Determination of the nucleotide sequence provides the amino acid sequences of protein 3A and viral coat protein. *Eur. J. Biochem.*, 126, 217, 1982.
33. **Jacrot, B., Chauvin, C., and Witz, J.,** Comparative neutron small-angle scattering study of small spherical RNA viruses, *Nature (London)*, 266, 417, 1977.
34. **Francki, R. I. B.,** Inactivation of cucumber mosaic virus (Q strain) nucleoprotein by pancreatic ribonuclease, *Virology*, 34, 694, 1968.
35. **Kaper, J. M. and Geelen, J. L. M. C.,** Studies on the stabilizing forces of simple RNA viruses. II. Stability, dissociation, and reassembly of cucumber mosaic virus, *J. Mol. Biol.*, 56, 277, 1971.
36. **Boatman, S. and Kaper, J. M.,** Molecular organization and stabilizing forces of simple RNA viruses. IV. Selective interference with protein-RNA interactions by use of sodium dodecyl sulfate, *Virology*, 70, 1, 1976.
37. **Habili, N. and Francki, R. I. B.,** unpublished results, 1973.
38. **Habili, N. and Francki, R. I. B.,** Comparative studies on tomato aspermy and cucumber mosaic viruses. II. Virus stability, *Virology*, 60, 29, 1974.
39. **Mink, G. I.,** Inactivation of peanut stunt virus by pancreatic ribonuclease, *Virology*, 63, 466, 1975.
40. **van Regenmortel, M. H. V.,** Biochemical and biophysical properties of cucumber mosaic virus, *Virology*, 31, 391, 1967.
41. **Dupont, G., Hern, P., Yot-Douthy, D., and Bové, J. M.,** Détermination de la masse moléculaire du virus de la mosaique du concombre, *C. R. Acad. Sci. Paris D*, 267, 1013, 1968.
42. **Diener, T. O., Scott, H. A., and Kaper, J. M.,** Highly infectious nucleic acid from crude and purified preparations of cucumber mosaic virus (Y strain), *Virology*, 22, 131, 1964.
43. **Kaper, J. M., Diener, T. O., and Scott, H. A.,** Some physical and chemical properties of cucumber mosaic virus (strain Y) and of its isolated ribonucleic acid, *Virology*, 27, 54, 1965.
44. **Peden, K. W. C. and Symons, R. H.,** Cucumber mosaic virus contains a functionally divided genome, *Virology*, 53, 487, 1973.
45. **Lot, H., Marchoux, G., Marrou, J., Kaper, J. M., West, C. K., van Vloten-Doting, L., and Hull, R.,** Evidence for three functional RNA species in several strains of cucumber mosaic virus, *J. Gen. Virol.*, 22, 81, 1974.
46. **Schwinghamer, M. W. and Symons, R. H.,** Fractionation of cucumber mosaic virus RNA and its translation in a wheat embryo cell-free system, *Virology*, 63, 252, 1975.
47. **Schwinghamer, M. W. and Symons, R. H.,** Translation of the four major RNA species of cucumber mosaic virus in plant and animal cell-free systems and in toad oocytes, *Virology*, 79, 88, 1977.
48. **Kohl, R. J. and Hall, T. C.,** Aminoacylation of RNA from several viruses: amino acid specificity and differential activity of plant, yeast and bacterial synthetases, *J. Gen. Virol.*, 25, 257, 1974.
49. **Symons, R. H.,** Extensive sequence homology at the 3′-termini of the four RNAs of cucumber mosaic virus, *Nucleic Acids Res.*, 7, 825, 1979.
50. **Wilson, P. A. and Symons, R. H.,** The RNAs of cucumoviruses: 3′ terminal sequence analysis of two strains of tomato aspermy virus, *Virology*, 112, 342, 1981.
51. **Alquist, P., Dasgupta, R., and Kaesberg, P.,** Near identity of 3′ RNA secondary structure of bromoviruses and cucumber mosaic virus, *Cell*, 23, 183, 1981.
52. **Kaper, J. M. and West, C. K.,** Polyacrylamide gel separation and molecular weight determination of the components of cucumber mosaic virus RNA, *Prep. Biochem.*, 2, 251, 1972.
53. **Marchoux, G., Douine, L., Lot, H., and Esvan, C.,** Identification et estimation du poids moléculaire des acides ribonucléiques du virus de la mosaique du concombre (VMC Souche D), *C. R. Acad. Sci. Paris D*, 277, 1409, 1973.
54. **Kaper, J. M. and Re, G. G.,** Redetermination of the RNA content and the limiting RNA size of three strains of cucumber mosaic virus, *Virology*, 60, 308, 1974.
55. **Fiers, W., Contreras, R., Duerinck, F., Haegeman, G., Iserentant, D., Merregaert, J., Min Jou, W., Molemans, F., Raeymaekers, A., van den Berghe, A., Volckaert, G., and Ysebaert, M.,** Complete nucleotide sequence of bacteriophage MS2 RNA: primary and secondary structure of the replicase gene, *Nature (London)*, 260, 500, 1976.
56. **Brosius, J., Palmer, M. L., Kennedy, P. J., and Noller, H. F.,** Complete nucleotide sequence of a 16S ribosomal RNA gene from *Escherichia coli*, *Proc. Natl. Acad. Sci. U.S.A.*, 75, 4801, 1978.
57. **Brosius, J., Dull, T. J., and Noller, H. F.,** Complete nucleotide sequence of a 23S ribosomal RNA gene from *Escherichia coli*, *Proc. Natl. Acad. Sci. U.S.A.*, 77, 201, 1980.
58. **Lot, H. and Kaper, J. M.,** Physical and chemical differentiation of three strains of cucumber mosaic virus and peanut stunt virus, *Virology*, 74, 209, 1967.
59. **Lot, H. and Kaper, J. M.,** Further studies on the RNA component distribution among the nucleoproteins of cucumber mosaic virus, *Virology*, 74, 223, 1976.

60. **Rao, A. L. N. and Francki, R. I. B.,** Comparative studies on tomato aspermy and cucumber mosaic viruses. VI. Partial compatibility of genome segments from the two viruses, *Virology*, 114, 573, 1981.
61. **Habili, N. and Francki, R. I. B.,** Comparative studies on tomato aspermy and cucumber mosaic viruses. III. Further studies on relationship and construction of a virus from parts of the two viral genomes, *Virology*, 61, 443, 1974.
62. **Marchoux, G., Maron, J., and Quiot, J. B.,** Complémentation entre ARN de différentes souches du virus de la mosaïque du concombre. Mise en évidence d'une interaction entre deux ARN pour determiner un type de symptôme, *C. R. Acad. Sci. Ser. D*, 278, 1943, 1974.
63. **Mossop, D. W. and Francki, R. I. B.,** Association of RNA 3 with aphid transmission of cucumber mosaic virus, *Virology*, 81, 177, 1977.
64. **Hanada, K. and Tochihara, H.,** Genetic analysis of cucumber mosaic, peanut stunt and chrysanthenum mild mottle viruses, *Ann. Phytopathol. Soc. Jpn.*, 46, 159, 1980.
65. **Rao, A. L. N. and Francki, R. I. B.,** Distribution of determinants for symptom production and host range of the three RNA components of cucumber mosaic virus, *J. Gen. Virol.*, 61, 197, 1982.
66. **Edwards, M. C., Gonsalves, D., and Provvidenti, R.,** Genetic analysis of cucumber mosaic virus in relation to host resistance: location of determinants for pathogenicity to certain legumes and *Lactuca saligna*, *Phytopathology*, 73, 269, 1983.
67. **Schwinghamer, M. W. and Symons, R. H.,** Fractionation of cucumber mosaic virus RNA and its translation in a wheat embryo cell-free system, *Virology*, 63, 252, 1975.
68. **Gonda, T. J. and Symons, R. H.,** Cucumber mosaic virus replication in cowpea protoplasts: time course of virus, coat protein and RNA synthesis, *J. Gen. Virol.*, 45, 723, 1979.
69. **Roberts, P. L. and Wood, K. R.,** Protein synthesis in cucumber mosaic virus-infected tobacco leaves, pre-infected cells and protoplasts, *Arch. Virol.*, 70, 115, 1981.
70. **Roberts, P. L. and Wood, K. R.,** Comparison of protein synthesis in tobacco systemically infected with a severe or a mild strain of cucumber mosaic virus, *Phytopathol. Z.*, 102, 257, 1981.
71. **Kaper, J. M., Tousignant, M. E., and Lot, H.,** A low molecular weight replicating RNA associated with a divided genome plant virus: defective or satellite RNA?, *Biochem. Biophys. Res. Commun.*, 72, 1237, 1976.
72. **Kaper, J. M., Tousignant, M. E., Diaz-Ruiz, J. R., and Tolin, S. A.,** Peanut stunt virus-associated RNA 5: second tripartite genome virus with an associated satellite-like replicating RNA, *Virology*, 88, 166, 1978.
73. **Takanami, Y.,** A striking change in symptoms on cucumber mosaic virus-infected tobacco plants induced by a satellite RNA, *Virology*, 109, 120, 1981.
74. **Kaper, J. M., Tousignant, M. E., and Thompson, S. M.,** Cucumber mosaic virus-associated RNA 5. VIII. Identification and partial characterization of a CARNA 5 incapable of inducing tomato necrosis, *Virology*, 114, 526, 1981.
75. **Gonsalves, D., Provvidenti, R., and Edwards, M. C.,** Tomato white leaf: the relation of an apparent satellite RNA and cucumber mosaic virus, *Phytopathology*, 72, 1533, 1982.
76. **Lot, H., Jonard, G., and Richards, K.,** Cucumber mosaic virus RNA 5. Partial characterization and evidence for no large sequence homologies with genomic RNAs, *FEBS Lett.*, 80, 395, 1977.
77. **Diaz-Ruiz, J. R. and Kaper, J. M.,** Cucumber mosaic virus-associated RNA 5. III. Little nucleotide sequence homology between CARNA 5 and helper RNA, *Virology*, 80, 204, 1977.
78. **Gould, A. R., Palukaitis, P., Symons, R. H., and Mossop, D. W.,** Characterization of a satellite RNA associated with cucumber mosaic virus, *Virology*, 84, 443, 1978.
79. **Mossop, D. W. and Francki, R. I. B.,** Survival of a satellite RNA *in vivo* and its dependence on cucumber mosaic virus for replication, *Virology*, 86, 562, 1978.
80. **Richards, K. E., Jonard, G., Jacquemond, M., and Lot, H.,** Nucleotide sequence of cucumber mosaic virus-associated RNA 5, *Virology*, 89, 395, 1978.
81. **Gordon, K. H. J. and Symons, R. H.,** Satellite RNA of cucumber mosaic virus forms a secondary structure with partial 3′ terminal homology to genomal RNAs, *Nucleic Acids Res.*, 11, 947, 1983.
82. **Mossop, D. W. and Francki, R. I. B.,** Comparative studies on two satellite RNAs of cucumber mosaic virus, *Virology*, 95, 395, 1979.
83. **Mossop, D. and Francki, R. I. B.,** The stability of satellite viral RNAs *in vivo* and *in vitro*, *Virology*, 94, 243, 1979.
84. **Waterworth, H. E., Kaper, J. M., and Tousignant, M. E.,** CARNA 5, the small cucumber mosaic virus-dependent replicating RNA, regulates disease expression, *Science*, 204, 845, 1979.
85. **Kaper, J. M. and Waterworth, H. E.,** Cucumber mosaic virus associated RNA 5: causal agent for tomato necrosis, *Science*, 196, 429, 1977.
86. **Matsui, C. and Yamaguchi, A.,** Some aspects of plant viruses *in situ*, *Adv. Virus Res.*, 12, 127, 1966.
87. **Gerola, F. M., Bassi, M., and Belli, G.,** Some observations on the shape and localization of different viruses in experimentally infected plants, and on the fine structure of host cells. II. *Nicotiana glutinosa* systemically infected with cucumber mosaic virus, *Caryologia*, 18, 567, 1965.

88. **Misawa, T. and Ehara, Y.,** Electron microscopic observation of host cells infected with cucumber mosaic virus. *Tohoku J. Agric. Res.*, 16, 159, 1965.
89. **Honda, Y. and Matsui, C.,** Electron microscopy of intracellular modifications of tobacco by mixed infections with cucumber mosaic and tobacco mosaic viruses, *Phytopathology*, 58, 1230, 1968.
90. **Ehara, Y. and Misawa, T.,** Studies on the infection of cucumber mosaic virus. IX. Crystalline virus aggregates in infected cells. *Tohoku J. Agric. Res.*, 25, 43, 1974.
91. **Honda, Y. and Matsui, C.,** Electron microscopy of cucumber mosaic virus-infected tobacco leaves showing mosaic symptoms. *Phytopathology*, 64, 534, 1974.
92. **Honda, Y., Matsui, C., Otsuki, Y., and Takebe, I.,** Ultrastructure of tobacco mesophyll protoplasts inoculated with cucumber mosaic virus. *Phytopathology*, 64, 30, 1974.
93. **Ehara, Y. and Misawa, T.,** Occurrence of abnormal chloroplasts in tobacco leaves infected systemically with the ordinary strain of cucumber mosaic virus. *Phytopathol. Z.*, 84, 233, 1975.
94. **Ehara, Y., Misawa, T., and Nagayama, H.,** Arrangement of cucumber mosaic virus particles in the virus aggregates. *Phytopathol. Z.*, 87, 28, 1976.
95. **Lawson, R. H. and Hearon, S.,** Subcellular localization of chrysanthemum aspermy virus in tobacco and chrysanthemum leaf tissue. *Virology*, 41, 30, 1970.
96. **Ehara, Y.,** Complex membranous organelles in the cells of the local lesion area on cowpea leaves infected with cucumber mosaic virus, *Ann. Phytopathol. Soc. Jpn.*, 45, 17, 1979.
97. **Kraev, V. G., Semernikova, L. I., Vutenko, S. I., and Porembskaya, N. V.,** Certain structural changes in cells of host plants induced by peanut stunt virus. *Microbiol. Zhurm. Kiev*, 37, 761, 1975.
98. **Hatta, T. and Francki, R. I. B.,** Enzyme cytochemical method for identification of cucumber mosaic virus particles in infected cells. *Virology*, 93, 265, 1979.
99. **Hatta, T. and Francki, R. I. B.,** Identification of small polyhedral virus particles in thin sections of plant cells by an enzyme cytochemical technique, *J. Ultrastruct. Res.*, 74, 116, 1981.
100. **Hatta, T. and Francki, R. I. B.,** unpublished data, 1980.
101. **Hatta, T. and Francki, R. I. B.,** Cytopathic structures associated with tonoplasts of plant cells infected with cucumber mosaic and tomato aspermy viruses. *J. Gen. Virol.*, 53, 343, 1981.
102. **Mossop, D. W., Francki, R. I. B., and Grivell, C. J.,** Comparative studies on tomato aspermy and cucumber mosaic viruses. V. Purification and properties of a cucumber mosaic virus inducing severe chlorosis, *Virology*, 74, 544, 1976.
103. **Takanami, Y., Kubo, S., and Imaizumi, S.,** Synthesis of single- and double-stranded cucumber mosaic virus RNAs in tobacco mesophyll protoplasts. *Virology*, 80, 376, 1977.
104. **Kaper, J. M. and Diaz-Ruiz, J. R.,** Molecular weights of the double-stranded RNAs of cucumber mosaic virus strain S and its associated RNA 5, *Virology*, 80, 214, 1977.
105. **Kumarasamy, R. and Symons, R. H.,** Extensive purification of the cucumber mosaic virus-induced RNA replicase, *Virology*, 96, 622, 1979.
106. **Gill, D. S., Kumarasamy, R., and Symons, R. H.,** Cucumber mosaic virus-induced RNA replicase: solubilization and partial purification of the particulate enzyme. *Virology*, 113, 1, 1981.
107. **Gordon, K. H. J., Gill, D. S., and Symons, R. H.,** Highly purified cucumber mosaic virus-induced RNA-dependent RNA polymerase does not contain any of the full length translation products of the genomic RNAs, *Virology*, 123, 284, 1982.

Chapter 6

BROMOVIRUS GROUP

The group derives its name from brome mosaic virus (BMV), the type member. The viruses are similar in many respects to the Cucumoviruses. Bromoviruses have polyhedral particles with T = 3 symmetry built from 180 identical polypeptides of M_r about 20 × 10^3. These particles encapsidate four species of single-stranded, positive-sense RNA with M_r from about 1.1 × 10^6 to 0.3 × 10^6. Packaging of the RNA is such that all particles sediment at similar rates.[1-3]

Bromoviruses are transmitted by mechanical inoculation, each to a limited number of host plants. They are also transmitted by beetles and there are reports that some may be transmitted by nematodes or aphids.[3,4] There is one report that broad bean mottle virus (BBMV) is seed transmitted.[5]

I. MEMBERS OF THE GROUP AND THEIR RELATIONSHIPS

Matthews[6] lists three members of the group, BMV, BBMV, and cowpea chlorotic mottle virus (CCMV). The recently described melandrium yellow fleck virus (MYFV) has also been considered to be a member of this group but has not been characterized extensively.[7] Matthews[6] has included it as a possible member.

BMV, BBMV, and CCMV are all very distantly related serologically, the coat proteins more closely than the intact virus particles.[8]

There are a number of striking similarities between members of the Bromovirus and Cucumovirus groups,[9] although the RNA genomes and coat protein subunits of the Bromoviruses are a little smaller than those of the Cucumoviruses[2,10] and the particles also have a slightly smaller diameter. The RNAs of viruses belonging to both groups have very similar tRNA-like 3′ termini which accept tyrosine and extensive 3′-terminal sequence and structure homologies.[11] On the other hand, they differ in some biological properties such as natural means of transmission. As far as is known there are no serological relationships between viruses of the two groups. Differences and similarities of viruses within the groups are such that logical arguments can be put forward either for retaining the two groups as separate or for amalgamating them into one.[9]

II. VIRUS STRUCTURE AND COMPOSITION

A. Particle Structure

In electron micrographs such as shown in Figure 1, the polyhedral Bromovirus particles are between 25 and 28 nm in diameter depending on the preparation and staining methods used.[12] When highly purified and concentrated Bromovirus particles are spread on cleaved mica in the presence of ammonium molybdate they can form two-dimensional crystalline arrays.[13] The center-to-center distances between BMV and CCMV particles measured from optical diffraction patterns of the arrays were found to be 26.0 nm. This is in good agreement with measurements from X-ray scattering data and conventional electron microscopy of all three Bromoviruses,[14-17] although White and Fischbach[18] reported a slightly larger diameter for BBMV.

The particles of Bromoviruses are sensitive to changes in pH. They exist in a stable, compact form at pH 3 to 6, but above pH 6.5, they swell[17-19] and are then sensitive to ribonucleases, trypsin, and chymotrypsin.[20-22]

Changes during swelling of BMV have been studied in some detail. Incardona and his

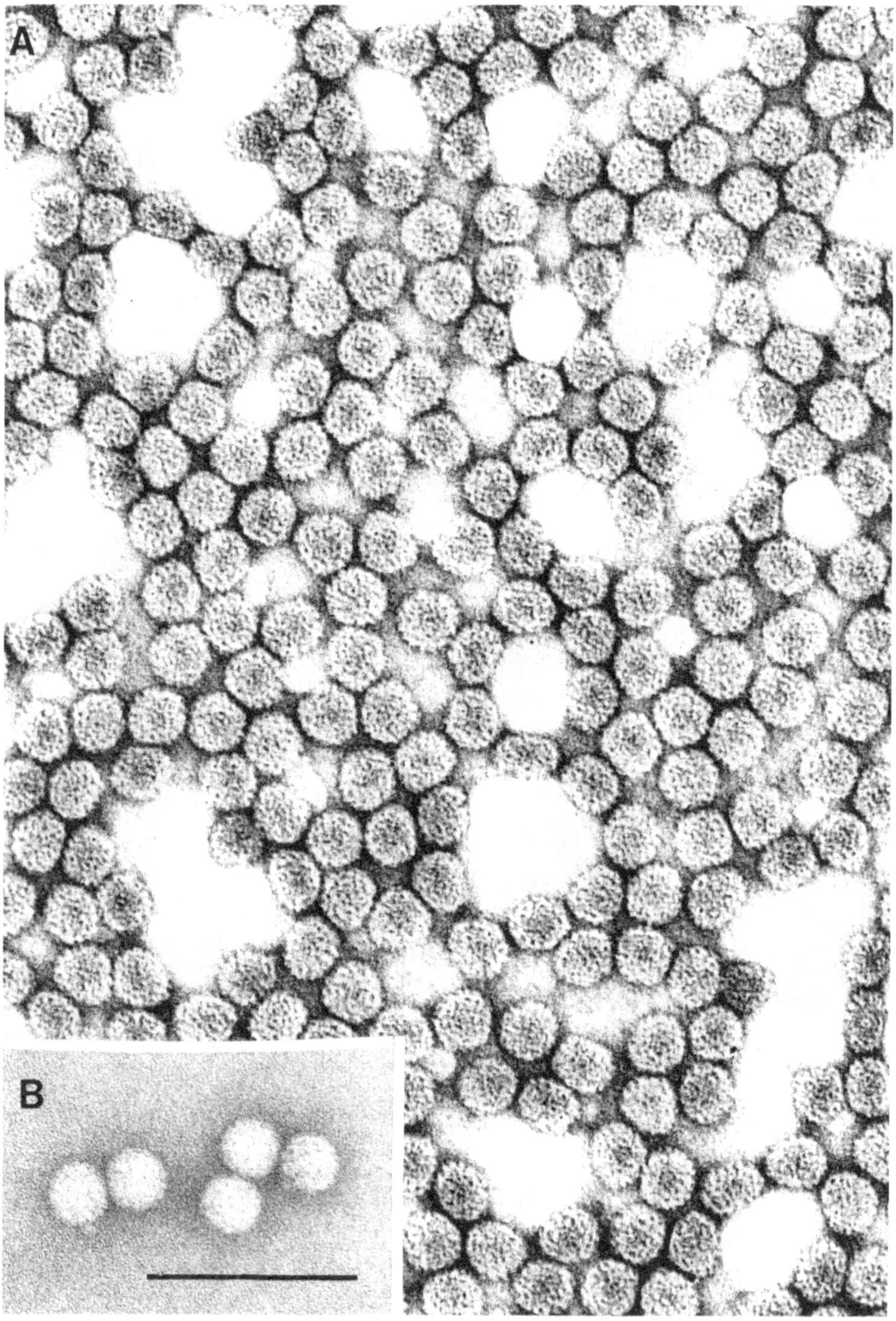

FIGURE 1. A, Purified particles of brome mosaic virus (BMV) and B, broad bean mottle virus (BBMV), both stained in uranyl acetate and photographed at the same magnification. Bar = 100 nm.

colleagues[23] demonstrated that the particles first underwent a pH-induced configuration change independent of temperature and that this was followed by temperature-dependent swelling which could be prevented by the addition of Mg^{2+}. Optically detectable changes in the encapsidated RNA indicate that Mg^{2+} prevents the thermal transition by intercalating with the RNA. Swollen particles which are stabilized with Mg^{2+} and other cations are resistant to ribonuclease indicating that the RNA becomes accessible to the enzyme only after the temperature-dependent swelling step.[24,25] It appears that BMV particles, and probably those of the other Bromoviruses, can exist in three different conformations; the "native" compact, the partially expanded Mg^{2+} stabilized, and the fully expanded forms.[24]

The protein shell of Bromoviruses is constructed from pentamer and hexamer clusters of 180 identical polypeptides of M_r about 20×10^3 to give 32 morphological subunits arranged with T = 3 symmetry. Computer image processing and optical diffraction studies have been applied to electron micrographs of tetragonal crystalline arrays of CCMV yielding a detailed three-dimensional model of the virus particle.[26] The structure of Bromovirus particles (Figure 2) appears to be very similar to that of the Cucumoviruses (see Chapter 5), except that the Cucumoviruses are not so prone to swelling.

Some negative stains penetrate Bromovirus particles indicating the presence of a central hole (Figures 1A, 2A, and 2B). A space about 11 nm in diameter was also detected by X-ray diffraction studies of BBMV[16] and by small-angle neutron diffraction of BMV.[27] The encapsidated RNA of Bromoviruses appears to line the inside of the protein shell leaving a central hole and partially interpenetrating the spaces between the coat protein subunits. The depth of penetration depends on the pH and presence or absence of Mg^{2+}.[27] Data from nuclear magnetic resonance studies indicate that the N-terminal region of the coat protein, which is susceptible to proteases when the particles are swollen, is the region which binds to the RNA.[28] This protein-RNA interaction does not appear to be altered during the pH-dependent swelling of the particles.[29]

Like the Cucumoviruses the RNA complement of Bromovirus particles consists of four distinct RNA species, RNAs 1 to 4 of M_r about 1.1, 1.0, 0.8, and 0.3×10^6, respectively. RNAs 1 and 2 are encapsidated separately whereas RNAs 3 and 4 are packed together in a single particle.[30] The three particle types have identical protein shells and encapsidate very similar amounts of RNA by weight, and hence are similar in size with M_r about 4.6×10^6, sedimentation coefficients of about 85S, and buoyant densities of about 1.35 g/cm^3 in CsCl and 1.31 g/cm^3 in Cs_2SO_4.[31]

Bromovirus particles can readily be dissociated by treating swollen particles with high concentrations of salts. The dissociation is reversible and particles of various polyhedral and tubular forms can be assembled in vitro.[17,32-37]

B. The Viral Genome

The Bromovirus genome is tripartite in that RNAs 1, 2, and 3 are all essential for infectivity.[30,38,39] RNA 4, which is also encapsidated, is a subgenomic mRNA for viral coat protein.[40] BMV RNAs 1 and 2 appear to be monocistronic mRNAs in that they have been shown to be translated in vitro, each to yield a protein of M_r about 120×10^3 and 110×10^3, respectively.[41] Although the coat protein cistron is present also on RNA 3 near the 3' terminus, the RNA is translated in vitro into only traces of coat protein and much larger amounts of a protein of M_r about 35×10^3 (3a protein) which is unrelated to the coat protein.[41] It is interesting that, in mixtures of RNAs 3 and 4, the latter is translated preferentially, apparently because of its preferential binding to ribosomes.[42] In vitro translation studies with CCMV have shown that CCMV RNA yields four proteins similar to those translated from BMV RNA.[43] These data indicate that the Bromovirus genome and its strategy (Figure 3) for translation is very similar to that of the Cucumoviruses (see Chapter 5, Figure 2).

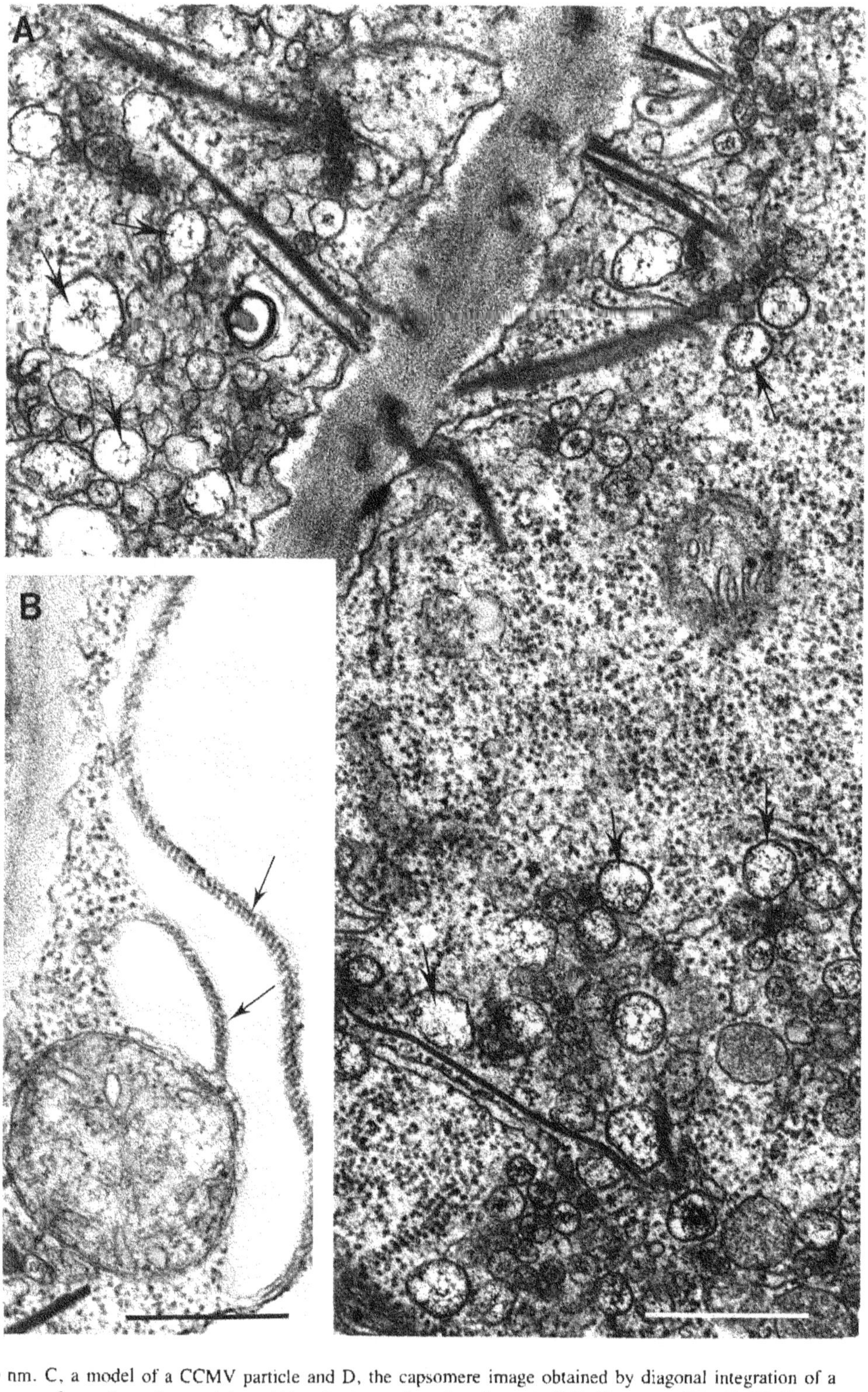

= 100 nm. C, a model of a CCMV particle and D, the capsomere image obtained by diagonal integration of a square array of superimposing particles within a lattice such as that shown in B.[12] (Courtesy of Dr. R. W. Horne, John Innes Institute, Norwich, U. K.)

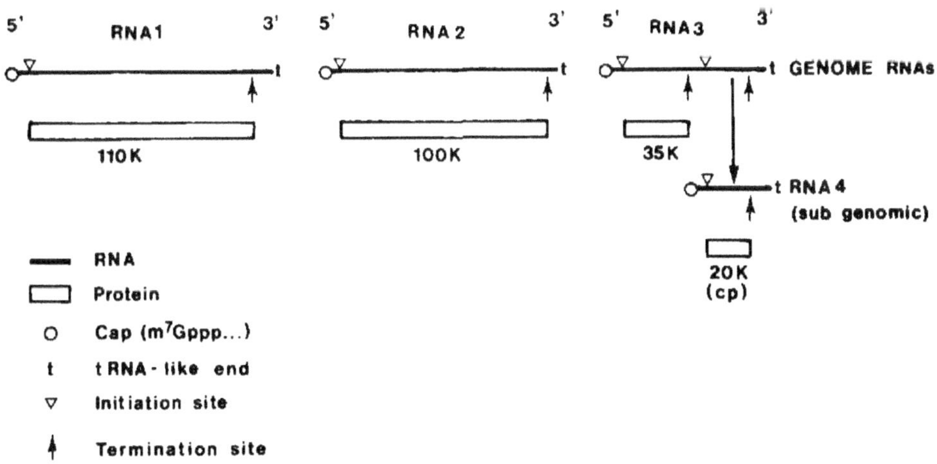

FIGURE 3. Physical map of the organization of the Bromovirus genome.

Four proteins specific to BMV infection of M_r about 107 × 10^3, 100 × 10^3, 35 × 10^3, and 20 × 10^3 (coat protein) have been detected in protoplasts from both host and nonhost plants of the virus.[44,45] They are probably the same as those detected in the in vitro translation system.[41]

The four BMV RNAs have very similar 3′ termini of about 200 bases[46] which accept tyrosine.[47] Modification of these termini results in a drastic loss of infectivity.[48] It has also been demonstrated that although encapsidated BMV RNA is not aminoacylated, that detected free of coat protein in infected plant cells is charged with tyrosine.[49] This suggests that the tRNA-like 3′ termini of the BMV RNAs have an important biological function. The 3′ termini of CCMV and BBMV RNAs also accept tyrosine[47] and have high degrees of base sequence homologies with those of BMV RNA as well as those of cucumber mosaic virus.[11]

At their 5′ termini, BMV RNAs are capped with $m^7G^{5'}ppp^{5'}Gp$.[50] The cap does not appear to be essential for in vitro translation, but its removal reduces the messenger activity and infectivity.[48,51]

The complete sequence of 2114 bases of BMV RNA 3 has been determined.[52] It was shown that the complete sequence of RNA 4 is encoded without interruption at the 3′ end of RNA 3 and that a third long open reading frame overlaps the cistrons for the coat and 3a proteins. There is a 16- to 22-base stretch of poly A, 20 bases from the coat protein cistron towards the 5′ terminus. It has been suggested that the poly A tract may help to identify the initiation of sequences encoding RNA 4.[52]

Only a limited amount of work has been done on the genetics of Bromoviruses by the in vitro construction of pseudorecombinants. This is because generally the genomes of the three Bromoviruses are incompatible and because strains of the viruses lack good genetic markers.[3] However, Bancroft[53] did obtain a pseudorecombinant from RNAs 1 and 2 of BMV and RNA 3 of CCMV although the reverse combination of RNAs did not appear to be viable. Nevertheless, pseudorecombinants from several Bromovirus strains and mutants have contributed some genetic information.[3,30,54-61] These studies established that RNA 3 has the genetic determinant for coat protein as well as affecting local lesion appearance, systemic disease symptoms, and temperature sensitivity to virus growth. RNA 2 also appears to affect lesion characteristics and RNA 1 can control temperature-sensitive growth and confer the ability to break the resistance of cowpea cultivars to CCMV.

III. CYTOPATHOLOGY

Bromovirus particles have been observed in the cytoplasm, vacuoles, and nuclei of infected cells.[62-68] The particles are usually most numerous in the cytoplasm where they sometimes occupy most of the space (Figures 4 and 5), and may form crystalline arrays[63,68] that probably correspond to the inclusions in BBMV-infected cells observed by light microscopy.[69]

The most extensive studies on the cytopathology of Bromoviruses have been done with BBMV. Autoradiographic experiments established that virus-dependent RNA synthesis occurred in the cytoplasm and was concentrated in small areas, often near golgi bodies.[70] These areas may correspond to the patches of vesicles containing fibrils suggestive of nucleic acid found in BBMV- (Figure 6) and CCMV-infected cells.[64,67] Similar vesicles have also been observed in the perinuclear spaces of infected cells (Figure 4A).[67] These observations indicate that Bromovirus RNA is replicated in the cytoplasm.

Double-stranded RNA replicative forms corresponding to the three genomic RNAs have been isolated from BMV-infected plants. However, there is some disagreement as to whether a double-stranded RNA corresponding to RNA 4 also exists in infected cells.[71,72] RNA-dependent RNA polymerase has also been detected in a membrane fraction from BMV-infected plants capable of synthesising virus-specific RNAs in vitro.[73] Similar enzymes have been detected in plants infected with BBMV[74] and CCMV.[75]

It was observed, using ^{125}I-labeled antibodies and autoradiography, that at early stages of infection, BBMV antigen appeared both in the nuclei and cytoplasm.[66] Thereafter, the concentration of antigen remained more or less constant in the nucleus but increased rapidly in the cytoplasm. On the basis of this study the authors suggested that viral protein is synthesized in the nucleus and is later transported to the cytoplasm where virus particles are assembled. However, biochemical evidence using cycloheximide and chloramphenicol to inhibit protein synthesis by cytoplasmic and chloroplast ribosomes, respectively, indicate that BBMV coat protein is synthesized on cytoplasmic ribosomes.[76]

Infection by Bromoviruses causes proliferation of the endoplasmic reticulum and induces numerous, amorphous, electron-dense inclusions.[64,66,67,77] The amorphous inclusions can often be seen among the virus-induced fibril-containing vesicles (Figures 5A and 6). They resemble inclusions induced by a number of viruses belonging to other groups such as tomato spotted wilt virus (Volume 1, Chapter 6) and some Potyviruses (Chapter 14 in this volume).

From the data available, it appears that most of the events leading to the synthesis of Bromovirus particles occur in the cytoplasm. The role of the nucleus remains enigmatic. However, at early stages of infection, the virus-specific vesicles in the perinuclear space may play an important role in virus replication.

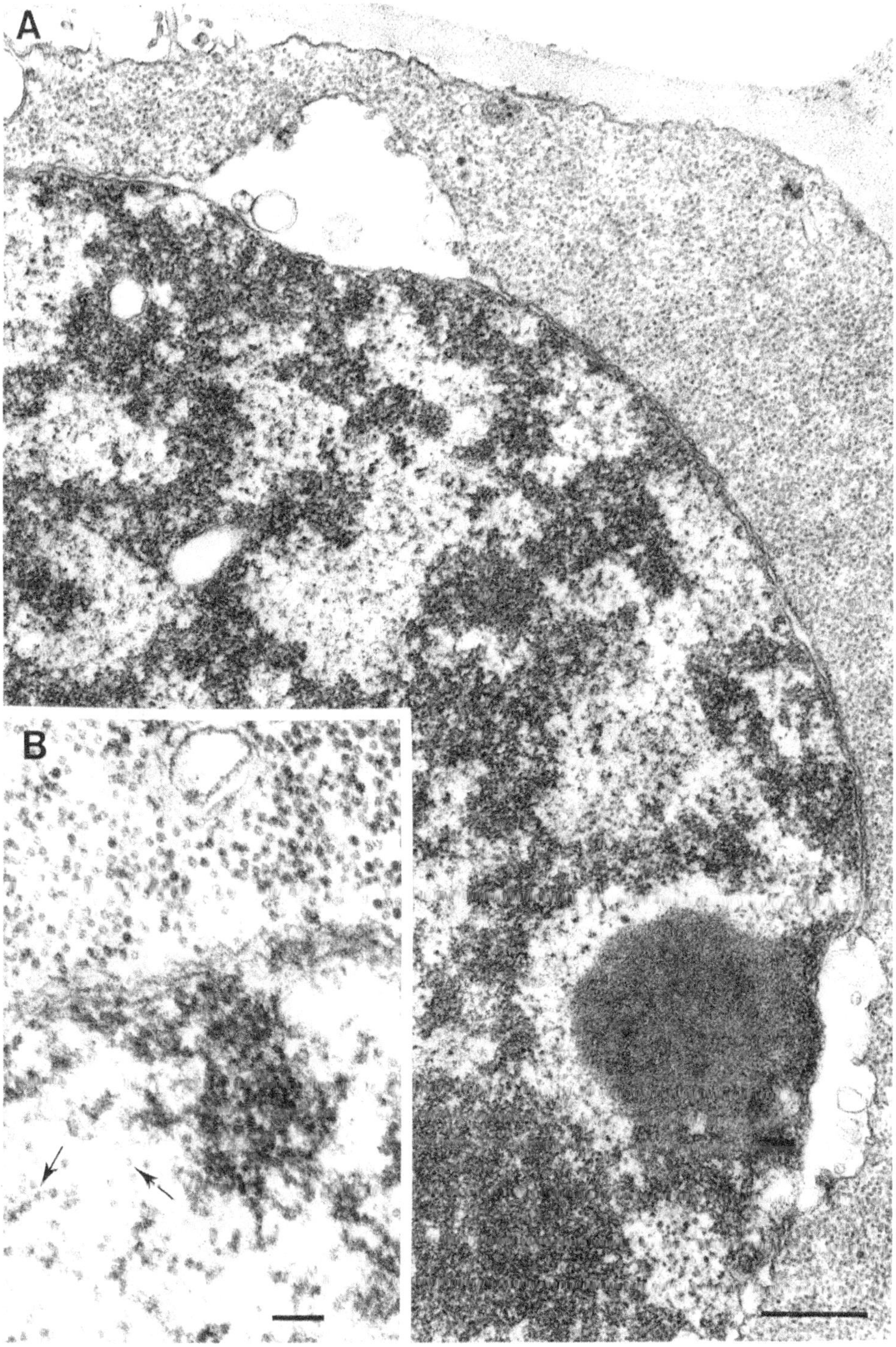

FIGURE 4. Thin sections of BBMV-infected *Vicia faba* (broad bean) leaf cells. (Ribosomes cleared by RNase treatment). A, Note the cytoplasm packed with virus particles, and the vesiculation of the outer nuclear membrane; bar = 500 nm. B, Note the virus particles in the nucleus (arrows) as well as the cytoplasm; bar = 100 nm.

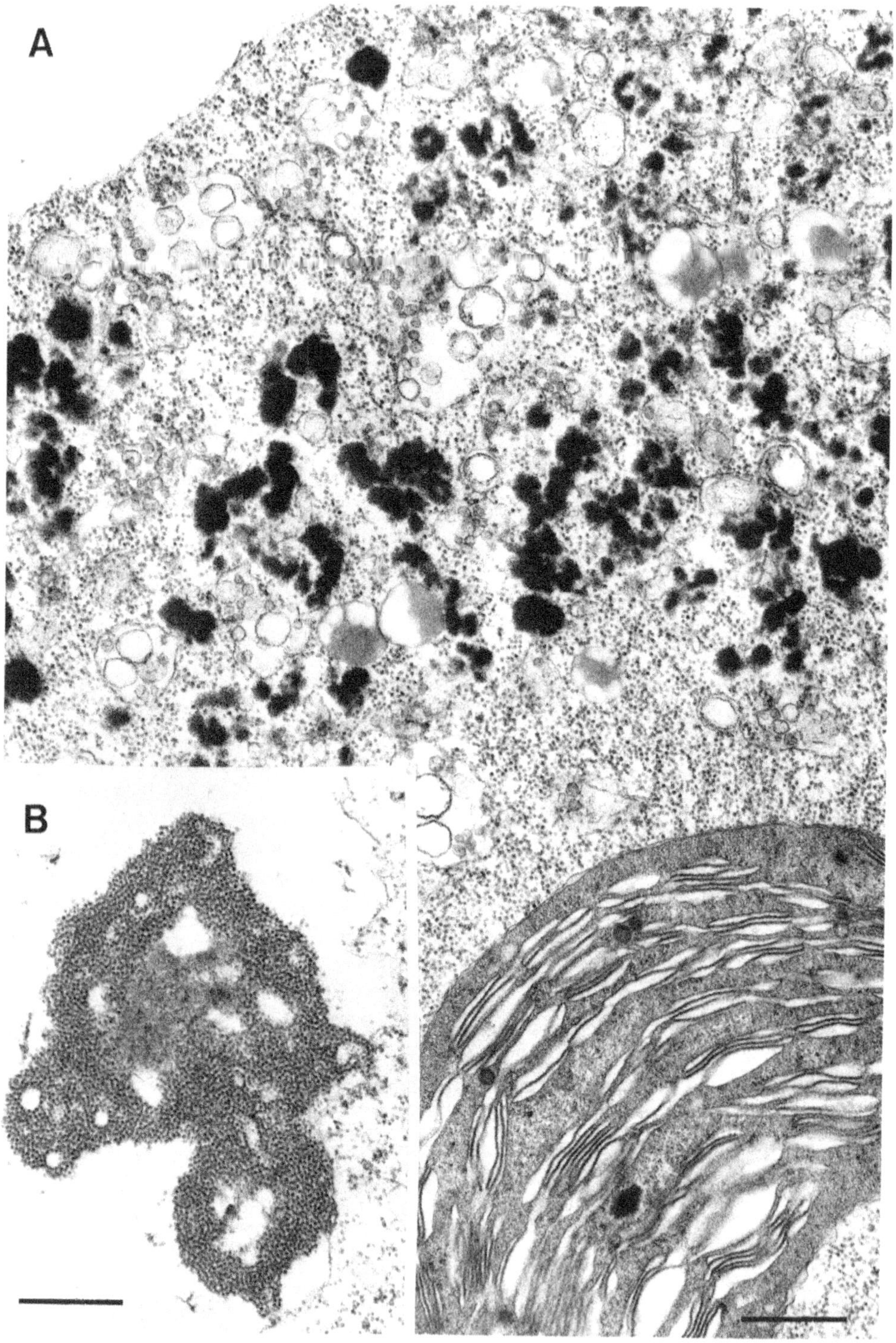

FIGURE 5. Material as in Figure 4. A. Note that the cytoplasm contains many virus particles, patches of amorphous material, and vesicles; the chloroplast lamellae are partly disrupted. B. An aggregate of virus particles in the cell vacuole. Bars = 500 nm.

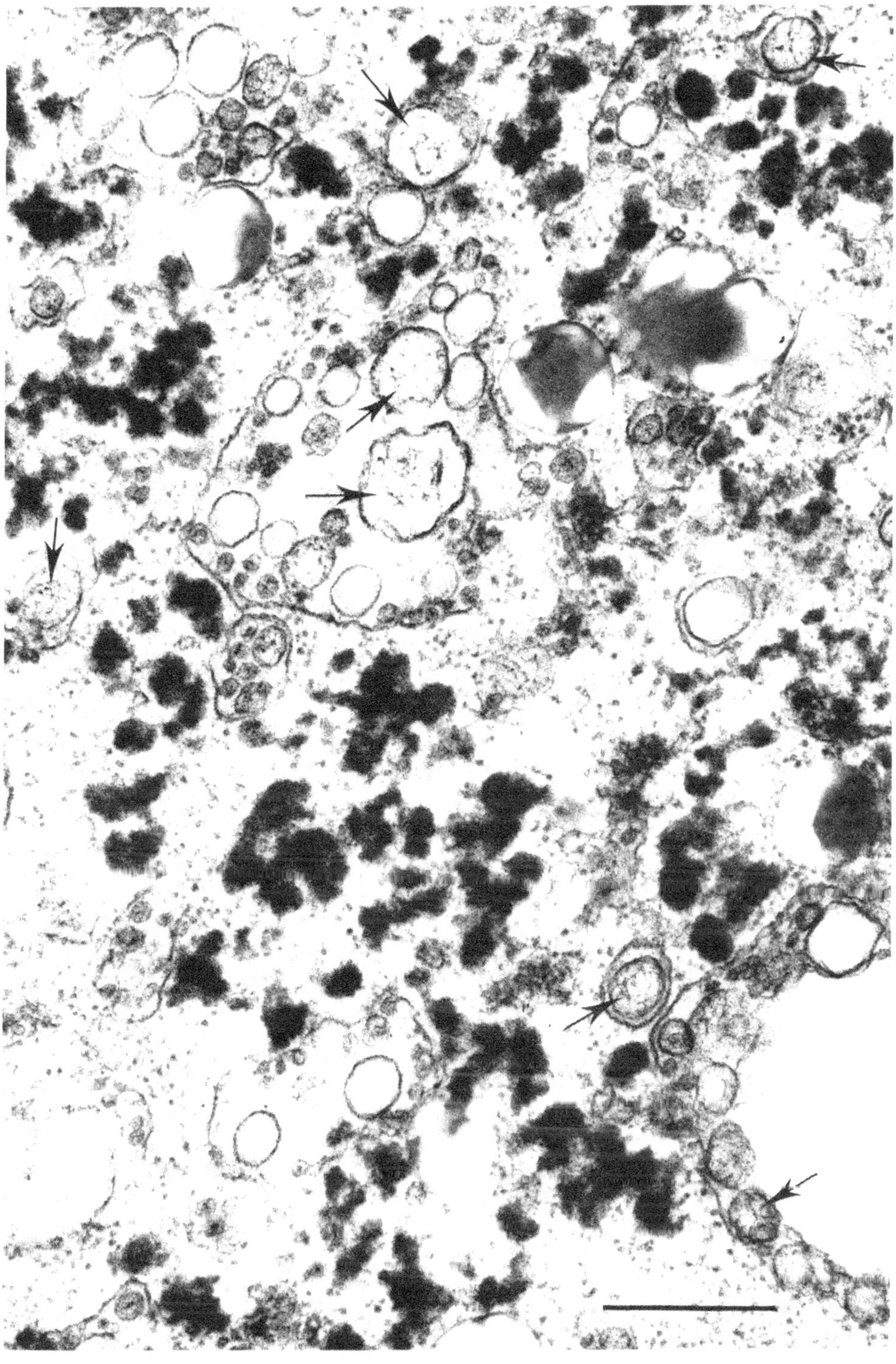

FIGURE 6. Material as in Figure 4. The cytoplasm contains amorphous densely stained material and fibril-containing vesicles (arrows). Bar = 500 nm.

REFERENCES

1. **Lane, L. C.,** The Bromoviruses, *Adv. Virus Res.*, 19, 151, 1974.
2. **Lane, L. C.,** Bromovirus group, *CMI/AAB Descriptions of Plant Viruses*, No. 215, 1979.
3. **Lane, L. C.,** Bromoviruses, in *Handbook of Plant Virus Infections and Comparative Diagnosis*, Kurstak, E., Ed. Elsevier, Amsterdam, 1981, 333.
4. **Rybicki, E. P. and von Wechmar, M. B.,** Characterisation of an aphid-transmitted virus disease of small grains, *Phytopathol. Z.*, 103, 306, 1982.
5. **Murant, A. F., Abu Salih, H. S., and Goold, R. A.,** Viruses from broad bean in the Sudan, *Rep. Scott. Hortic. Res. Inst. for 1973*, p. 67, 1974.
6. **Matthews, R. E. F.,** Classification and nomenclature of viruses. Fourth report of the International Committee on Taxonomy of Viruses, *Intervirology*, 17, 1, 1982.
7. **Hollings, M. and Horváth, T.,** Melandrium yellow fleck virus, *CMI/AAB Descriptions of Plant Viruses*, No. 236, 1981.
8. **Rybicki, E. P. and von Wechmar, M. B.,** The serology of the bromoviruses. I. Serological interrelationships of the bromoviruses, *Virology*, 109, 391, 1981.
9. **van Vloten-Doting, L., Francki, R. I. B., Fulton, R. W., Kaper, J. M., and Lane, L. C.,** Tricornaviridae — a proposed family of plant viruses with tripartite, single-stranded RNA genomes, *Intervirology*, 15, 198, 1981.
10. **Francki, R. I. B., Mossop, D. W., and Hatta, T.,** Cucumber mosaic virus, *CMI/AAB Descriptions of Plant Viruses*, No. 213, 1979.
11. **Ahlquist, P., Dasgupta, R., and Kaesberg, P.,** New identity of 3′ RNA secondary structure in bromoviruses and cucumber mosaic virus, *Cell*, 23, 183, 1981.
12. **Bancroft, J. B. and Horne, R. W.,** Bromovirus (brome mosaic virus) group, in *The Atlas of Insect and Plant Viruses*, Maramorosch, K., Ed., Academic Press, New York, 1977, 287.
13. **Horne, R. W. and Pasquali-Ronchetti, I.,** A negative staining-carbon film technique for studying viruses in the electron microscope. I. Preparative procedures for examining icosahedral and filamentous viruses, *J. Ultrastruct. Res.*, 47, 361, 1974.
14. **Anderegg, J. W., Wright, M., and Kaesberg, P.,** An X-ray scattering study of bromegrass mosaic virus, *Biophys. J.*, 3, 175, 1963.
15. **Finch, J. T. and Klug, A.,** Structure of broad bean mottle virus. I. Analysis of electron micrographs and comparison with turnip yellow mosaic virus and its top component, *J. Mol. Biol.*, 24, 289, 1967.
16. **Finch, J. T., Leberman, R., and Berger, J. E.,** Structure of broad bean mottle virus. II. X-ray diffraction studies, *J. Mol. Biol.*, 27, 17, 1967.
17. **Bancroft, J. B., Hills, G. J., and Markham, R.,** A study of the self-assembly process in a small spherical virus. Formation of organized structures from protein subunits in vitro, *Virology*, 31, 354, 1967.
18. **White, R. A. and Fischbach, F. A.,** An X-ray scattering investigation of broad bean mottle virus in solutions of various electron densities, *J. Mol. Biol.*, 75, 549, 1973.
19. **Incardona, N. L. and Kaesberg, P.,** A pH-induced structural change in bromegrass mosaic virus, *Biophys. J.*, 4, 11, 1964.
20. **Kassanis, B. and Lebeurier, G.,** The behaviour of tomato bushy stunt virus and bromegrass mosaic virus at different temperatures *in vivo* and *in vitro*, *J. Gen. Virol.*, 4, 385, 1969.
21. **Agrawal, H. O. and Tremaine, J. H.,** Proteins of cowpea chlorotic mottle, broad bean mottle, and brome mosaic viruses, *Virology*, 47, 8, 1972.
22. **Pfeiffer, P. and Hirth, L.,** The effect of conformational changes in brome mosaic virus upon its sensitivity to trypsin, chymotrypsin and ribonuclease, *FEBS Lett.* 56, 144, 1975.
23. **Incardona, N. L., McKee, S., and Flanegan, J. B.,** Noncovalent interactions in viruses: characterization of their role in the pH and thermally induced conformational changes in bromegrass mosaic virus, *Virology*, 53, 204, 1973.
24. **Pfeiffer, P., Herzog, M., and Hirth, L.,** Stabilization of brome mosaic virus, *Phil. Trans. R. Soc. London Ser. B*, 276, 99, 1976.
25. **Pfeiffer, P.,** Changes in the organization of bromegrass mosaic virus in response to cation binding as probed by changes in susceptibility to degradative enzymes, *Virology*, 102, 54, 1980.
26. **Steven, A. C., Smith, P. R., and Horne, R. W.,** Capsid fine structure of cowpea chlorotic mottle virus: from a computer analysis of negatively stained virus arrays, *J. Ultrastruct. Res.*, 64, 63, 1978.
27. **Jacrot, B., Pfeiffer, P., and Witz, J.,** The structure of a spherical plant virus (bromegrass mosaic virus) established by neutron diffraction, *Phil. Trans. R. Soc. London Ser. B*, 276, 109, 1976.
28. **Vriend, G., Hemminga, M. A., Verduin, B. J. M., de Wit, J. L., and Schaafsma, T. J.,** Segmental mobility involved in protein-RNA interaction in cowpea chlorotic mottle virus, *FEBS Lett.*, 134, 167, 1981.
29. **Vriend, G., Hemminga, M. A., Verduin, B. J. M., and Schaafsma, T. J.,** Swelling of cowpea chlorotic mottle virus studied by proton nuclear magnetic resonance, *FEBS Lett.*, 146, 319, 1982.

30. **Lane, L. C. and Kaesberg, P.,** Multiple genetic components in bromegrass mosaic virus, *Nature New Biol.*, 232, 40, 1971.
31. **Hull, R.,** The behavior of salt-labile plant viruses in gradients of cesium sulphate, *Virology*, 75, 18, 1976.
32. **Bancroft, J. B., Bracker, C. E., and Wagner, G. W.,** Structure derived from cowpea chlorotic mottle and brome mosaic virus protein, *Virology*, 38, 324, 1969.
33. **Bancroft, J. B., Hiebert, E., and Bracker, C. E.,** The effects of various polyanions on shell formation of some spherical viruses, *Virology*, 39, 924, 1969.
34. **Bancroft, J. B.,** The self-assembly of spherical plant viruses, *Adv. Virus Res.*, 16, 99, 1970.
35. **Adolph, K. W. and Butler, P. J. G.,** Studies on the assembly of a spherical plant virus. I. States of aggregation of the isolated protein. *J. Mol. Biol.*, 88, 327, 1974.
36. **Pfeiffer, P. and Hirth, L.,** Aggregation states of brome mosaic virus protein. *Virology*, 61, 160, 1974.
37. **Cuillel, M., Jacrot, B., and Zulauf, M.,** A T = 1 capsid formed by protein of brome mosaic virus in the presence of trypsin, *Virology*, 110, 63, 1981.
38. **Bancroft, J. B.,** The significance of the multicomponent nature of cowpea chlorotic mottle virus RNA, *Virology*, 45, 830, 1971.
39. **Hull, R.,** The multicomponent nature of broad bean mottle virus and its nucleic acid, *J. Gen. Virol.*, 17, 111, 1972.
40. **Shih, D. S. and Kaesberg, P.,**Translation of brome mosaic viral ribonucleic acid in a cell-free system derived from wheat embryo, *Proc. Natl. Acad. Sci. U.S.A.*, 70, 1799, 1973.
41. **Shih, D. S. and Kaesberg, P.,** Translation of the RNAs of brome mosaic virus: the monocistronic nature of RNA1 and RNA2, *J. Mol. Biol.*, 103, 77, 1976.
42. **Pyne J. W. and Hall, T. C.,** Efficient ribosome binding of brome mosaic virus (BMV) RNA4 contributes to its ability to outcompete the other BMV RNAs for translation, *Intervirology*, 11, 23, 1979.
43. **Davies, J. W. and Verduin, B. J. M.,** *In vitro* synthesis of cowpea chlorotic mottle virus polypeptides, *J. Gen. Virol.*, 44, 545, 1979.
44. **Sakai, F., Dawson, J. R. O., and Watts, J. W.,** Synthesis of proteins in tobacco protoplasts infected with brome mosaic virus, *J. Gen. Virol.*, 42, 323, 1979.
45. **Okuno, T. and Furusawa, I.,** RNA polymerase activity and protein synthesis in brome mosaic virus-infected protoplasts, *Virology*, 99, 218, 1979.
46. **Dasgupta, R. and Kaesberg, P.,** Sequence of an oligonucleotide derived from the 3′ end of each of the four brome mosaic viral RNAs, *Proc. Natl. Acad. Sci. U.S.A.*, 74, 4900, 1977.
47. **Kohl, R. J. and Hall, T. C.,** Aminoacylation of RNA from several viruses: amino acid specificity and differential activity of plant, yeast and bacterial synthetases, *J. Gen. Virol.*, 25, 257, 1974.
48. **Kohl, R. J. and Hall, T. C.,** Loss of infectivity of brome mosaic virus RNA after chemical modification of the 3′ or 5′ terminus, *Proc. Natl. Acad. Sci. U.S.A.*, 74, 2682, 1977.
49. **Loesch-Fries, L. S. and Hall, T. C.,** *In vivo* aminoacylation of brome mosaic and barley stripe mosaic virus RNAs, *Nature (London)*, 298, 771, 1982.
50. **Dasgupta, R., Harada, F., and Kaesberg, P.,** Blocked 5′ termini in brome mosaic virus RNA, *J. Virol.*, 18, 260, 1976.
51. **Shih, D. S., Dasgupta, R., and Kaesberg, P.,** 7-Methyl-guanosine and efficiency of RNA translation, *J. Virol.*, 19, 637, 1976.
52. **Ahlquist, P., Luckow, V., and Kaesberg, P.,** Complete nucleotide sequence of brome mosaic virus RNA 3, *J. Mol. Biol.*, 153, 23, 1981.
53. **Bancroft, J. B.,** A virus made from parts of the genomes of brome mosaic and cowpea chlorotic mottle viruses, *J. Gen. Virol.*, 14, 223, 1972.
54. **Bancroft, J. B. and Lane, L. C.,** Genetic analysis of cowpea chlorotic mottle and brome mosaic viruses, *J. Gen. Virol.*, 19, 381, 1973.
55. **Bancroft, J. B., McLean, G. D., Rees, M. W., and Short, M. N.,** The effect of an arginyl to a cysteinyl replacement on the uncoating behaviour of a spherical plant virus, *Virology*, 45, 707, 1971.
56. **Bancroft, J. B., Rees, M. W., Dawson, J. R. O., McLean, G. D., and Short, M. N.,** Some properties of a temperature-sensitive mutant of cowpea chlorotic mottle virus, *J. Gen. Virol.*, 16, 69, 1972.
57. **Bancroft, J. B., Rees, M. W., Johnson, M. W., and Dawson, J. R. O.,** A salt-stable mutant of cowpea chlorotic mottle virus, *J. Gen. Virol.*, 21, 507, 1973.
58. **Bancroft, J. B., McDonald, J. G., and Rees, M. W.,** A mutant of cowpea chlorotic mottle virus with a perturbed assembly mechanism, *Virology*, 75, 293, 1976.
59. **Dawson, W. O.,** Isolation and mapping of replication-deficient, temperature-sensitive mutants of cowpea chlorotic mottle virus, *Virology*, 90, 112, 1978.
60. **Kuhn, C. W. and Wyatt, S. D.,** A variant of cowpea chlorotic mottle virus obtained by passage through beans, *Phytopathology*, 69, 621, 1979.
61. **Kuhn, C. W. and Wyatt, S. D.,** Derivation of strains of cowpea chlorotic mottle virus, *Phytopathology*, 69, 1035, 1979.

62. **Hills, G. J. and Plaskitt, A.,** A protein stain for the electron microscopy of small isometric plant virus particles, *J. Ultrastruct. Res.*, 25, 323, 1968.
63. **Paliwal, Y. C.,** Electron microscopy of bromegrass mosaic virus in infected leaves, *J. Ultrastruct. Res.*, 30, 491, 1970.
64. **Burgess, J., Motoyoshi, F., and Fleming, E. N.,** Structural changes accompanying infection of tobacco protoplasts with two spherical viruses, *Planta*, 117, 133, 1974.
65. **Moline, H. E. and Ford, R. E.,** Bromegrass mosaic virus infection of seedling roots of *Zea mays, Triticum aestivum, Avena sativa and Hordeum vulgare*, *Physiol. Plant Pathol.*, 4, 209, 1974.
66. **Lastra, J. R. and Schlegel, D. E.,** Viral protein synthesis in plants infected with broad bean mottle virus, *Virology*, 65, 16, 1975.
67. **Kim, K. S.,** An ultrastructural study of inclusions and disease development in plant cells infected by cowpea chlorotic mottle virus, *J. Gen. Virol.*, 35, 535, 1977.
68. **de Zoeten, G. A. and Schlegel, D. E.,** Broad bean mottle virus in leaf tissue, *Virology*, 31, 173, 1967.
69. **Rubio, M. and van Slogteren, D. H. M.,** Light and electron microscopy of X-bodies associated with broad-bean mottle virus and phaseolus virus 2, *Phytopathology*, 46, 401, 1956.
70. **de Zoeten, G. A. and Schlegel, D. E.,** Nucleolar and cytoplasmic uridine-^{3}H incorporation in virus-infected plants, *Virology*, 32, 416, 1967.
71. **Philipps, G., Gigot, C., and Hirth, L.,** Replicative forms and viral RNA synthesis in leaves infected with brome mosaic virus, *Virology*, 60, 370, 1974.
72. **Bastin, M. and Kaesberg, P.,** A possible replicative form of brome mosaic virus RNA 4, *Virology*, 72, 536, 1976.
73. **Semal, J., Kummert, J., and Dekegel, D.,** Localization of viral RNA polymerase activity and its products in extracts of barley leaves infected with bromegrass mosaic virus, *Phytopathology*, 64, 446, 1974.
74. **Jacquemin, J. M. and Lopez, M.,** RNAs labelled *in vitro* by polymerase from leaves infected with broad bean mottle virus, *Intervirology*, 4, 45, 1974.
75. **White, J. L. and Dawson, W. O.,** Characterization of RNA-dependent RNA polymerases in uninfected and cowpea chlorotic mottle virus-infected cowpea leaves: selective removal of host RNA polymerase from membranes containing CCMV RNA replicase, *Virology*, 88, 33, 1978.
76. **Gibbs, A. and MacDonald, P. W.,** Preliminary experiments to determine which host ribosomes are used by some further plant viruses, *Intervirology*, 4, 52, 1974.
77. **Motoyoshi, F., Bancroft, J. B., Watts, J. W., and Burgess, J.,** The infection of tobacco protoplasts with cowpea chlorotic mottle virus and its RNA, *J. Gen. Virol.*, 20, 177, 1973.

Chapter 7

ILARVIRUS GROUP

This group, whose type member is tobacco streak virus (TSV), is named from the description "isometric labile ringspot".[1] Although most viruses in this group are labile in vitro and often produce ringspot or line-pattern symptoms, the virus particles are not truly isometric. In negative stain, they appear characteristically quasi-spherical of slightly pleomorphic and have a range of sizes from 20 to 32 nm in diameter. In addition, some members yield a proportion of bacilliform particles up to 73 nm in length that approach the morphology of alfalfa mosaic virus (AMV) particles (see Chapter 8). The particles of most Ilarviruses separate into three or four sedimenting components ranging from 78 to 114S but of similar buoyant densities[2,3] The capsids are built from a single species of protein of M_r between 24 and 30 $\times$ 10^3. There are four classes of positive-sense, single-stranded RNA; all four segments are required for infection, except that the coat protein can be substituted for the smallest RNA in activating the infectivity of the otherwise noninfectious mixture of the three largest (genomic) RNAs.[4-8]

Ilarviruses frequently, though not always, have woody natural hosts. Their experimental host ranges are wide and include many herbaceous species. No specific natural vectors are known, but there are often high rates of transmission through seed, and transmission to both the progeny and the parent plant may occur by way of the pollen from infected plants.[4,5] There are some unconfirmed reports of experimental vector-transmission of Ilarviruses. Thus TSV is reportedly vectored by the thrips *Frankliniella* and prunus necrotic ringspot virus (PNRSV) by a mite *(Vasates fockeui)* and a nematode *(Longidorus macrosoma)*.[4]

I. MEMBERS OF THE GROUP AND THEIR RELATIONSHIPS

As described in Chapter 8 of this volume, AMV has labile anisometric particles that package single-stranded RNAs of four different sizes; for infectivity, the three largest (RNAs 1 to 3) are required plus either the smallest RNA (RNA 4) or the coat protein. This description exactly fits the Ilarviruses, and in fact "protein activation" of the RNAs of certain Ilarviruses has been achieved with AMV coat protein and vice versa.[9-14] For these reasons it has been proposed that the Ilarviruses and AMV may be related and should be considered as a "supergroup" of protein- (or RNA 4-) dependent multicomponent plant viruses packaging their RNAs in quasi-isometric capsids of different size but similar density.[7,15] AMV is aphid transmitted while Ilarviruses are not, but this does not affect the apparently fundamental nature of the relationship. It has further been suggested that a family of plant viruses be constituted, consisting of viruses possessing tripartite single-stranded RNA genomes.[16] The proposals suggest the inclusion of Bromoviruses, Cucumoviruses, Ilarviruses and AMV, under the Tricornaviridae. The merits of the various aspects of this proposal are still under discussion by the ICTV.

The Ilarvirus group was formerly divided into subgroups A and B[8,17,18] based on the idea that some but not all members of each subgroup were serologically distantly related, whereas they were unrelated to members of the other subgroup. However, as Fulton[4] has pointed out, this serological grouping is not clear cut and does not correlate well with other differences such as those in host range, stability in vitro, or particle morphology. In addition, TSV, the type member, was not placed by Matthews[18] in either subgroup A or B. In the latest official ICTV report, the A and B subgrouping has been abandoned.[1]

The serological data do nevertheless appear to indicate clustering of some kind within the Ilarviruses. We have, therefore, in listing some properties of the known Ilarviruses in Table

Table 1
ILARVIRUSES AND SOME OF THEIR PROPERTIES

Serological Subgroup	Virus	Particle sizes (diameter, nm)	Sedimentation coeff (S_{20w})	Size of RNAs ($M_r \times 10^6$)	Size of coat protein ($M_r \times 10^3$)	Ref.
1	Tobacco streak virus (TSV)	29.5 + 27.5 + 24.6 (strain M)	113 + 96 + 84 + 78 (strain M)	1.04 + 0.98 + 0.78 + smaller	28.5—30.00 (strains GV and BRAZ)	2,3,15,20
		35 + 30 + 28 (strain GV)	109 + 93 + 78 (strain GV)	ND[a]	ND	
			114 + 98 + 90 (strain RN)	ND	ND	
	Black raspberry latent virus (BRLV)	~26 ~28.5	98 + 89 + 81 (pH 5) 93 + 88 + 78 (pH 7)	1.35 + 1.10 + 0.85 (+0.4)	28	21,22
	Asparagus stunt virus (ASV)	~30	98 + 85 (+ 2 minor)	ND	ND	23
2	Tulare apple mosaic virus (TAMV)	31 + 29.5 + 27.5 + bacilliform	91 + 5 114 + 108 + 93	1.01 + 0.92 + 0.74	19	15,24,25
	Citrus leaf rugose virus (CiLRV)	32 + 31 + 26 + 25	105 + 98 + 89 + 79	1.1 + 1.0 + 0.7 + 0.3 (+ trace of 5th)	26	11—13,26—28
	Citrus variegation virus (CVV)	33 + 31 + 28	110 + 93 + 83 + 79	1.0 + 1.0 + 0.7 + 0.3 (+ trace of 5th)	26	11,12,29
	Citrus crinkly leaf virus (CiCLV)	ND	ND	ND	ND	27
	Elm mottle virus (EMoV)	30—25	101 + 88 + 83	1.3 + 1.15 + 0.82 (+ 2 minor, 0.39 + 0.30)	25	30,31
	Asparagus virus II (AV II)	32 + 28 + 26	104 + 95 + 90	ND	ND	32,33
3	Prunus necrotic ringspot virus (PNRSV)	~23 + bacilliform	95 + 90 + 72, 119—107	1.3 + 0.89 + 0.69 + 0.31	25	8,24,34—37

	Cherry rugose mosaic virus[b]	ND	ND	ND	ND	4,38
	Rose mosaic virus (some isolates)	ND	ND	ND	ND	39,40
	Hop virus C[c]	ND	ND	ND	ND	41
	Apple mosaic virus (ApMV)	29 + 25 + bacilliform 32.5 + 30 + 26.6	117 + 88 + third intermed.	1.23 + 1.00 + 0.69 + 0.31	ND	15,42,43,
	Danish plum line pattern virus		ND	ND	ND	34
	Rose mosaic virus (some isolates)	23—29 + bacilliform	ND	ND	25	14,39,40,44
	Hop virus A	31 + 25	107 + 79	ND	ND	41
4	Prune dwarf virus (PDV)	~22 + bacilliform up to 73 nm long	113 + 98 + 85 + 81 + 75	1.26 + 0.95 + 0.76	24	45,46
5	American plum line pattern virus (APLPV)		ND	ND	ND	8,47
6[d]	Spinach latent virus (SPLV)	~27 + bacilliform	108 + 98 + 87	1.30 + 1.18 + 0.91 + 0.35 + 0.27	28	48
7[d]	Lilac ring mottle virus (LRMV)	~27	98 + 83	1.18 + 1.13 + 0.90 + 0.37	27.5	49,50

[a] Not determined.
[b] Viruses appearing as indented have been reported as very closely related serologically to the viruses listed above them; we consider them as isolates of those viruses.
[c] Indicated in Matthews[1] as hop virus B.
[d] Tentative serological subgroups only.

1, divided the group into seven serological subgroups at least two of which are tentative for lack of complete data. Each subgroup contains a virus or viruses appearing at present to be unrelated serologically to viruses in other sections; viruses within sections are related. The scheme suggested in Table 1 agrees closely with that independently proposed by Uyeda and Mink.[19]

The comparative serology of the Ilarviruses is, however, not yet satisfactory, partly because the viruses are unpromising subjects for purification and raising of good antisera. Another difficulty expressed with regard to TSV[24] and PNRSV[36] is that different isolates of the viruses present wide and variable spectra of antigens. Thus what is a "strain" and what a "virus" becomes unusually difficult to pin down. Pseudorecombinant viruses that have exchanged RNA segments have been artificially constructed,[9,17,51-54] and if they also occur naturally, may well contribute to this variation.

Within serotype 3 (Table 1) there is particular confusion as to the naming of viruses and strains. One sensible suggestion is that the term "rose mosaic" be reserved for a disease (caused by one or more viruses, not necessarily Ilarviruses) and that the terms "PNRSV-cherry serotype" or "apple serotype" be used to cover what we have listed as PNRSV and apple mosaic virus (ApMV) and their respective strains.[39,40]

Apart from serological and host-range differences, there appear to be no data that consistently separate the Ilarviruses into different clusters. The figures for coat protein and genome molecular sizes are as yet not precise or consistent enough for this purpose, and there are no reports on comparisons of amino acid or nucleotide sequences among Ilarviruses.

There are two viruses that we list as possible Ilarviruses (Table 2) though further work may show that they do not belong in this group. Pelargonium zonate spot virus (PZSV)[55,56] has rather unstable quasi-spherical particles that are disrupted in neutral PTA and that sediment as three components; there is one coat protein of M_r about 23 × 10^3, and the virus is seed transmitted and may also be pollen transmitted. No vector has been found. In these respects PZSV would qualify as an Ilarvirus. However, it appears that the 118S sedimenting component may not be a separate particle type but a dimer of the 80S component. Furthermore, there seem to be only two RNA species; a mixture of these is infective and the infectivity is not enhanced by the addition of coat protein.

The second virus, raspberry bushy dwarf (RBDV), has quasi-spherical particles of typical Ilarvirus morphology that are rather unstable and are disrupted in neutral PTA. All the biological properties (type of symptoms, wide host range, easy sap transmission, seed and pollen transmission, lack of vector) and the M_r of the coat protein are also consistent with those of Ilarviruses, but RBDV differs from these viruses in sedimentation behavior and in RNA composition (Table 2). Neither PZSV nor RBDV appear to be serologically related to any of the Ilarviruses tested.

II. VIRUS STRUCTURE AND COMPOSITION

A. Particle Structure

No elements of symmetry have been detected in Ilarvirus particles, and they do not appear to possess strictly icosahedral structures, though the possibility was considered by Ghabrial and Lister.[3] The particles may be built from a single protein species in a manner similar to the smaller particles of AMV (Chapter 8). They are unstable in neutral PTA unless previously fixed, but survive intact in uranyl acetate (Figures 1 to 3). As seen from Table 1, some Ilarviruses have been shown to have particles of discrete size and sedimentation rate, but this is unlikely to be true for all members of the group. The proportions of the different nucleoprotein components obtained vary considerably and can be influenced by the host and method of extraction.[58]

Usually, each of the three largest RNAs is separately packaged, while the slowest sedi-

Table 2
POSSIBLE ILARVIRUSES AND SOME OF THEIR PROPERTIES

Virus	Particle sizes (diameter, nm)	Sedimentation coeff (S_{20w})	Size of RNAs ($M_r \times 10^6$)	Size of coat protein ($M_r \times 10^3$)	Ref.
Pelargonium zonate spot virus (PZSV)	35—25	118 + 90 + 80	1.25 + 0.95	23	55,56
Raspberry bushy dwarf virus (RBDV)	~33	115 + minor 111	2.0 + 0.8 + 0.3	29	57

menting class of particle contains two molecules of the smallest RNA. If this were always so, it should mean that individual well separated and purified nucleoprotein components would not be infective. As Lane[9] has pointed out, some residual infectivity often remains due to the presence, in the bottom component for example, of "pseudo-bottom" particles packaging several molecules of the smaller RNAs.

Protein shells devoid of nucleic acid are not found with the Ilarviruses; furthermore, the virus particles disrupt under high-salt conditions, are stabilized by EDTA, and are sensitive to ribonuclease. These are indications that the particles are held together by protein-RNA salt linkages.[7] They are also degraded by low concentrations of SDS,[15] a further indication of structural stability being due to protein-RNA interactions.[59]

B. The Viral Genome

Ilarviruses have four molecules of positive-sense, single-stranded RNA (RNAs 1 to 4) of M_r about 1.1, 0.9, 0.7, and 0.3 $\times$ 10^6, respectively (Table 1). Sometimes additional RNAs of unknown significance have been detected (Table 1). In common with the Bromoviruses, Cucumoviruses, and AMV, the three largest RNAs carry all the genetic information of the virus. The nucleotide sequence specifying the coat protein, contained in part of RNA 3, is also carried on RNA 4, the so-called subgenomic RNA.

In vitro, RNA 4 of TSV functions as an efficient coat-protein messenger; RNAs 1 and 2 are each translated into polypeptides of M_r about 100 $\times$ 10^3 which corresponds to almost the total coding capacities of the RNAs, while from RNA 3 only the 5′ proximal portion is translated, into a product of M_r about 35 $\times$ 10^3. The coat-protein sequence, located toward the 3′ end of RNA 3 is not translated in vitro.[7,8]

Like AMV, Ilarviruses require RNAs 1 to 3 plus either RNA 4 or the coat protein for infectivity. If RNAs 1 to 3 plus coat protein are supplied, RNA 4 is regenerated as a separate entity when the virus multiplies. The Bromoviruses and Cucumoviruses, in contrast, require only RNAs 1 to 3 for infectivity, but RNA 4 is synthesized in infected cells and becomes encapsidated (see Chapters 5 and 6 of this volume).

Studies on the activation of Ilarvirus RNAs 1 to 3 by coat protein caused some surprise when it was found that the coat proteins of several Ilarviruses and of AMV were interchangeable in this respect.[10,11,13,14] This occurs despite the fact that the coat proteins of Ilarviruses are not serologically related to those of AMV, and that genome segments 1 to 3 cannot be interchanged between Ilarviruses and AMV to form viable pseudorecombinants. It is interesting in this context that the RNAs of both Ilarviruses and AMV have a strong affinity in vitro for their own or each other's coat protein. Indeed, the RNAs will strip protein subunits from intact particles.[10,14]

Koper-Zwarthoff and Bol[60] reported evidence that each of the four RNAs of both TSV and AMV has the same or a very similar nucleotide sequence at the 3′ terminus, whose function appears to be to recognize and attach to the coat protein. They postulated that the 3′ termini contain multiple occurrences of the sequence AUGC embedded in a series of

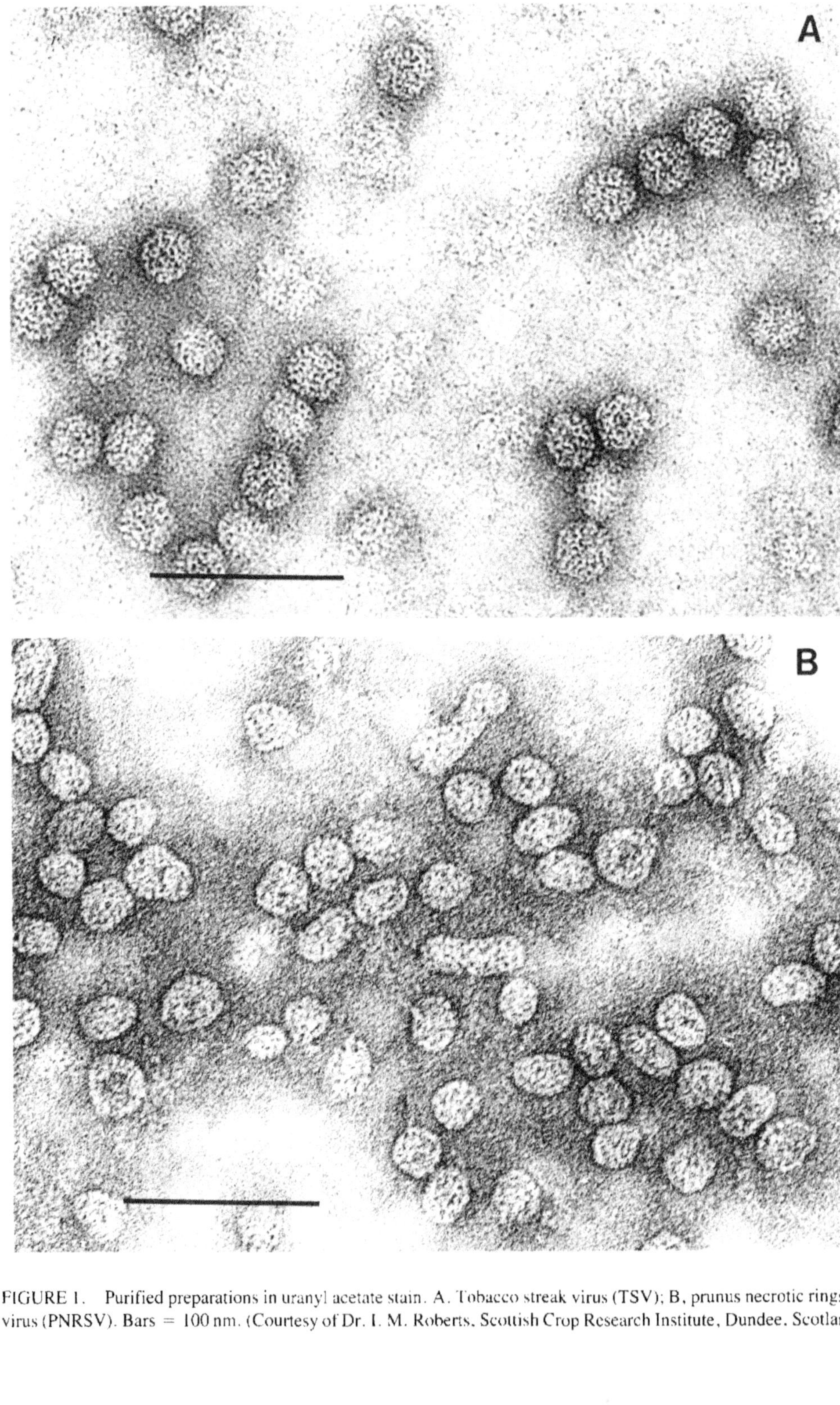

FIGURE 1. Purified preparations in uranyl acetate stain. A. Tobacco streak virus (TSV); B, prunus necrotic ringspot virus (PNRSV). Bars = 100 nm. (Courtesy of Dr. I. M. Roberts, Scottish Crop Research Institute, Dundee, Scotland.)

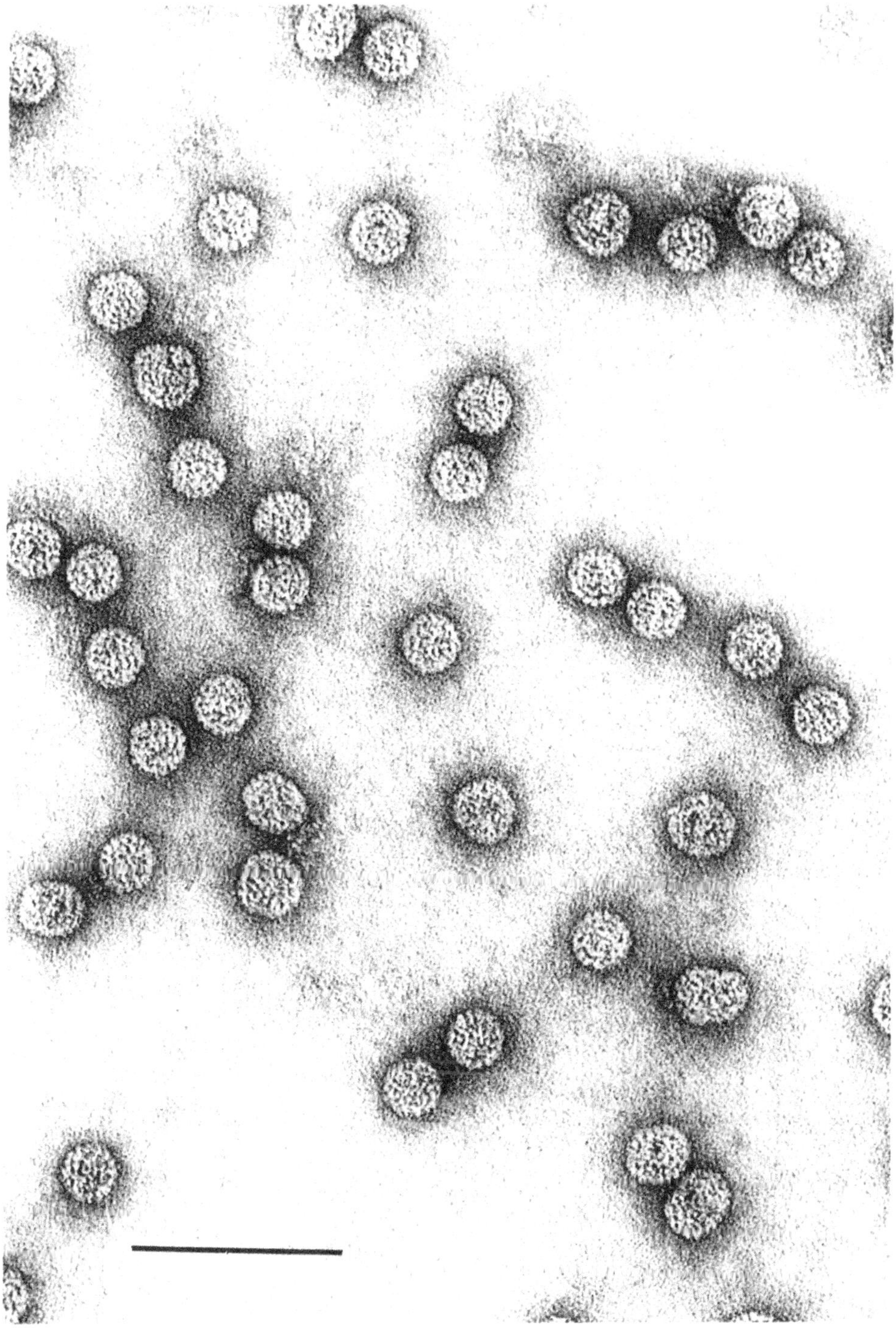

FIGURE 2. Purified preparation of elm mottle virus (EMoV) in uranyl acetate stain. Bar = 100 nm. (Courtesy of Dr. I. M. Roberts, Scottish Crop Research Institute, Dundee, Scotland.)

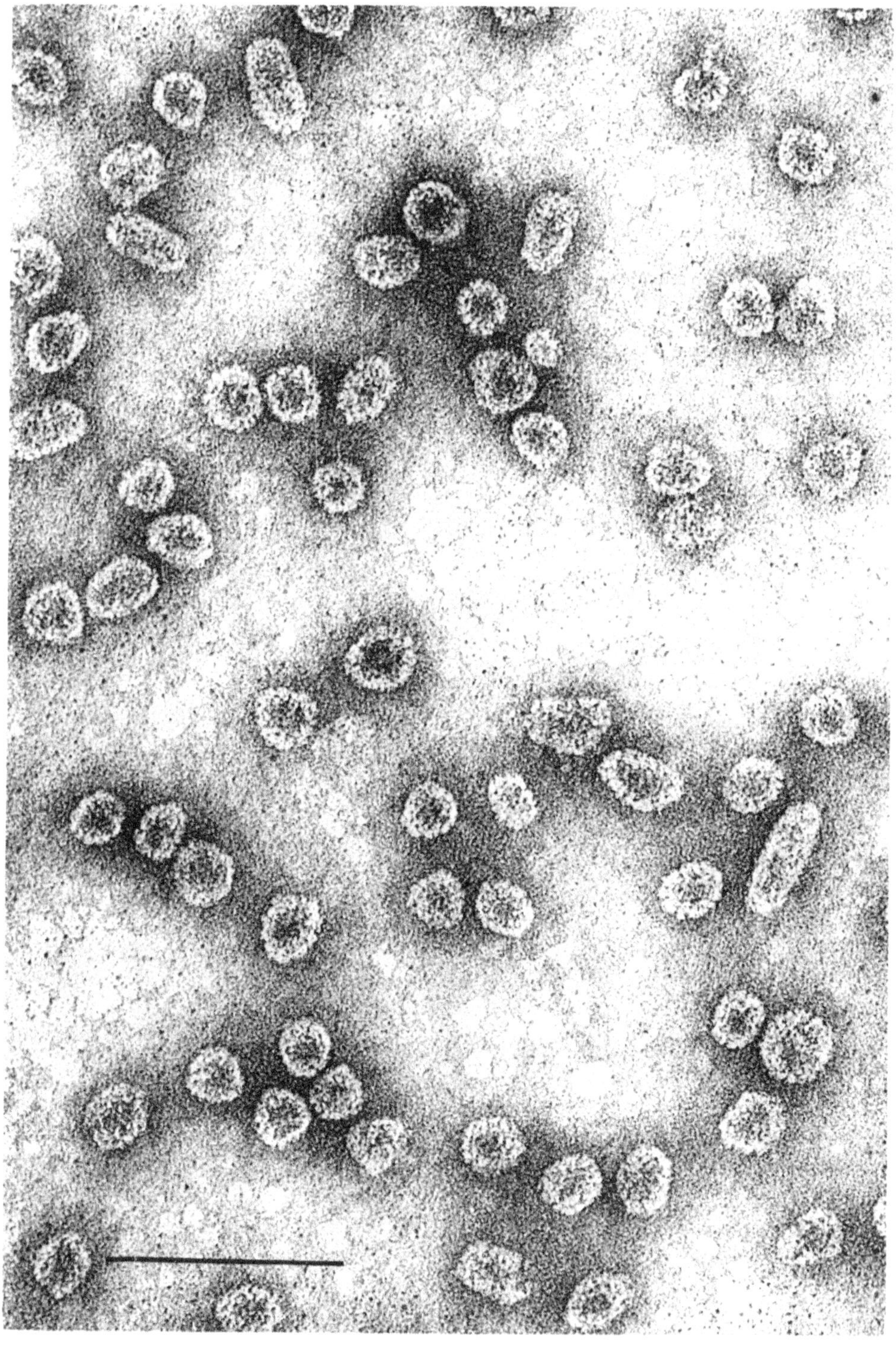

FIGURE 3. Purified preparation of prune dwarf virus (PDV) in uranyl acetate stain. Bar = 100 nm. (Courtesy of Dr. J. M. Roberts, Scottish Crop Research Institute, Dundee, Scotland.)

hairpin structures, and that these sequences would interact cooperatively to bind the coat protein.

Fulton[51-54] attempted to assign functions to each segment of the TSV genome by constructing pseudorecombinants between strains bearing different markers. While progeny with recombined characteristics were certainly obtained, interpreting the results is "intellectually taxing" because of various technical problems.[7,9] As most Ilarviruses are deficient in RNA 4, the purified RNAs, even when unfractionated, often have little infectivity. Fulton therefore used carefully separated nucleoprotein components as recombinant units. As we have seen, however, each centrifugally homogeneous component is likely to be contaminated with small amounts of "pseudo-components" containing different RNA molecules. The recombinant units used by Fulton were therefore probably not genetically pure. Lane[9] has interpreted these results to indicate that the determinants for lesion size are located on RNA 3, lesion color on RNA 2, and antigenic specificity on RNA 1. However, as we have noted, the coat protein sequence appears to be on RNA 3, and certainly is found there with viruses belonging to the Cucumovirus, Bromovirus, and AMV groups (see Chapters 5, 6, and 8).

III. CYTOPATHOLOGY

We have found only one report on the cytopathology of Ilarvirus infections.[61] This may be because few attempts have been made to study them. It seems more likely, however, that studies have been made but have turned up nothing dramatic or specific enough to publish.

An examination of reports of AMV infections (Chapter 8) suggests that, in thin sections, AMV particles can be clearly detected only when they appear bacilliform or form distinct aggregates. Ilarvirus particles might therefore easily pass undetected or be indistinguishable from ribosomes, especially if not numerous. Furthermore, AMV does not apparently induce specific alterations of organelles or the formation of inclusion bodies, and it is quite possible that Ilarviruses behave similarly.

The above speculations receive some support from studies of a strain of TSV in thin sections of infected *Chenopodium quinoa*.[62] These results indicate that virus particles may be present in the cytoplasm but are not clearly identifiable, that no recognizable virus aggregates (such as crystalline arrays) are formed, and that organelles such as nuclei, chloroplasts, mitochondria, microbodies, and cell walls are not altered in characteristic or specific ways.

Classical embedding and thin sectioning, it is clear, are inadequate for studying Ilarvirus cytopathology, and newer techniques such as enzyme cytochemistry and immune labeling must be tried before useful results are obtained.

The cytopathology of PZSV has been examined in thin sections[63] and virus particles were found both in the cytoplasm and the nucleus. In the cytoplasm, particles were sometimes aligned inside tubes reminiscent of the structures found with, for example, Nepoviruses and Comoviruses (see Chapters 1 and 2). Chloroplasts were deranged, the nuclear membranes became vesiculated, and wall thickenings and outgrowths were commonly induced. The authors noted that the taxonomic position of PZSV is uncertain, and that the cytopathology of PZSV may not at all represent that of the Ilarviruses.

REFERENCES

1. **Matthews, R. E. F.,** Classification and nomenclature of viruses. Fourth report of the International Committee on Taxonomy of Viruses, *Intervirology*, 17, 1, 1982.
2. **Lister, R. M., Ghabrial, S. A., and Saksena, K. N.,** Evidence that particle size heterogeneity is the cause of centrifugal heterogeneity in tobacco streak virus, *Virology*, 49, 290, 1972.
3. **Ghabrial, S. A. and Lister, R. M.,** Chemical and physicochemical properties of two strains of tobacco streak virus, *Virology*, 57, 1, 1974.
4. **Fulton, R. W.,** Ilarviruses, In *Handbook of Plant Virus Infections and Comparative Diagnosis*, Kurstak, E., Ed., Elsevier, Amsterdam, 1981, 377.
5. **Fulton, R. W.,** Ilarvirus group, *CMI/AAB Descriptions of Plant Viruses*, No. 275, 1983.
6. **Jaspars, E. M. J.,** Plant viruses with multipartite genome, *Adv. Virus Res.*, 19, 37, 1974.
7. **van Vloten-Doting, L. and Jaspars, E. M. J.,** Plant covirus systems: three-component systems, in *Comprehensive Virology*, Vol. 11, Fraenkel-Conrat, H. and Wagner, R. R., Eds., Plenum Press, New York, 1977, 1.
8. **Loesch-Fries, L. S., Halk, E. L., and Fulton, R. W.,** Ilarvirus (Necrotic ringspot) group, subgroup B, in *The Atlas of Insect and Plant Viruses*, Maramorosch, K., Ed., Academic Press, New York, 1977, 371.
9. **Lane, L. C.,** The nucleic acids of multipartite, defective, and satellite plant viruses, in *Nucleic Acids in Plants*, Vol. 2, Hall, T. C. and Davies J. W., Eds., CRC Press, Boca Raton, Fla., 1979, 65.
10. **van Vloten-Doting, L.,** Coat protein is required for infectivity of tobacco streak virus: biological equivalence of the coat proteins of tobacco streak and alfalfa mosaic viruses, *Virology*, 65, 215, 1975.
11. **Gonsalves, D. and Garnsey, S. M.,** Functional equivalence of an RNA component and coat protein for infectivity of citrus leaf rugose virus, *Virology*, 64, 23, 1975.
12. **Gonsalves, D. and Garnsey, S. M.,** Nucleic acid components of citrus variegation virus and their activation by coat protein, *Virology*, 67, 311, 1975.
13. **Gonsalves, D. and Garnsey, S. M.,** Infectivity of heterologous RNA-protein mixtures from alfalfa mosaic, citrus leaf rugose, citrus variegation, and tobacco streak viruses, *Virology*, 67, 319, 1975.
14. **Gonsalves, D. and Fulton, R. W.,** Activation of Prunus necrotic ringspot virus and rose mosaic virus by RNA 4 components of some ilarviruses, *Virology*, 81, 398, 1977.
15. **Lister, R. M. and Saksena, K. N.,** Some properties of tulare apple mosaic and ILAR viruses suggesting grouping with tobacco streak virus, *Virology*, 70, 440, 1976.
16. **van Vloten-Doting, L., Francki, R. I. B., Fulton, R. W., Kaper, J. M., and Lane, L. C.,** Tricornaviridae — a proposed family of plant viruses with tripartite, single-stranded RNA genomes, *Intervirology*, 15, 198, 1981.
17. **Shepherd, R. J., Francki, R. I. B., Hirth, L., Hollings, M., Inouye, T., MacLeod, R., Purcifull, D. E., Sinha, R. C., Tremaine, J. H., Valenta, V., and Wetter, C.,** New groups of plant viruses approved by the International Committee on Taxonomy of Viruses, September, 1975, *Intervirology*, 6, 181, 1975/76.
18. **Matthews, R. E. F.,** Classification and nomenclature of viruses. Third report of the International Committee on Taxonomy of Viruses, *Intervirology*, 12, 137, 1979.
19. **Uyeda, I. and Mink, G. I.,** Relationships among some ilarviruses: proposed revision of subgroup A, *Phytopathology*, 73, 47, 1983.
20. **Fulton, R. W.,** Tobacco streak virus, *CMI/AAB Descriptions of Plant Viruses*, No. 44, 1971.
21. **Lister, R. M. and Converse, R. H.,** Black raspberry latent virus, *CMI/AAB Descriptions of Plant Viruses*, No. 106, 1972.
22. **Jones, A. T. and Mayo, M. A.,** Further properties of black raspberry latent virus, and evidence for its relationship to tobacco streak virus, *Ann. Appl. Biol.*, 79, 297, 1975.
23. **Brunt, A. A. and Paludan, N.,** The serological relationship between "asparagus stunt" and tobacco streak viruses, *Phytopathol. Z.*, 69, 277, 1970.
24. **Barnett, O. W. and Fulton, R. W.,** Some chemical properties of Prunus necrotic ringspot and Tulare apple mosaic viruses, *Virology*, 39, 556, 1969.
25. **Fulton, R. W.,** Tulare apple mosaic virus, *CMI/AAB Descriptions of Plant Viruses*, No. 42, 1971.
26. **Gonsalves, D. and Garnsey, S. M.,** Infectivity of the multiple nucleoprotein and RNA components of citrus leaf rugose virus, *Virology*, 61, 343, 1974.
27. **Garnsey, S. M.,** Purification and properties of citrus-leaf-rugose virus, *Phytopathology*, 65, 50, 1975.
28. **Garnsey, S. M. and Gonsalves, D.,** Citrus leaf rugose virus, *CMI/AAB Descriptions of Plant Viruses*, No. 164, 1976.
29. **Corbett, M. K. and Grant, T. J.,** Purification of citrus variegation virus, *Phytopathology*, 57, 137, 1967.
30. **Jones, A. T. and Mayo, M. A.,** Purification and properties of elm mottle virus, *Ann. Appl. Biol.*, 75, 347, 1973.
31. **Jones, A. T.,** Elm mottle virus, *CMI/AAB Descriptions of Plant Viruses*, No. 139, 1974.

32. **Weissenfels, M., Schmidt, H. B., and Schmelzer, K.,** Partikelgestalt des Spargel-Virus 2 (asparagus virus 2), *Arch. Phytopathol. Pflanzenschutz*, 12, 63, 1976.
33. **Uyeda, I. and Mink, G. I.,** Properties of asparagus virus II, a new member of the ilarvirus group, *Phytopathology*, 71, 1264, 1981.
34. **Fulton, R. W.,** Prunus necrotic ringspot virus, *CMI/AAB Descriptions of Plant Viruses*, No. 5, 1970.
35. **Fulton, R. W.,** Serology of viruses causing cherry necrotic ringspot, plum line pattern, rose mosaic, and apple mosaic, *Phytopathology*, 58, 635, 1968.
36. **Casper, R.,** Serological properties of prunus necrotic ringspot and apple mosaic virus isolates from rose, *Phytopathology*, 63, 238, 1973.
37. **Loesch, L. S. and Fulton, R. W.,** Prunus necrotic ringspot virus as a multicomponent system, *Virology*, 68, 71, 1975.
38. **Nyland, G. and Lowe, S. K.,** The relation of cherry rugose mosaic and almond calico viruses to prunus ringspot virus, *Phytopathology*, 54, 1435, 1964.
39. **Thomas, B. J.,** The detection by serological methods of viruses infecting the rose, *Ann. Appl. Biol.*, 94, 91, 1980.
40. **Thomas, B. J.,** Studies on rose mosaic disease in field-grown roses produced in the United Kingdom, *Ann. Appl. Biol.*, 98, 419, 1981.
41. **Bock, K. R.,** Strains of *Prunus* necrotic ringspot virus in hop (*Humulus lupulus* L.), *Ann. Appl. Biol.*, 59, 437, 1967.
42. **De Sequeira, O. A.,** Purification and serology of an apple mosaic virus, *Virology*, 31, 314, 1966.
43. **Fulton, R. W.,** Apple mosaic virus, *CMI/AAB Descriptions of Plant Viruses*, No. 83, 1972.
44. **Basit, A. A. and Francki, R. I. B.,** Some properties of rose mosaic virus from South Australia, *Aust. J. Biol. Sci.*, 23, 1197, 1970.
45. **Fulton, R. W.,** Prune dwarf virus, *CMI/AAB Descriptions of Plant Viruses*, No. 19, 1970.
46. **Halk, E. L. and Fulton, R. W.,** Stabilization and particle morphology of prune dwarf virus, *Virology*, 91, 434, 1978.
47. **Fulton, R. W.,** Ilar-like characteristics of American plum line pattern virus and its serological detection in *Prunus*, *Phytopathology*, 72, 1345, 1982.
48. **Bos, L., Huttinga, H., and Maat, D. Z.,** Spinach latent virus, a new ilarvirus seed-born in *Spinacia oleracea*, *Neth. J. Plant Pathol.*, 86, 79, 1980.
49. **Huttinga, H. and Mosch, W. H. M.,** Lilac ring mottle virus: a coat protein-dependent virus with a tripartite genome, *Acta Hortic.*, 59, 113, 1976.
50. **van der Meer, F. A. and Huttinga, H.,** Lilac ring mottle virus, *CMI/AAB Descriptions of Plant Viruses*, No. 201, 1979.
51. **Fulton, R. W.,** The role of particle heterogeneity in infection by tobacco streak virus, *Virology*, 41, 288, 1970.
52. **Fulton, R. W.,** Inheritance and recombination of strain-specific characters in tobacco streak virus, *Virology*, 50, 810, 1972.
53. **Fulton, R. W.,** The role of top particles in recombination of some characters of tobacco streak virus, *Virology*, 67, 188, 1975.
54. **Fulton, R. W.,** Superinfection by strains of tobacco streak virus, *Virology*, 85, 1, 1978.
55. **Quacquarelli, A. and Gallitelli, D.,** Tre virosi del geranio in Puglia, *Phytopathol. Mediterr.*, 18, 61, 1979.
56. **Gallitelli, D.,** Properties of a tomato isolate of Pelargonium zonate spot virus, *Ann. Appl. Biol.*, 100, 457, 1982.
57. **Murant, A. F.,** Raspberry bushy dwarf virus, *CMI/AAB Descriptions of Plant Viruses*, No. 165, 1976.
58. **Lister, R. M. and Bancroft, J. B.,** Alteration of tobacco streak virus component ratios as influenced by host and extraction procedure, *Phytopathology*, 60, 689, 1970.
59. **Kaper, J. M.,** Arrangement and identification of simple isometric viruses according to their dominating stabilizing interactions, *Virology*, 55, 299, 1973.
60. **Koper-Zwarthoff, E. C. and Bol, J. F.,** Nucleotide sequence of the putative recognition site for the coat protein in the RNAs of alfalfa mosaic virus and tobacco streak virus, *Nucleic Acids Res.*, 8, 3307, 1980.
61. **Gerola, F. M., Lombardo, G., and Catara, A.,** Histological localisation of citrus infectious variegation virus (CCVV) in *Phaseolus vulgaris*, *Protoplasma*, 67, 319, 1969.
62. **Masenga, V., Caciagli, P., and Milne, R. G.,** unpublished results, 1980.
63. **Castellano, M. A. and Martelli, G. P.,** Electron microscopy of Pelargonium zonate spot virus in host tissues, *Phytopathol. Mediterr.*, 20, 64, 1981.

Chapter 8

ALFALFA MOSAIC VIRUS GROUP

This is a monotypic group with alfalfa mosaic virus (AMV) as its sole member, although the virus does occur as numerous distinct strains. AMV particles are built from a single species of polypeptide of M_r 24.3 × 10^3 to form quasi-spherical particles about 18 nm in diameter, plus a series of bacilliform particles of the same width but differing lengths. The particles encapsidate four major species of single-stranded, positive-sense RNA with M_r from 1.04 × 10^6 to 0.28 × 10^6.[1-6]

AMV has been transmitted by mechanical inoculation and by aphids in a nonpersistent manner to some 430 plant species in 51 dicotyledonous families. It is also seed transmitted in some species. In nature, the virus is widespread and can cause serious disease in many crops, though it is often symptomless.[1-3]

AMV has close affinities with the Ilarvirus group (Chapter 7) and is perhaps more distantly related to the Bromoviruses and Cucumoviruses. A higher taxon such as the proposed family Tricornaviridae[7] to include all four groups may eventually be approved. There are some convincing arguments for inclusion of AMV in the same taxon as the Ilarviruses.[7]

I. VIRUS STRUCTURE AND COMPOSITION

A. Particle Structure

Electron micrographs of AMV preparations (Figure 1) reveal a variety of particles ranging from polyhedra about 18 nm in diameter to bacilliform structures of the same diameter but various lengths.[1,4,8] Preparations of all AMV strains can be separated into at least four particle types, the bottom (B), middle (M), top b (Tb) and top a (Ta) components which sediment at about 94, 82, 73, and 66S, respectively.[5] B, M, Tb, and Ta particles are approximately 56, 43, 35, and 30 nm long, respectively, encapsidating RNAs of different lengths. Some strains of AMV may also have a number of other particles encapsidating a variety of RNA molecules singly or in combination of two or more. In preparations of one particular strain, as many as seventeen particle types were detected by electrophoresis in agarose gels.[9] Strains VRU and 15/64 contain small amounts of very long particles (up to 400 nm). However, such particles only encapsidate several RNA molecules of normal size, usually smaller than RNA 1.[10,11]

The bacilliform particles of AMV are cylindrical structures with rounded caps, built from many copies of a single polypeptide of M_r 24.3 × 10^3. The complete coat protein amino acid sequences have been determined for AMV strains 425, S, and VRU, and consist of 220, 219, and 217 residues, respectively.[12-15] Image reconstruction of electron micrographs and small-angle neutron scattering data show that the polypeptides are grouped in dimers forming a hexagonal lattice in the cylindrical portion of the particle. This arrangement transforms into a pentameric pattern at the ends of the particles to produce caps with icosahedral symmetry.[8,16,17] The particles of the four major components represent a series of structures with 60 + (n × 18) subunits, n being 10, 7, 5, and 4 for components B, M, Tb, and Ta, respectively.[18] Data from optical diffraction and computer processing of electron microscopic data revealed that AMV particles can have two kinds of structure — a stacked and a helical type.[4] In the stacked form the unit cells are arranged in staggered rings so that two rings comprise a repeat of the structure every 4.4 nm. In the helical form, these rings of unit cells are transformed into turns of a double helix.

AMV particles are stabilized mainly by protein-RNA interactions, being readily dissociated in low concentrations of sodium dodecyl sulfate[19] or high concentrations of salt.[20] Dissociation at high ionic strengths is reversible and AMV particles have been reconstituted from

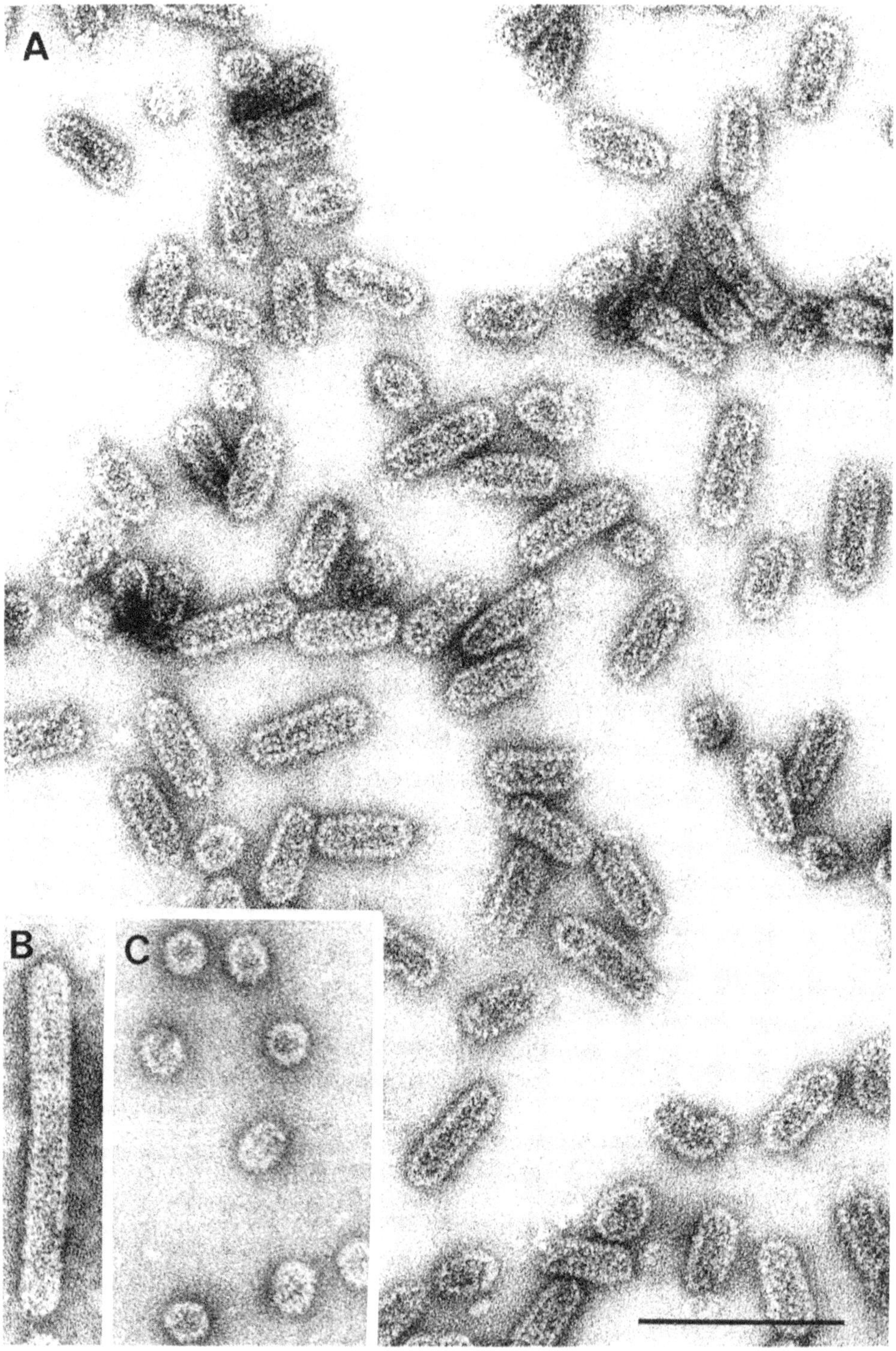

FIGURE 1. Purified alfalfa mosaic virus (AMV) in uranyl acetate stain. A, Unfractionated preparation (note the variety of particle sizes); B, particle of unusual length; C, small particles after caesium sulfate density gradient fractionation; note their resemblance to Ilarvirus particles (see Chapter 7, Figures 1 through 3). All micrographs are at the same magnification. Bar = 100 nm.

isolated RNA and protein.[20-21] AMV RNA has a very high affinity for coat protein subunits, to the extent that free RNA is capable of removing protein subunits from intact particles.[22]

AMV particles are unstable in the presence of ribonuclease which digests the RNA into fragments and this results in the degradation of the particles into smaller structures.[23] It seems that the enzyme may gain access to the encapsidated RNA through the large spaces between the subunits of the capsid.[8]

The N-terminal region of AMV coat protein is very basic and can be cleaved with trypsin.[24] On removal of this 26 residue end fragment, the protein can be assembled into isometric particles with T = 1 symmetry[24] which can then by crystallized.[25] However, it cannot assemble into elongated particles in the presence of RNA, indicating that the histone-like N-terminal region is essential for RNA-protein interaction during assembly.[25] Recent proton magnetic resonance studies have revealed that the N-terminal 36 or 50 amino acid residues form a flexible random coil structure.[26]

B. The Viral Genome

Like the Ilarviruses, AMV has a coat protein-dependent tripartite genome[5] consisting of RNAs 1, 2, and 3 of M_r 1.04×10^6, 0.73×10^6, and 0.62×10^6, all of which are encapsidated separately in B, M, and Tb particles, respectively. The subgenomic mRNA for coat protein (RNA 4) of M_r 0.25×10^6 is also encapsidated, two molecules per Ta particle.[11,27,28] When present, minor RNA species intermediate in size between RNAs 3 and 4 (X-RNAs) or smaller than RNA 4 (Z-RNAs) are packaged in additional nucleoprotein particle types as well as in some of the Ta and Tb particles.[9]

Although RNAs 1, 2, and 3 represent the entire AMV genome, they are not infective unless encapsidated. Alternatively, infective inoculum must contain RNAs 1, 2, 3, and 4 or a mixture of RNAs 1, 2, and 3 with viral coat protein added.[29,30] Because RNA 4 is the mRNA for coat protein,[27,28] it is logical to assume that on inoculation of plant cells with all four RNAs of AMV, RNA 4 is translated into coat protein which then activates the genome.[29] It has been estimated that after inoculation, the penetration and translation of RNA 4 takes from 15 to 30 min to be completed.[31] Coat protein activation of genomic RNAs is not very specific. It has been demonstrated that the coat protein of the Ilarvirus tobacco streak virus, which is quite distinct from that of AMV both chemically and serologically, can activate the AMV genomic RNAs.[32] However, coat proteins of some Cucumoviruses, Bromoviruses, and Tobamoviruses were unable to activate the genomes of AMV or several Ilarviruses.[29,32,33]

At present the role of the coat protein in the activation of the AMV genome is obscure. However, it has been observed that the activation is more efficient when the coat protein is added to all three genomic RNAs than when it is preincubated with only one or two of the RNAs before the complete inoculum is mixed; this suggests that the coat protein interacts with all three genomic RNAs.[34] It also seems likely that the coat protein acts by changing the conformation of the RNAs, enabling them to take part in some essential infection event such as replicase recognition.[35-37] Such changes in conformation upon binding of coat protein have been detected by studies involving calorimetry[38] as well as fluorometry and binding of ethidium bromide.[37] It has been demonstrated that the coat protein binds preferentially at sites near the 3′ end of RNA 4[35] and because the other three AMV RNAs share at least 80% sequence homology in the last 150 3′ terminal nucleotides,[39-41] the affinity for coat protein is probably in that region. This suggestion is supported by studies of coat protein binding sites on AMV RNA 1.[42]

All four AMV RNAs are capped by $m^7G^{5'}ppp$[43] but lack poly A tails at their 3′ ends and also the ability to accept amino acids.[44]

The complete sequences of RNAs 1 to 4 of AMV strain 425 have been determined and consist of 3644, 2593, 2037, and 881 nucleotides, respectively.[45-48] The sequence of RNA

1 reveals a long open reading frame for a protein of M_r 125.7 × 10^3 flanked by 5′ and 3′ noncoding regions of 100 and 163 nucleotides, respectively.[45] RNA 2 contains a reading frame for a protein of M_r 89.75 × 10^3 flanked by 54 and 166 noncoding nucleotides at the 5′ and 3′ ends, respectively.[46]

RNA 3 of AMV is dicistronic. At its 5′ end there is a 240 nucleotide noncoding sequence followed by 903 nucleotides of the cistron coding for a protein of M_r 32.40 × 10^3.[47] This is followed by an intracistronic region of 49 nucleotides, 36 of which are homologous to the noncoding region of RNA 4 at its 5′ end.[47] The coat protein cistron of 666 nucleotides is followed by 179 noncoding nucleotides at the 3′ end of RNA 3.[47]

The 881 nucleotide sequence of AMV RNA 4 is homologous with that of RNA 3 at its 3′ end.[47-48] The 39 nucleotide leader sequence at the 5′ end is the same in the RNA 4s of seven different strains of AMV, except at position 26, where A, G, or U residues were found.[49] The order and number of amino acids in the coat protein, as derived from the RNA data,[48] is in agreement with that found by directly sequencing the protein.[13]

In vitro, RNA 4 is an efficient monocistronic mRNA for coat protein.[27] Although RNA 3 is dicistronic, its in vitro translation results almost entirely in the M_r 32.4 × 10^3 protein and only traces of coat protein have been detected in some systems.[27,50,51]

In most in vitro systems, AMV RNAs 1 and 2 are translated into a large number of polypeptides.[52,53] The largest products have M_r of about 100 ± 20 × 10^3, which would correspond to the translation of almost the entire molecules of either RNAs 1 or 2. In the rabbit reticulocyte cell-free system, RNAs 2, 3, and 4 direct the synthesis, almost exclusively, of polypeptides of M_r about 100 × 10^3, 35 × 10^3, and 24 × 10^3 (coat protein), respectively.[54] Synthesis of the smaller products can be suppressed by a tRNA which can be charged with glutamine.[55] On the other hand, polypeptide synthesis directed by RNA 1 depends on the concentration of the RNA in the reaction mixture.[54] At low concentrations, RNA 1 directs the synthesis of a polypeptide of M_r 115 × 10^3, whereas at high concentrations a mixture of two proteins is produced with M_r about 58 × 10^3 and 62 × 10^3. Furthermore, at RNA 1 concentrations directing optimum polypeptide synthesis, all three proteins are produced. They represent overlapping peptide chains with identical N-termini. Thus, the presence of what appear to be leaky termination codons in RNA 1 allows it to direct the translation of three different polypeptides starting from a single initiation site.[55] However, the leaky terminators are not UGA as no such sequences in the appropriate positions can be located in RNA 1.[45] Current knowledge about the translation strategy of AMV is summarized in Figure 2.

Several pseudorecombinants between naturally occurring AMV strains, and artifically induced mutants have been constructed in vitro and inoculated to plants to determine the genetic information carried by each of the three genomic components. The only character so far found to be controlled by RNA 1 is the thermoresistance of the virus in vivo.[56] RNA 2 has been shown to control the ability of AMV strain YCMV to infect beans systemically.[57] The majority of the genetic characters studied have been shown to be controlled by RNA 3. These include the serological properties of the particles, a number of symptom reactions on plants, virus sensitivity to cycloheximide, particle component ratios, and poor virus multiplication at low temperatures.[56-59]

II. CYTOPATHOLOGY

A number of authors have observed AMV particles in thin sections in the cytoplasm of infected plant cells.[60-67] All the observations indicate that in cells of fully infected leaves, masses of AMV particles accumulate in the cytoplasm (Figures 3 and 4). In most instances the particles are randomly distributed throughout the cytoplasm but in others, prominent aggregates of regularly arranged particles have been seen (Figure 3B). Both free and aggregated particles have also been observed in the vacuoles of infected cells.[60,61,63,64]

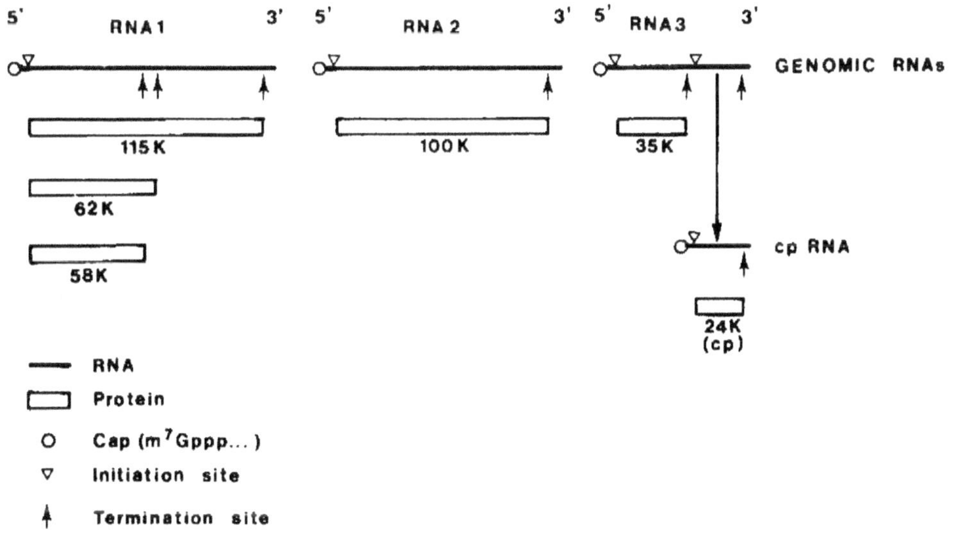

FIGURE 2. Physical map of the organization of the AMV genome.

Hull and colleagues[60] examined cells of plants infected with 24 different strains of AMV. Eleven of the strains produced only randomly distributed particles in the cytoplasm. The other 13 strains, in addition to randomly distributed particles, formed particle aggregates which they classified into four morphological types. Cells of plants infected by only certain strains contained particles in the vacuoles. Many such particles formed vacuolar inclusions like those in the cytoplasm. The vacuolar aggregates were shown to be similar to those induced in vitro by agents such as divalent cations, ribonuclease, and low pH.[1] It therefore seems likely that the vacuolar aggregates were induced by the ionic conditions in the vacuolar sap;[1] a similar mechanism could also be involved in the formation of cytoplasmic inclusions at least of some types.[1]

Although there is no convincing evidence of AMV particles invading any organelles, some authors have reported the presence of long tubular structures reminiscent of virus particles in the nuclei of cells infected by some strains of the virus.[60,65] Highly convoluted nuclei have been seen in some infected cells (Figure 4). It has also been observed, using electron microscopic autoradiography, that leaf cells infected with AMV and treated with actinomycin D to inhibit host-directed RNA synthesis, incorporate ^{3}H-labeled uridine mainly into the nuclei and cytoplasm.[68] This suggests involvement of the nucleus in AMV RNA synthesis. However, it is not clear if labeling in the cytoplasm represented RNA synthesised there or the accumulation of RNA previously synthesized in the nucleus.[68]

Double-stranded RNAs which appeared to be duplexes of each of the three AMV genomic RNAs have been isolated from infected plants.[69] However, their location in the infected cell has not been determined. Similarly, the location of RNA-dependent RNA polymerase associated with AMV infection remains unknown.[70]

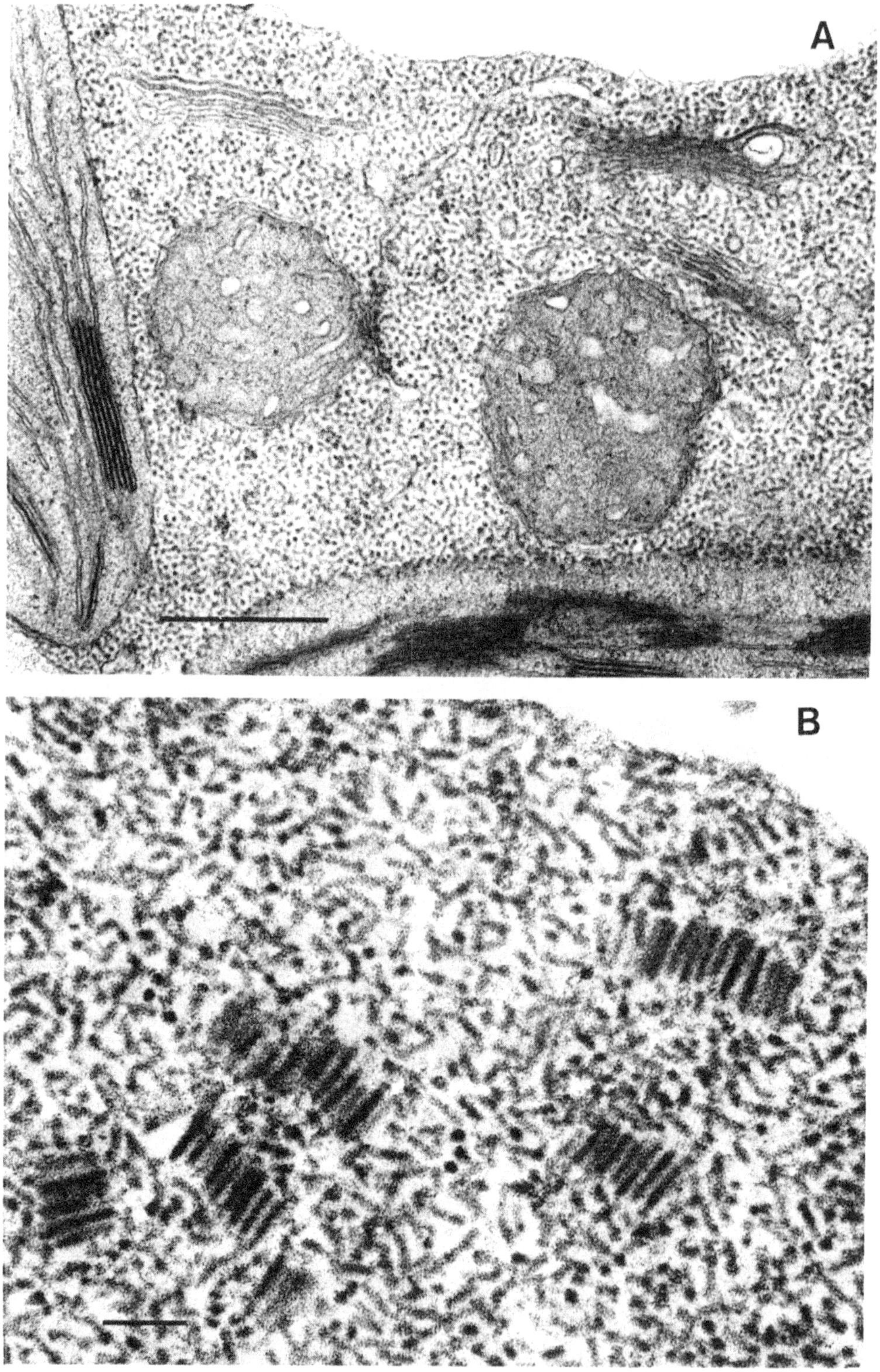

FIGURE 3. Thin sections of *Nicotiana clevelandii* leaf cells infected with AMV (tissue cleared of ribosomes by RNase treatment). A, Cytoplasm is filled with virus particles in random orientation (bar = 500 nm); B, some of the longer particles have formed regular arrays; bar = 100 nm.

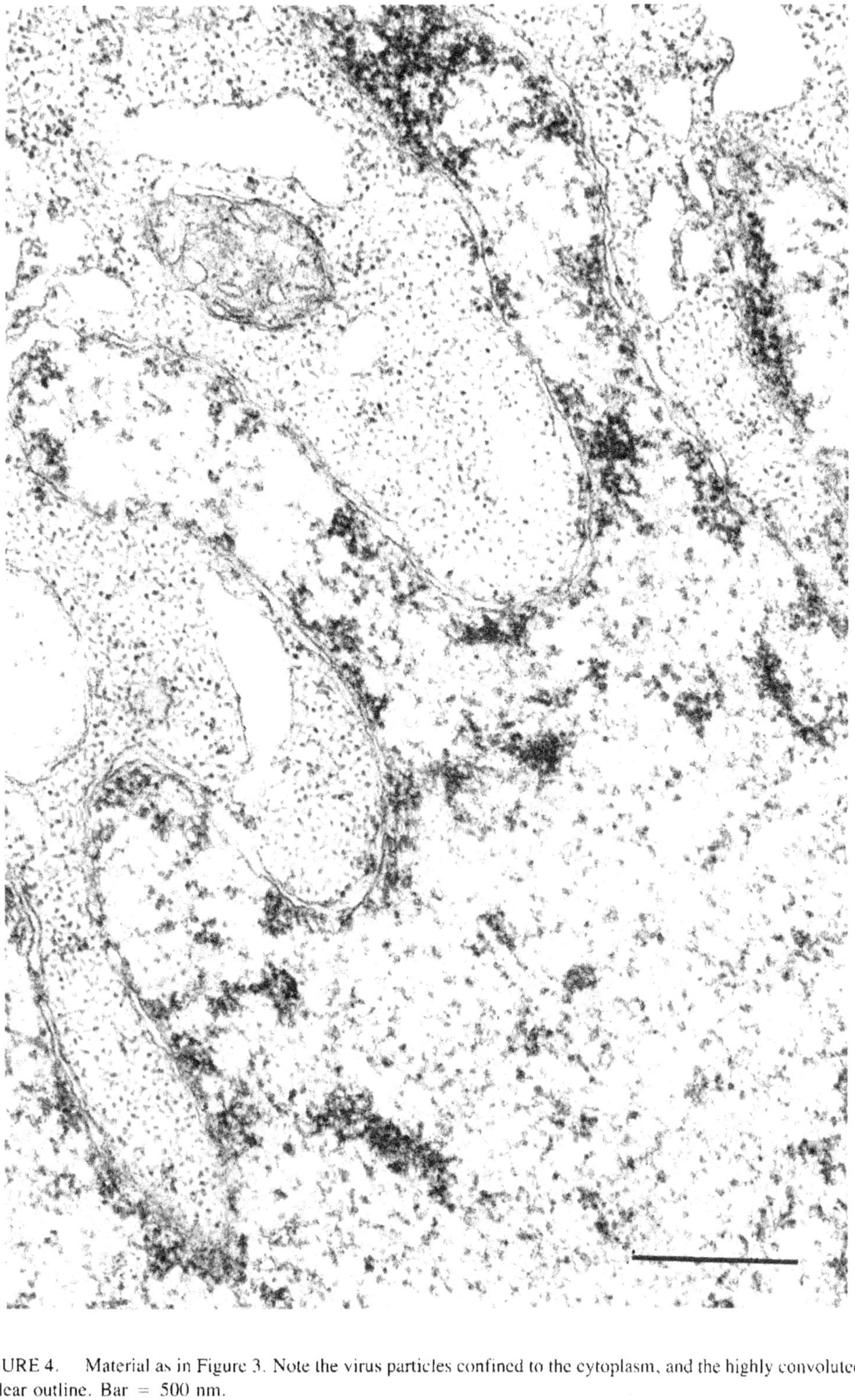

FIGURE 4. Material as in Figure 3. Note the virus particles confined to the cytoplasm, and the highly convoluted nuclear outline. Bar = 500 nm.

REFERENCES

1. **Hull, R.,** Alfalfa mosaic virus, *Adv. Virus Res.*, 15, 365, 1969.
2. **Crill, P., Hagedorn, D. J., and Hanson, E. W.,** Alfalfa mosaic, the disease and its virus incitant, *Univ. Wis. Res. Bull.*, No. 280, 1970. 40 pp.
3. **Jaspars, E. M. J. and Bos, L.,** Alfalfa mosaic virus, *CMI/AAB Descriptions of Plant Viruses*, No. 229, 1980.
4. **Mellema, J. E. and van den Berg, H. J. N.,** The quaternary structure of alfalfa mosaic virus, *J. Supramolec. Struc.*, 2, 17, 1974.
5. **Jaspars, E. M. J.,** Plant viruses with a multipartite genome, *Adv. Virus Res.*, 19, 37, 1974.
6. **van Regenmortel, M. H. V. and Pinck, L.,** Alfalfa mosaic virus, in *Handbook of Plant Virus Infections and Comparative Diagnosis*, Kurstak, E., Ed, Elsevier, Amsterdam, 1981, 415.
7. **van Vloten-Doting, L., Francki, R. I. B., Fulton, R. W., Kaper, J. M., and Lane, L. C.,** Tricornaviridae — a proposed family of plant viruses with tripartite, single-stranded RNA genomes, *Intervirology*, 15, 198, 1981.
8. **Mellema, J. E.,** Model for the capsid structure of alfalfa mosaic virus, *J. Mol. Biol.*, 94, 643, 1975.
9. **Bol, J. F. and Lak-Kaashoek, M.,** Composition of alfalfa mosaic virus nucleoproteins, *Virology*, 60, 476, 1974.
10. **Hull, R.,** Studies on alfalfa mosaic virus. IV. An unusual strain, *Virology*, 42, 283, 1970.
11. **Heijtink, R. A. and Jaspars, E. M. J.,** RNA contents of abnormally long particles of certain strains of alfalfa mosaic virus, *Virology*, 59, 371, 1974.
12. **Kraal, B., van Beynum, G. M. A., de Graaf, J. M., Castel, A., and Bosch, L.,** The primary structure of the coat protein of alfalfa mosaic virus, *Virology*, 74, 232, 1976.
13. **van Beynum, G. M. A., de Graaf, J. M., Castel, A., Kraal, B., and Bosch, L.,** Structural studies on the coat protein of alfalfa mosaic virus, *Eur. J. Biochem.*, 72, 63, 1977.
14. **Collot, D., Peter, R., Das, B., Wolff, B., and Duranton, H.,** Primary structure of alfalfa mosaic virus coat protein (Strain S), *Virology*, 74, 236, 1976.
15. **Castel, A., Kraal, B., de Graaf, J. M., and Bosch, L.,** The primary structure of the coat protein of alfalfa mosaic virus strain VRU, *Eur. J. Biochem.*, 102, 125, 1979.
16. **Cusack, S., Miller, A., Krijgsman, P. C. J., and Mellema, J. E.,** An investigation of the structure of alfalfa mosaic virus by small-angle neutron scattering, *J. Mol. Biol.*, 145, 525, 1981.
17. **Cremers, A. F. M., Oostergetel, G. T., Schilstra, M. J., and Mellema, J. E.,** An electron microscopic investigation of the structure of alfalfa mosaic virus, *J. Mol. Biol.*, 145, 545, 1981.
18. **Heijtink, R. A., Houwing, C. J., and Jaspars, E. M. J.,** Molecular weights of particles and RNAs of alfalfa mosaic virus. Number of subunits in protein capsids. *Biochemistry*, 16, 4684, 1977.
19. **Kaper, J. M.,** Arrangement and identification of simple isometric viruses according to their domination stabilizing interactions. *Virology*, 55, 299, 1973.
20. **Bol, J. F. and Kruseman, J.,** The reversible dissociation of alfalfa mosaic virus, *Virology*, 37, 485, 1969.
21. **Hull, R.,** Studies on alfalfa mosaic virus. III. Reversible dissociation and reconstitution studies, *Virology*, 40, 34, 1970.
22. **Verhagen, W., van Boxsel, J. A. M., Bol, J. F., van Vloten-Doting, L., and Jaspars, E. M. J.,** RNA-protein interactions in alfalfa mosaic virus, *Ann. Microbiol. (Paris)*, 127a,165, 1976.
23. **Bol, J. F. and Veldstra, H.,** Degradation of alfalfa mosaic virus by pancreatic ribonuclease, *Virology*, 37, 74, 1969.
24. **Bol, J. F., Kraal, B., and Brederode, F. Th.,** Limited proteolysis of alfalfa mosaic virus: influence on the structural and biological function of the coat protein, *Virology*, 58, 101, 1974.
25. **Fukuyama, K., Abdel-Meguid, S. S., and Rossmann, M. G.,** Crystallization of alfalfa mosaic virus coat protein as a T = 1 aggregate, *J. Mol. Biol.*, 150, 33, 1981.
26. **Kan, J. H., Andree, P-J., Kouijzer, L. C., and Mellema, J. E.,** Proton-magnetic-resonance studies on the coat protein of alfalfa mosaic virus, *Eur. J. Biochem.*, 126, 29, 1982.
27. **Gerlinger, P., Mohier, E., Le Meur, M. A., and Hirth, L.,** Monocistronic translation of alfalfa mosaic virus RNAs, *Nucleic Acids Res.*, 4, 813, 1977.
28. **Castel, A., Kraal, B., Kerklaan, P. R. M., Klok, J., and Bosch, L.,** Initiation of polypeptide synthesis with various NH_2-blocked aminoacyl-tRNAs under the direction of alfalfa mosaic virus RNA4, *Proc. Natl. Acad. Sci. U.S.A.*, 74, 5509, 1977.
29. **Bol, J. F., van Vloten-Doting, L., and Jaspars, E. M. J.,** A functional equivalence of top component *a* RNA and coat protein in the initiation of infection by alfalfa mosaic virus, *Virology*, 46, 73, 1971.
30. **Alblas, F. and Bol, J. F.,** Coat protein is required for infection of cowpea protoplasts with alfalfa mosaic virus, *J. Gen. Virol.*, 41, 653, 1978.
31. **van Vloten-Doting, L.,** Early events in the infection of tobacco with alfalfa mosaic virus, *J. Gen. Virol.*, 41, 649, 1978.

32. **van Vloten-Doting, L.,** Coat protein is required for infectivity of tobacco streak virus: biological equivalence of the coat proteins of tobacco streak and alfalfa mosaic viruses, *Virology*, 65, 215, 1975.
33. **Gonsalves, D. and Garnsey, S. M.,** Infectivity of heterologous RNA-protein mixtures from alfalfa mosaic, citrus leaf ugose, citrus variegation, and tobacco streak viruses, *Virology*, 67, 319, 1975.
34. **Smit, C. H. and Jaspars, E. M. J.,** Activation of the genome of alfalfa mosaic virus is enhanced by the presence of the coat protein on all three genome parts, *Virology*, 104, 454, 1980.
35. **Houwing, C. J. and Jaspars, E. M. J.,** Coat protein binds to the 3'-terminal part of RNA 4 of alfalfa mosaic virus, *Biochemistry*, 17, 2927, 1978.
36. **Koper-Zwarthoff, E. C. and Bol, J. F.,** Nucleotide sequence of the putative recognition site for coat protein in the RNAs of alfalfa mosaic virus and tobacco streak virus, *Nucleic Acids Res.*, 8, 3307, 1980.
37. **Srinivasan, S. and Jaspars, E. M. J.,** Alterations of the conformation of the RNAs of alfalfa mosaic virus upon binding of a few coat protein molecules, *Biochim. Biophys. Acta*, 696, 260, 1982.
38. **Srinivasan, S. and Jaspars, E. M. J.,** Colorimetric studies on the interaction of RNA and coat protein of alfalfa mosaic virus, *FEBS Lett.*, 80, 288, 1977.
39. **Pinck, L. and Pinck, M.,** Sequence homology at the 3' ends of alfalfa mosaic virus RNAs, *FEBS Lett.*, 107, 61, 1979.
40. **Koper-Zwarthoff, E. C., Brederode, F.Th., Walstra, P., and Bol, J. F.,** Nucleotide sequence of the 3'-noncoding region of alfalfa mosaic virus RNA4 and its homology with the genomic RNAs, *Nucleic Acids Res.*, 7, 1887, 1979.
41. **Gunn, M. R. and Symons, R. H.,** Sequence homology at the 3'-termini of the four RNAs of alfalfa mosaic virus, *FEBS Lett.*, 109, 145, 1980.
42. **Zuidema, D., Brerhuizeu, M. F. A., Cornelissen, B. J. C., Bol, J. F., and Jaspars, E. M. J.,** Coat protein binding sites on RNA 1 of alfalfa mosaic virus, *Virology*, 125, 361, 1983.
43. **Pinck, L.,** The 5'-end groups of alfalfa mosaic virus RNAs are $m^7G^{5'}ppp^{5'}Gp$, *FEBS Lett.*, 59, 24, 1975.
44. **van Vloten-Doting, L. and Jaspars, E. M. J.,** Plant covirus systems: three component systems, *Compr. Virol.* 11, 1, 1977.
45. **Cornelissen, B. J. C., Brederode, F.Th., Moorman, R. J. M., and Bol, J. F.,** Complete nucleotide sequence of alfalfa mosaic virus RNA1, *Nucleic Acids Res.*, 5, 1253, 1983.
46. **Cornelissen, B. J. C., Brederode, F. Th., Veeneman, G. H., van Boom, J. H., and Bol, J. F.,** Complete nucleotide sequence of alfalfa mosaic virus RNA2, *Nucleic Acids Res.*, 11, 3019, 1983.
47. **Barker, R. F., Jarvis, N. P., Thompson, D. V., Loesch-Fries, L. S., and Hall, T. C.,** Complete nucleotide sequence of alfalfa mosaic virus RNA 3, *Nucleic Acids Res.*, 11, 2881, 1983.
48. **Brederode, F.Th., Koper-Zwarthoff, E. C., and Bol, J. F.,** Complete nucleotide sequence of alfalfa mosaic virus RNA4, *Nucleic Acids Res.*, 8, 2213, 1980.
49. **Swinkels, P. P. H. and Bol, J. F.,** Limited sequence variation in the leader sequence of RNA4 from several strains of alfalfa mosaic virus, *Virology*, 106, 145, 1980.
50. **Mohier, E., Hirth, L., Le Meur, M-A., and Gerlinger, P.,** Analysis of alfalfa mosaic virus 17S RNA translational products, *Virology*, 71, 615, 1976.
51. **Neeleman, L., Rutgers, T., and van Vloten-Doting, L.,** Internal initiation of protein synthesis on RNA of eukaryotic virus?, in *Translation of Natural and Synthetic Polynucleotides*, Legocki, A. B., Ed., University of Agriculture, Pozmari, Poland, 1977, 292.
52. **van Vloten-Doting, L., Bol, J., Neeleman, O., Rutgers, T., van Dalen, D., Castel, A., Bosch, L., Marbaix, G., Huez, G., Hubert, E., and Clenter, Y.,** *In vivo* and *in vitro* translation of the RNAs of alfalfa mosaic virus, in *Nucleic Acids and Protein Synthesis in Plants*, Bogoard, L. and Wiel, J. H., Eds., NATO Advanced Studies Institute Series, Plenum Press, New York, 1977, 387.
53. **Davies, J. W.,** Translation of plant virus ribonucleic acids in extracts from eukaryotic cells, in *Nucleic Acids in Plants*, Vol. 2, Hall, T. C. and Davies, J. W., Eds., CRC Press, Boca Raton, Fla., 1979, 111.
54. **van Tol, R. G. L. and van Vloten-Doting, L.,** Translation of alfalfa mosaic virus RNA1 in the mRNA dependent translation system from rabbit reticulocyte lysates, *Eur. J. Biochem.*, 93, 461, 1979.
55. **van Tol, R. G. L., van Gemeren, R., and van Vloten-Doting, L.,** Two leaky termination codons in AMV RNA 1, *FEBS Lett.*, 118, 67, 1980.
56. **Franck, A. and Hirth, L.,** Temperature-resistant strains of alfalfa mosaic virus, *Virology*, 70, 283, 1976.
57. **Dingjan-Versteegh, A., van Vloten-Doting, L., and Jaspars, E. M. J.,** Alfalfa mosaic virus hybrids constructed by exchanging nucleoprotein components, *Virology*, 49, 716, 1972.
58. **Dingjan-Versteegh, A., van Vloten-Doting, L., and Jaspars, E. M. J.,** Confirmation of the constitution of alfalfa mosaic virus hybrid genomes by backcross experiments, *Virology*, 59, 328, 1974.
59. **Hartmann, D., Mohier, E., Leroy, C., and Hirth, L.,** Genetic analysis of alfalfa mosaic virus mutants, *Virology*, 74, 470, 1976.
60. **Hull, R., Hills, G. J., and Plaskitt, A.,** Electron microscopy on *in vivo* aggregation forms of a strain of alfalfa mosaic virus, *J. Ultrastruct. Res.*, 26, 465, 1969.
61. **Hull, R., Hills, G. J., and Plaskitt, A.,** The *in vivo* behaviour of twenty-four strains of alfalfa mosaic virus, *Virology*, 42, 753, 1970.

62. **Gerola, F. M., Bassi, M., and Betto, E.,** A sub-microscopical study of leaves of alfalfa, basil and tobacco experimentally infected with lucerne mosaic virus, *Protoplasma*, 67, 307, 1969.
63. **de Zoeten, G. A. and Gaard, G.,** Distribution and appearance of alfalfa mosaic virus in infected plant cells, *Virology*, 39, 768, 1969.
64. **Hull, R. and Plaskitt, A.,** Electron microscopy on the behaviour of two strains of alfalfa mosaic virus in mixed infections, *Virology*, 42, 773, 1970.
65. **Conti, G. G., Favali, M. A., and Vegetti, G.,** The behaviour of yellow spot mosaic virus, a strain of alfalfa mosaic virus, in different host plants. Symptomatology and ultrastructural observations, *Rev. Patol. Veg.*, 8, 323, 1972.
66. **Wilcoxson, R. D., Frosheiser, F. I., and Johnson, L. B.,** The appearance of eight strains of alfalfa mosaic virus in alfalfa leaves and their effect on ultrastructure of infected cells, *Can. J. Bot.*, 52, 979, 1974.
67. **Hatta, T. and Francki, R. I. B.,** Identification of small polyhedral virus particles in thin sections of plant cells by an enzyme cytochemical technique, *J. Ultrastruct. Res.*, 74, 116, 1981.
68. **Bassi, M., Favali, M. A., Conti, G. G., and Betto, E.,** Uridine-^{3}H incorporation in leaf cells infected with lucerne mosaic virus. A quantitative electron microscopic autoradiographic study, *Phytopathol. Z.*, 69, 247, 1970.
69. **Mohier, E., Pinck, L., and Hirth, L.,** Replication of alfalfa mosaic virus RNAs, *Virology*, 58, 9, 1974.
70. **Linthorst, H. J. M., Bol, J. F., and Jaspars, E. M. J.,** Lack of temperature sensitivity of RNA-dependent RNA polymerases isolated from tobacco plants infected with *ts* mutants of alfalfa mosaic virus, *J. Gen. Virol.*, 46, 511, 1980.

Chapter 9

TOBAMOVIRUS GROUP

The name Tobamovirus is derived from the type member of the group, tobacco mosaic virus (TMV). The virus particles are rigid tubes 18 nm in diameter, and generally have one modal length of 300 nm. They are built from a single species of protein subunit of M_r about 17.6×10^3 arranged in a helix; each particle contains a single molecule of positive-sense, single-stranded RNA of M_r about 2×10^6.[1-3]

Definite members of the group have very stable particles that usually occur in high concentrations in their host plants. They are easily mechanically transmissible, and are naturally transmitted by contact and incidental wounding. There seem to be no efficient natural vectors. Seed transmission sometimes occurs.[1,4-6]

I. MEMBERS OF THE GROUP AND THEIR RELATIONSHIPS

A. True Tobamoviruses

The number of isolates and strains of Tobamoviruses that have been studied and named is very large.[1,3,7-9] Many of the viruses have been described under different names and some of those now considered to be distinct viruses (Table 1) have in the past been referred to as strains of TMV. All the Tobamoviruses have very similar physical properties but vary in the primary structure of their RNA[10-13] and coat protein[1,14-17] as well as in their host range.[1,18] At present there appear to be at least eleven distinct Tobamoviruses (Table 1).

Siegel and Wildman[18] compared 8 Tobamovirus isolates labeled U_1 to U_8, of which U_1 was type or "vulgare" TMV and U_3, U_4, U_6, and U_7 resembled it. U_2 (T_2MV) is now recognized as a separate virus, while U_5 has affinities with both TMV and T_2MV.[1,11,18] U_8 was derived from the virus isolated originally by Holmes from ribgrass (*Plantago lanceolata*)[26] and is listed in Table 1 as ribgrass mosaic virus (RMV).[22] Two Tobamovirus isolates have frequently been worked on in Japan, the common or OM[27] and cowpea "strain"[28] of TMV. It seems that OM is very similar to U_1TMV and should be considered as an isolate of TMV[19] while the cowpea "strain" is almost certainly an isolate of sunn-hemp mosaic virus (SHMV).[5]

Two strains of tomato mosaic virus (ToMV), dahlemense and Y-TAMV (yellow tomato atypical mosaic virus), are well known. However, G- (green) TAMV is not a strain of ToMV, but is related to T_2MV.[14,29-31] The name *para*-tobacco mosaic has been used for strains of T_2MV.

Cucumber virus 3 (CV3) is now named cucumber green mottle mosaic virus (CGMMV), and cucumber virus 4 (CV4) has been included as a strain of this virus,[4] but is now recognized as distinct.[9]

Frangipani mosaic virus (FMV) is a distinct virus found only in frangipani,[20] odontoglossum ringspot virus (ORSV) is one of the several Tobamoviruses found in orchids,[21] Sammons' opuntia virus (SOV) was isolated from a prickly pear cactus,[23] and SHMV represents a group of several Tobamovirus strains infecting legumes.[5] Youcai mosaic virus (YMV) is a distinct though not officially recognised Tobamovirus infecting crucifers and having affinities with RMV.[25]

Recently the host reactions and serological relationships of Tobamoviruses infecting peppers (*Capsicum* spp.), eggplant (*Solanum melongena*), and tobacco were compared.[8] The viruses fell into two serological groups, containing strains A1, F0, P8, P14, and SL (group 1), and P11 and O6 (group 2), respectively. The strains were distantly related serologically to TMV, and some to ToMV, but no comparisons were made to other recognized Tobamoviruses.[8]

Table 1
MEMBERS OF THE TOBAMOVIRUS GROUP[a]

Virus	Natural hosts	Ref.
Tobacco mosaic virus (TMV)	Tobaccos	19
Cucumber green mottle mosaic virus (CGMMV) (= cucumber virus 3)	Cucurbits	4
Cucumber virus 4 (CV4)	Cucurbits	1
Frangipani mosaic virus (FMV)	Frangipani	20
Odontoglossum ringspot virus (ORSV)	Orchids	21
Ribgrass mosaic virus (RMV)	*Plantago* species	22
Sammons' opuntia virus (SOV)	Cacti	23,24
Sunn-hemp mosaic virus (SHMV) (= cowpea strain of TMV)	Leguminosae	5
Tomato mosaic virus (ToMV)	Tomatoes	6
Tobacco mosaic U_2 virus (T_2MV) (= para-tobacco mosaic virus)	Solanaceae	1
Youcai mosaic virus (YMV) (= Chinese rape mosaic virus)	Cruciferae	25

[a] Data summarized from Gibbs[1] and Matthews.[9]

Quantifiable indicators of affinity among the Tobamoviruses are (1) serological relationships between the coat proteins,[3,14,15,30-32] (2) amino acid composition of the coat proteins,[1,16,17] and (3) the extent of nucleotide sequence homology of the RNAs.[10-13] Serology and amino acid analysis reveal relationships over a wide range of variants whereas nucleotide sequence homology analysis at present permits only close relationships to be detected. Nevertheless, this last technique should in principle be a better approach as it takes the whole genome into consideration and not only one gene product. The entire genome of type TMV has recently been sequenced[33] so that direct comparisons of entire Tobamovirus genomes will soon be possible.

Figures 1 and 2 summarize the results obtained from serology, amino acid analysis, and nucleotide sequence homology studies of certain Tobamoviruses. The correlation between the results of these approaches is close, except that ORSV falls near TMV, ToMV, and especially T_2MV according to amino acid composition, whereas according to sequence homology it is quite distinct. It is also interesting that CGMMV and CV_4 appear closely related in sequence homology, but are rather far apart (6.2 SDI units) serologically. Both these results are, however, compatible with the amino acid composition dendrogram (Figure 2).

B. Possible Tobamoviruses

Certain other viruses (Table 2) have affinities with the Tobamoviruses and are at present considered as possible members of the group.[9] They are in fact a rather heterogeneous collection whose taxonomic positions are far from clear. This is because their morphology, serology, and physicochemical properties have generally not been carefully studied and compared.

Chara corallina virus (CCV)[34,35] stands alone in having a green alga (*Chara corallina*, Characeae) as its host. It has straight tubular particles 18nm in diameter like true Tobamoviruses, but the pitch of the helix was found to be 2.75 nm (2.66 nm after correction of possible error), and the particles have a modal length of 532 nm. The RNA base ratio fits within the range found among the Tobamoviruses, but the coat protein amino acid composition is very distinct. The virus was found to be serologically distantly related to TMV and ORSV,[1,34,35] though this result needs careful reinvestigation.

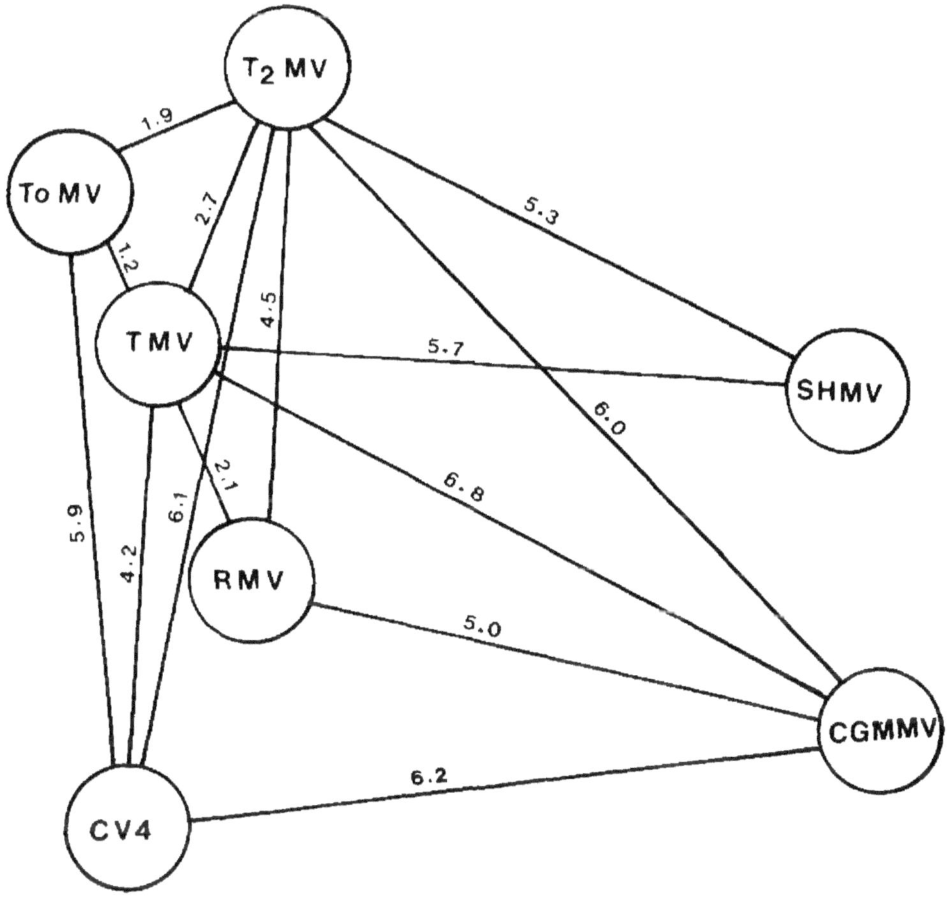

FIGURE 1. Serological relationships among some of the Tobamoviruses (see Table 1 for explanation of acronyms). Serological differentiation indices (SDI) between pairs of viruses are shown. Data after van Regenmortel.[14,15,30,31]

CCV apart, the other possible Tobamoviruses have particle morphology somewhat similar to that of true Tobamoviruses, but the electron microscopy has often not been good enough to say exactly how similar. The particles of some of the viruses appear to have a significantly larger diameter and helix pitch. All the viruses have particles of two or more modal lengths, and the particles are generally rather unstable. The host ranges of the viruses are relatively narrow and some of them are vectored by plasmodiophoromycete root-infecting fungi (Table 2).

Gibbs[1] has suggested that beet necrotic yellow vein virus (BNYVV), potato mop top virus (PMTV), soil-borne wheat mosaic virus (SBWMV), and probably also peanut clump virus (PCV) and nicotiana velutina mosaic virus (NVMV) may be closely related to one another, and Thouvenal and Fauquet[45] have suggested that PCV and related viruses may eventually be constituted as a separate group intermediate between the Tobamoviruses and the Hordeiviruses (see Chapter 10). SBWMV and PMTV were found to be serologically related,[55] and there is evidence of their distant serological relationship to TMV.[49,51] This work should be repeated carefully, in view of the finding that at least SBWMV resembles the Hordeiviruses rather than the Tobamoviruses in morphology[54] (Figures 3 and 4). BNYVV is very distinct from the definitive Tobamoviruses in coat protein amino acid composition and is now known, like the Hordeiviruses but unlike the Tobamoviruses, to have a divided genome.[53]

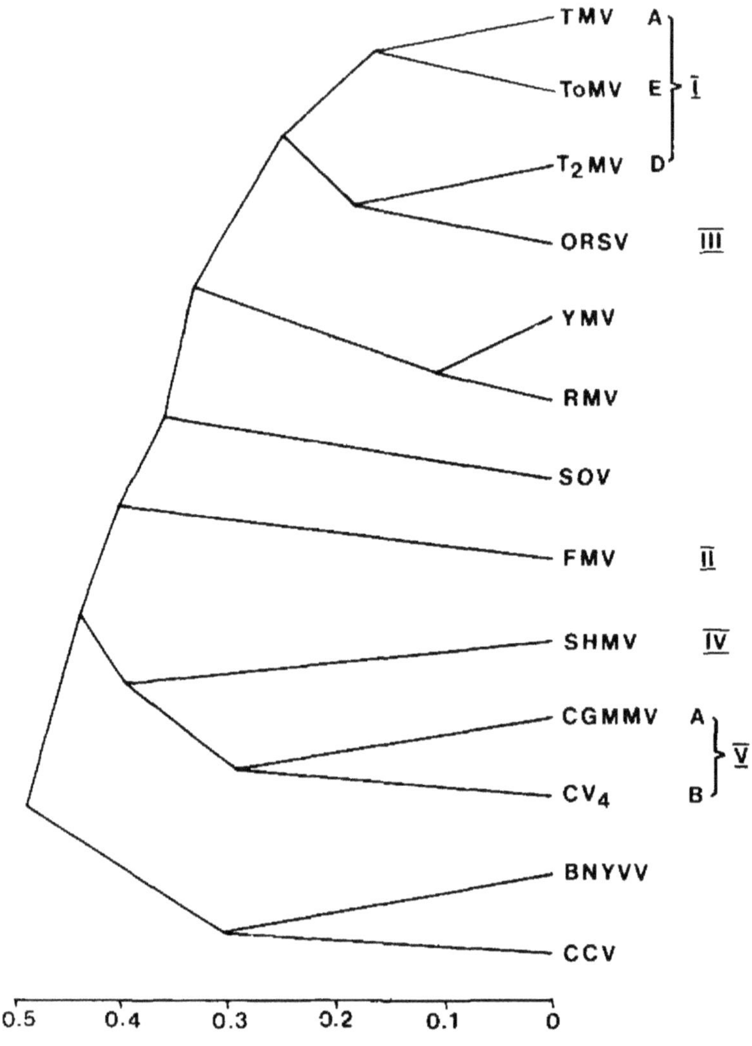

FIGURE 2. Relationships among some Tobamoviruses based on their coat protein amino acid composition[1,17] and nucleotide sequence homology of their RNAs.[11] The dendrogram indicates relationships computed from the viral coat protein amino acid compositions using an agglomerative hierarchial classification,[1,17] with the horizontal scale being a quantitative measure of dissimilarity at which the clusters fuse. Roman numbers refer to groups and capital letters to subgroups delimited by viral RNA nucleotide sequence homology.[11]

It appears that PCV is not serologically related to TMV, BNYVV, PMTV, or NVMV, or to the Hordeivirus, barley stripe mosaic virus (BSMV), but the particle buoyant density and sedimentation values of PCV are close to those of BSMV.[45] Indian peanut clump virus (IPCV) resembles PCV (which comes from West Africa) in the properties compared, but does not appear to be serologically related.[46]

Hypochoeris mosaic virus (HyMV) seems to have properties similar to those of other possible Tobamoviruses, except that it forms characteristic crystalline inclusions in infected cells; it was found to be serologically unrelated to BNYVV, broad bean necrosis virus (BBNV), BSMV, PCV, PMTV, NVMV, SBWMV, and defective strains of TMV.[42,43]

Table 2
SOME PROPERTIES OF POSSBILE MEMBERS OF THE TOBAMOVIRUS GROUP

Possible members	Particle length(s) (nm)	Particle width (nm)	M_r of RNA ($\times 10^6$)	M_r of coat protein ($\times 10^3$)	Pitch of particle helix (nm)	Cytopathology[a]	Host range	Natural vector	Ref.
Chara corallina virus (CCV)	332	18	3.6	17.5	2.66 2.75	ND[b]	*C. corallina* (Characeae, Algae)	Unknown	1,34,35
Beet necrotic yellow vein virus (BNYVV)	390; 270; 65—105 or 390; 270	~20	2.3; 1.8; 0.7; (0.6)[c]	2[illegible].0	2.6	C,A,M	Chenopodiaceae + narrow experimental	*Polymyxa betae*	1,36—40
Broad bean necrosis virus (BBNV)	~250; 150	~25	ND	ND	ND	C,M	Leguminosae + narrow experimental	Soil borne	41
Hypochoeris mosaic virus (HyMV)	~250 ~130	21—22.5 (up to 25.7)	ND	24.5	2.58	C,I	Moderately wide	ND	42,43
Nicotiana velutina mosaic virus (NVMV)	No clear modal length; 100—175 +, some longer	18—19	2.3	21.4	2.86	ND	Solanaceae + narrow experimental	High seed trans. other unknown	1,44
Peanut clump virus (PCV)	245; 190; + some shorter	21	2.1; 1.7	~23	ND	C,A	Wide	Probably *P. graminis*	1,45
Indian peanut clump virus (IPCV)	240—250; 70—185	20—24	ND	ND	2.5	ND	Leguminosae (not tested extensively)	Soil borne	46
Potato mop-top virus (PMTV)	250—300; 100—150	18—20	ND	19.8, 18.5—20.0	2.4—2.5	C,P	Solanaceae + narrow experimental	*Spongospora subterranea*	1,47—49
Soil-borne wheat mosaic virus (SBWMV)	~300; 100—160	~20	2.0; 1.0	ND	2.6	C	Gramineae + narrow experimental	*P. graminis*	1,50—54

[a] Abbreviations: C, virus particles present in cytoplasm; A, presence of angled-layer aggregates; M, accumulation of membranous vesicles; P, virus particles found in parallel arrays; I, presence of nonviral cytoplasmic crystalline inclusions.

[b] ND, not determined.

[c] The French but not the Japanese isolate has this fourth RNA.

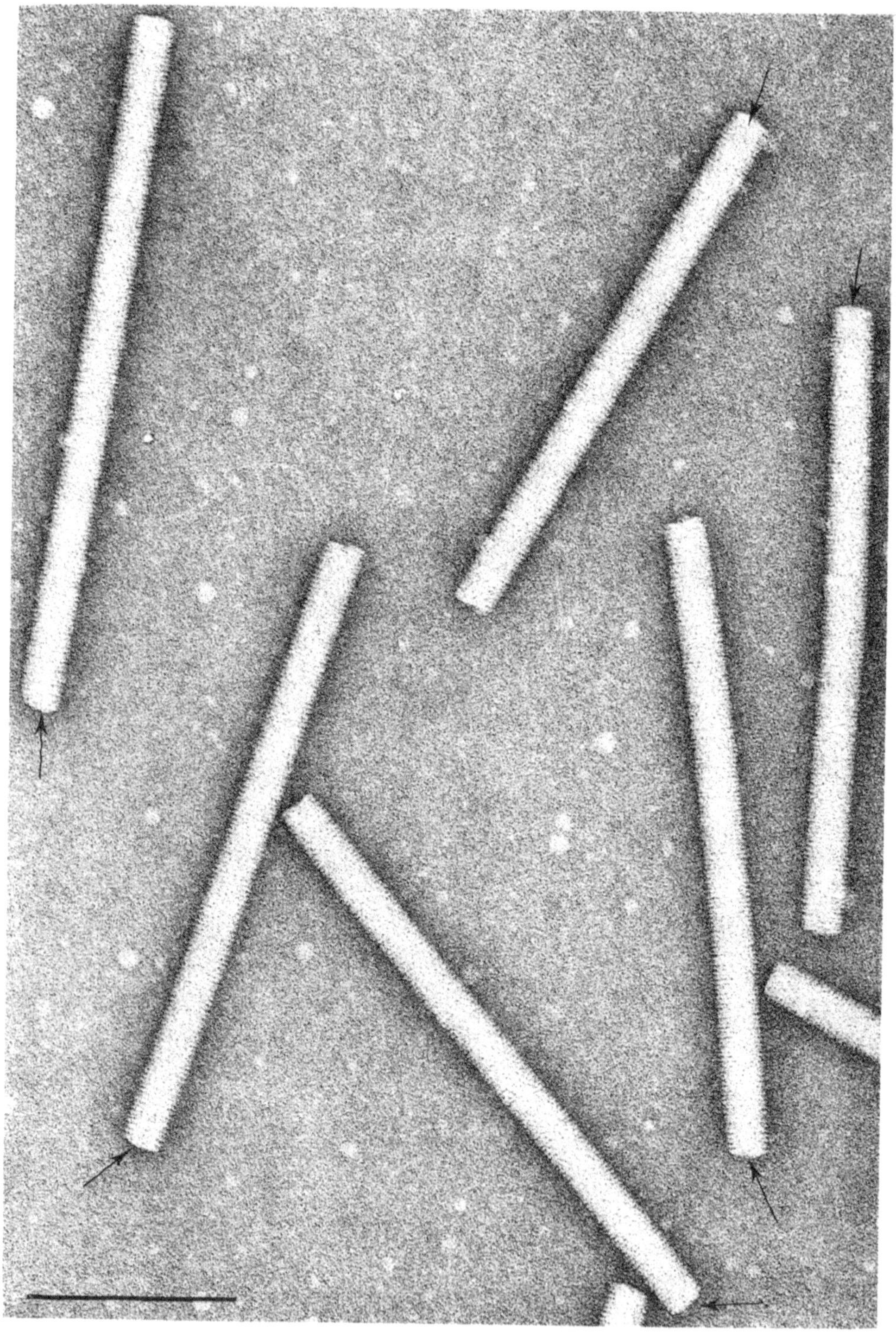

FIGURE 3. Tobacco mosaic virus (TMV) particles in uranyl formate stain. The axial canal is not visible in this particular micrograph. Note that each particle has a convex end (arrows) and a concave or flat end. The 3′ terminus of the RNA lies at the convex end (see text). Bar = 100 nm.

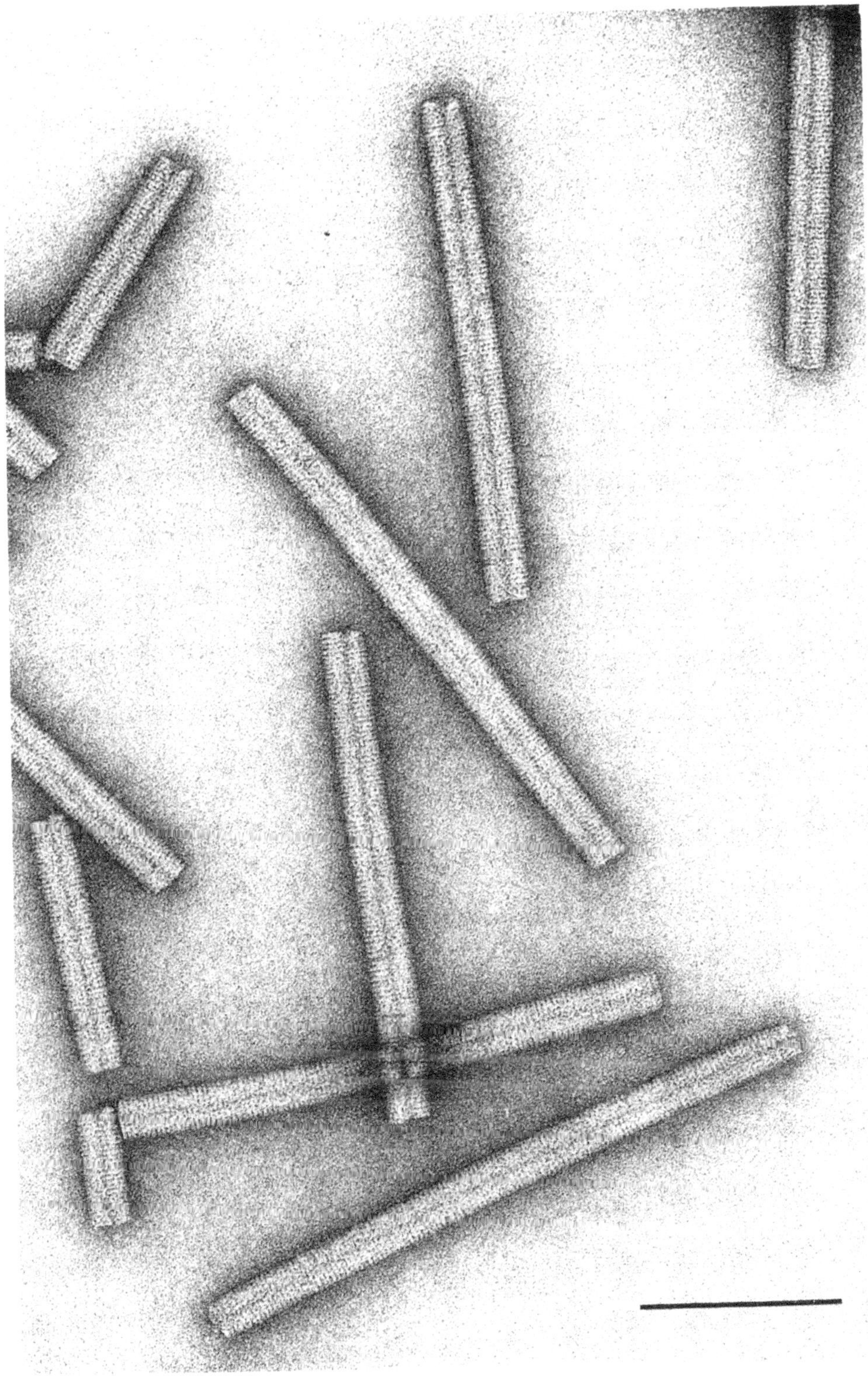

FIGURE 4. *Soil*-borne wheat mosaic virus (SBWMV) particles in uranyl formate stain. Bar = 100 nm.

II. VIRUS STRUCTURE AND COMPOSITION

A. Particle Structure

1. True Tobamoviruses

The particles of all true Tobamoviruses are indistinguishable in the electron microscope. They are rigid, helically constructed nucleoprotein tubes having external and internal diameters of 18 and 4 nm, respectively and a modal length of 300 nm (Figure 3).

SOV and some Tobamovirus isolates from orchids reportedly have a modal length near 320 nm.[24,56,57] SHMV has, in addition to the 300-nm particles which alone are infectious, a population of 40-nm particles the RNA of which is the coat protein mRNA.[58,59] FMV preparations also generally contain many shorter particles of various lengths as well as the 300-nm particles.[20]

Electron micrographs of TMV particles negatively stained in neutral sodium or potassium phosphotungstate, uranyl acetate, or uranyl formate show a pitch of 2.3 nm that can be directly measured on good photographs (Figure 3) such as those published by a number of workers.[2,60-63] Optical diffractograms of such photographs also indicate this repeat of 2.3 nm, and show a layer line corresponding to a further repeat every 6.9 nm, i.e., after every three turns of the helix.[64] Data obtained from optical linear integration of electron micrographs[65] and from X-ray diffraction by oriented gels of wet virus particles[66] corroborate these data (Figure 5).

The TMV particle is built from a single-start right-handed helix[67] of 2130 identical protein subunits (Figure 3). Each subunit of the type strain contains 158 amino acid residues whose complete sequence has been determined experimentally[68,69] and by deduction from the RNA sequence.[33] This can be compared with CGMMV coat protein, which contains 160 residues and whose sequence has been deduced from the RNA sequence.[70] Each particle encloses one molecule of single-stranded RNA at a radius of 4 nm, as established by X-ray diffraction[66] and confirmed by electron microscopy.[71] Three nucleotides are bonded to each successive protein subunit.

TMV particles in negative stain appear to have a convex end and a flat or concave end (Figure 3). It has been established that the 3′ terminus of the RNA lies at the convex end and the 5′ terminus at the concave end of the particle.[72,73]

TMV coat protein can be isolated from virus particle preparations by treatment with acetic acid[74] or with alkali.[75] This "alkalischer Protein",[76] or A protein as it is called in its lowest state of aggregation, shows a tendency to polymerize into tubular structures resembling TMV particles. In the presence of the viral RNA it can be reassembled in vitro into complete particles indistinguishable from the native virus. Progress in understanding the mechanism of in vitro assembly, which is probably the same as that operating in vivo, has been reviewed recently.[77] The reader is referred to this review for the extensive bibliography not repeated here.

Briefly, it appears that all the information needed to specify the final structure resides within the building materials themselves. The process begins with the formation of a two-layered protein disc containing 17 subunits in each layer (Figure 6). A particular hairpin loop of the RNA containing the sequence 5′-AAGAAG-3′ recognizes and binds to a site between the layers of the disc, and is referred to as the assembly origin.[33,78-80] The assembly origin is about 925 nucleotides from the 3′ terminus of TMV RNA.[33]

Binding of the RNA to the disc causes the latter to dislocate and form a helical structure which then elongates by adding protein and entraining the RNA. During this process both of the as yet unencapsidated ends of the RNA trail from the 3′ end of the growing rod (Figure 6).[81-83] Elongation towards the 5′ end of the RNA proceeds rapidly, by addition of monomeric subunits, whereas the slower growth towards the 3′ end is apparently due to addition of discs.[84,85]

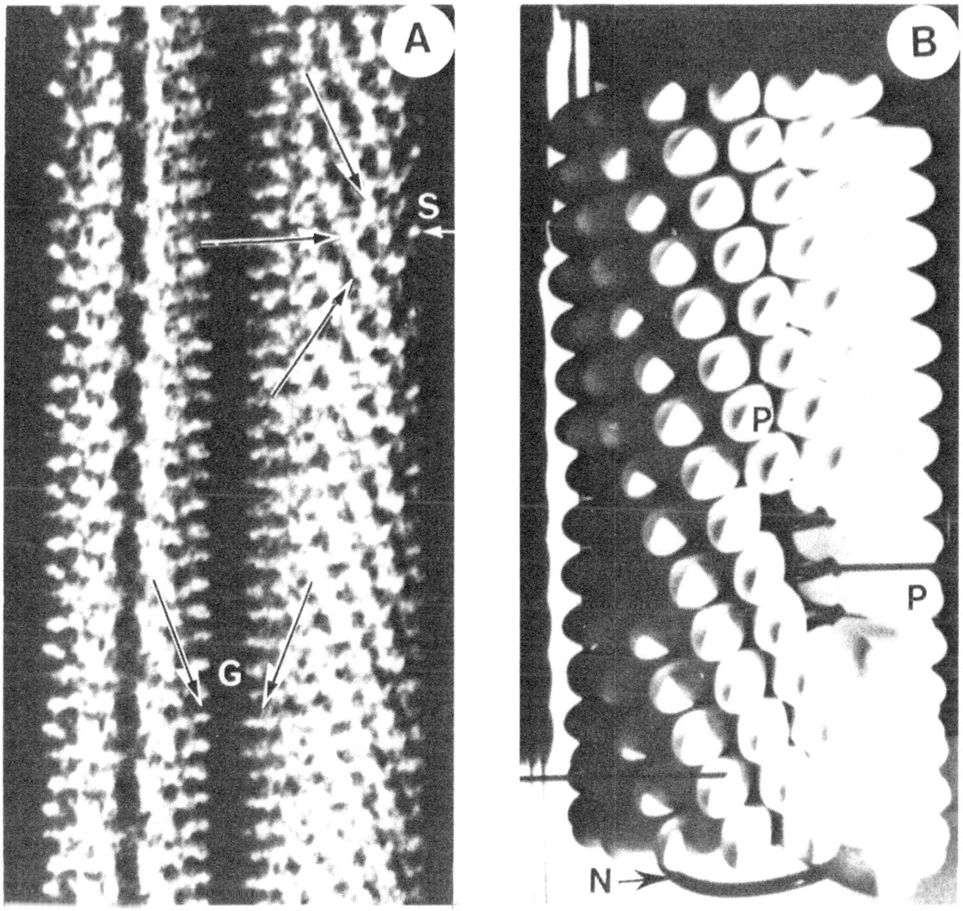

FIGURE 5. Images of two TMV rods (A) show essentially the same structural features as illustrated in the model (B). The original micrograph of the negatively stained particles was subjected to linear integration.[61] The grooves of the primary protein helix (G) and the individual protein subunits (S) are visible. The model of TMV (B) was constructed from data derived from X-ray analysis and biochemical studies.[67] The model shows the RNA (N) embedded in the helix of protein monomers (P). (Courtesy of Dr. A. Klug, Medical Research Council Laboratory of Molecular Biology, Cambridge, U. K.)

An interesting consequence of the existence of the assembly origin is that any RNA sequence containing it becomes encapsidated, whereas other sequences do not. Thus the subgenomic RNA coding for the coat protein of type TMV, though present in infected plants, is not encapsidated since it is at the 3′ end of the genome and the assembly origin is some 230 nucleotides further along towards the 5′ end. However, with the cowpea strain of TMV (presumably SHMV) and with CGMMV, the assembly origin is only about 320 nucleotides from the 3′ end of the viral RNA. This means that it is within the coat protein cistron and hence the coat protein mRNA is encapsidated as a short particle.[70,86,87] Similarly, any other subgenomic RNAs generated in infected cells, provided they include the assembly origin, will form a population of particles shorter than the standard length of 300 nm. The short particles that are a feature of SHMV and FMV (noted above) can thus be seen in context.

It can also be suggested that, with the coat protein mRNA routinely and independently encapsidated, a series of mutations could lead to the formation of a bipartite genome such as exists with the Tobraviruses, Hordeiviruses, and at least some possible Tobamoviruses.

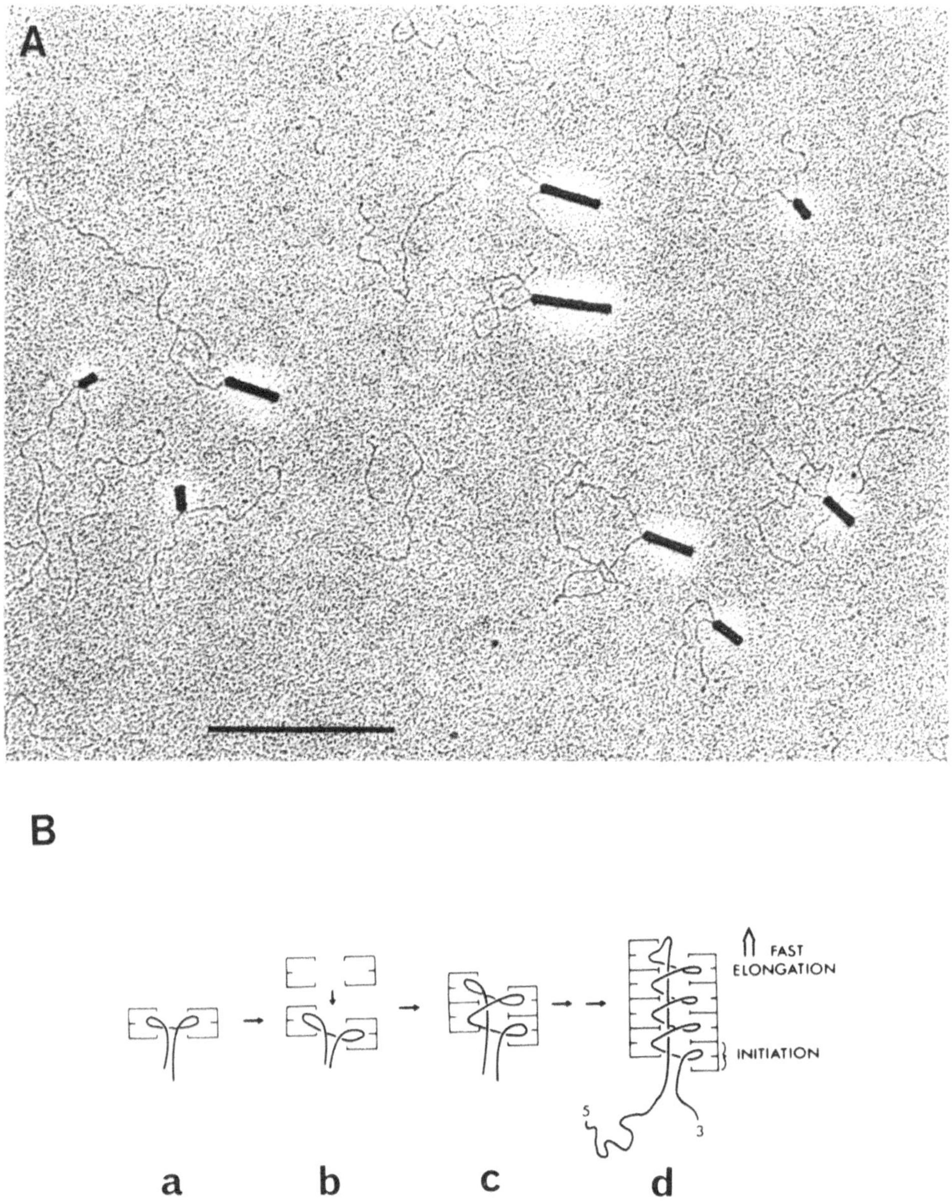

FIGURE 6. The assembly of TMV particles. A, Electron micrograph of shadowed partially assembled TMV rods, showing two RNA tails at one end of each particle. Bar = 500 nm. B, Model proposed by Hirth and Richards [77] for the assembly of TMV particles. For explanation, see text. (Courtesy of Drs. A. Nicolaieff and L. Hirth, Institute of Molecular and Cell Biology, University of Strasbourg, Strasbourg, France)

2. *Possible Tobamoviruses*

The particle structure of CCV has already been discussed (see Section I.A), and those of other possible Tobamoviruses are not well established apart from the fact that all the particles are rigid but fragile rods of at least two modal lengths (Figures 4 and 7, Table 2). Reported particle diameters range from 18.5 to 24.5 nm for different viruses (Table 2). Measurements on individual viruses have also varied; for example, IPCV particles reportedly have a diameter of 24 nm in uranyl acetate but 20 nm in neutral phosphotungstate,[46] while the particles of HyMV have a diameter of 21 to 22.5 nm and, exceptionally, up to 25.7 nm.[42,43] These data

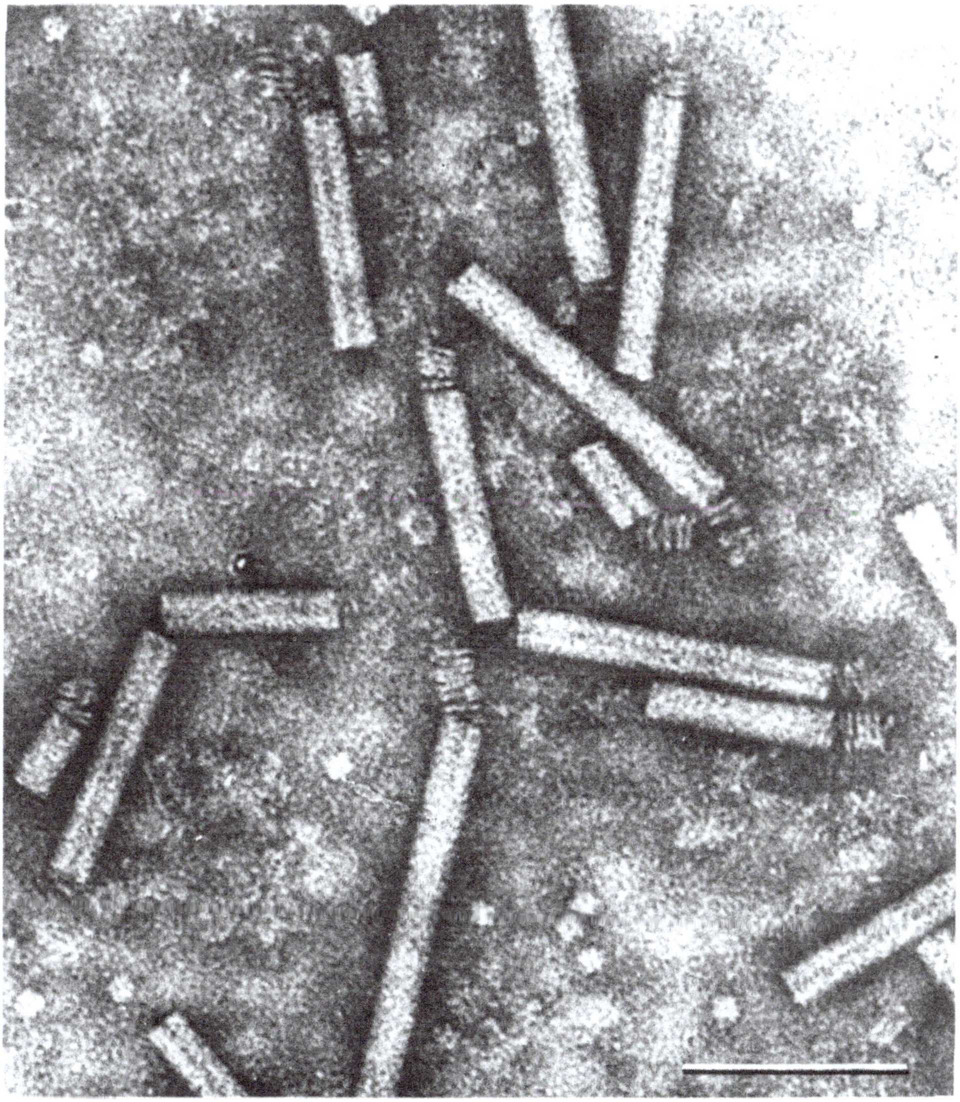

FIGURE 7. Potato mop-top virus (PMTV) in neutral sodium phosphotungstate stain. Note the characteristic loose helical structure at one end of some particles. Bar = 100 nm. (Courtesy of Dr. R. D. Woods, Rothamsted Experiment Station, Harpenden, England.)

and general experience suggest that the particles, under varying conditions of the electron microscope grid, are prone to both shrinkage and flattening (see Figures 1 and 2 in Chapter 10), so that reliable comparisons will be hard to make unless different viruses are co-imaged on the same grid.

The pitch of the helix of most of the possible Tobamoviruses is 2.4 to 2.6 nm (Table 2). The pitch of NVMV particles is reported as 2.86 nm, which is exceptionally large; that of CCV, measured as 2.75 nm, could be as low as 2.66 nm after correction, and there appear to be no reported data for PCV and BBNV.

The BNYVV particle is the most carefully studied among the possible Tobamoviruses.[88] It is apparently constructed as a single-start right-handed helix that has four bases associated with each protein subunit. This is interesting because, according to the calculations of

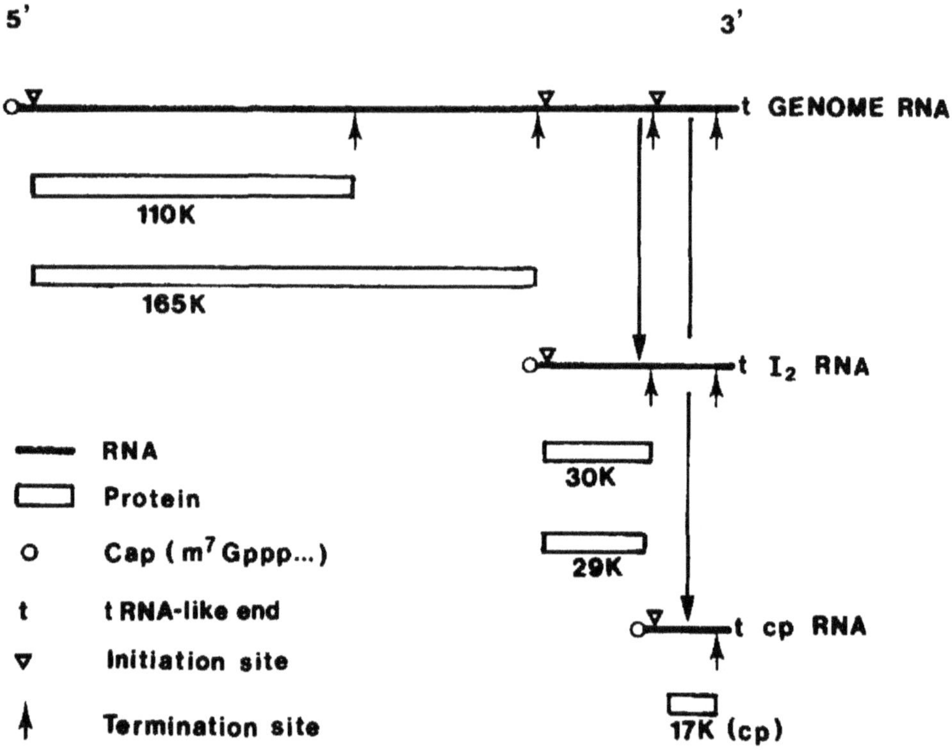

FIGURE 8. A physical map of the TMV RNA genome.

Veerisetty,[89] the Tobraviruses also have four bases per subunit in their particles, whereas SBWMV and the Hordeivirus BSMV appear to have three, like the true Tobamoviruses. This is one more problem that should be resolved in a concerted attack on the properties of the possible Tobamoviruses.

B. The Viral Genome

1. True Tobamoviruses

The Tobamovirus genome (Figure 8) consists of a single molecule of positive-sense, single-stranded RNA 6395 nucleotides long,[33] giving it an M_r about 2.1×10^6. The RNA is capped at the 5′ end with $m^7G^{5'}ppp$, and the 3′ end has a noncoding tRNA-like structure that can bind histidine[90-92] and valine in the case of the cowpea strain of TMV (presumably SHMV).[93]

TMV genomic and subgenomic RNAs have been translated in vitro.[94-97] With TMV RNA, after a leader sequence of 69 nucleotides from the 5′ end[33,98] there is an open reading frame for a protein of M_r 125,941 (which has been referred to as the 110K or 130K protein).[99] This protein is terminated by a UAG[100] leaky "amber" stop signal. Readthrough of this signal produces a second protein of M_r 183,253 (the so-called 165K protein) that is terminated by UAA. The end of this reading frame overlaps by five codons a reading frame, in a second phase, that codes for a protein of M_r 29,987 (the 30K protein),[101] also terminated by UAA. This codon, in turn, occurs two nucleotides before the initiator codon for the coat protein.

The 3′ portion of the sequence of CGMMV RNA is also known,[70] and comparison with that of type TMV reveals some interesting points. First, as already noted in Section II.A.1, the assembly origin is outside the coat protein cistron of type TMV but within that of CGMMV. Second, as also noted in Section II.A.1, the coat protein of type TMV has 158 residues and an M_r of 17,604, compared with the reported 160 residues and an M_r of 17,261 for CGMMV. Third, the cistrons for the 30K and coat proteins overlap in CGMMV but not, as we have seen, in type TMV. Nevertheless, the 3′ noncoding region of CGMMV is highly homologous with that of type TMV. In contrast, the cowpea or Cc strain of TMV (presumably SHMV)[102] resembles CGMMV in the location of the assembly origin and in the sequence of the cistrons for the 30K and coat proteins. Thus it is curious that the 3′ noncoding region of Cc shows less than 50% homology with that of CGMMV—data that lead Meshi and colleagues[70] to remark that relationships among the Tobamoviruses are "not simple."

The TMV genome appears to have the capacity to produce four proteins: the 165K, the 130K, the 30K, and the coat protein. Of these, only the coat protein has a known function, though the 30K protein may somehow assist in cell-to-cell movement of the virus.[103]

In infected plants, and particularly in infected protoplasts, the coat protein is easily detectable, and the 165K, 130K, and 30K proteins have also been found.[104-108] The 30K protein has recently been shown to appear transiently early in infection.[109] However, attempts to translate the TMV genome in vitro have succeeded in producing only the 165K and 130K proteins.[94,98,100,108] The 30K and coat proteins are apparently not translated because only the cistrons adjacent to the 5′ end of the RNA can be read, and internal starts are not possible.[109] How then are the 30K and coat proteins produced?

Many workers have detected a family of as many as eight subgenomic RNAs in infected plants (including the 30K protein and coat protein mRNAs), together with their double-stranded replicative forms.[109-112] These RNAs are co-terminal at the 3′ end and extend in the 5′ direction for various distances up to and beyond the cistrons for the 30K and coat proteins. A proposed mechanism for their synthesis is that the genomic positive strand is incompletely transcribed into a series of shorter 5′ co-terminal negative strands that in turn generate the subgenomic mRNAs that give rise to the 30K and coat proteins, and possibly other proteins also.[111] Within the genome, hypothetical start codons and ribosome binding sites for at least some of these subgenomic messengers have been identified.[111]

Several lines of evidence suggest that at least one viral-coded protein, an RNA-dependent RNA polymerase or replicase, is essential for TMV replication, and that the host plant cannot provide this function.[77] However, so far, no in vitro translation product of TMV RNA or its subgenomic RNAs has been shown to possess replicase activity.

2. Possible Tobamoviruses

The reported sizes of the RNAs of these viruses are listed in Table 2 where known. BNYVV, PCV, and SBWMV seem to possess at least two RNAs each, and the French isolate of BNYVV has four RNAs while the Japanese isolate probably has three.[88] It is not yet clear what sequences these RNAs may have in common, and only with SBWMV is it established that both the RNAs are needed for infection, and therefore that the genome is bipartite.[33,113] However, it seems likely that all the viruses listed in Table 2 except CCV will turn out to have bipartite (or multipartite) genomes.

It is interesting that all four RNAs of BNYVV are reported to be polyadenylated at the 3′ end[88,114] in contrast to the RNAs of Tobamoviruses, which possess a tRNA-like 3′ terminal structure instead. By way of comparison, it appears that Hordeiviruses have a 3′ terminal tRNA-like structure, upstream from which is a tract of up to 30 poly A residues (see Chapter 10).

BNYVV RNAs have m^7GpppNp cap structures at all their 5' termini. However, the cap with Tobamovirus RNAs is m^7GpppGp, whereas BNYVV RNAs do not appear to have G in position one.[114]

III. CYTOPATHOLOGY

A. True Tobamoviruses

Cells of different hosts infected with various Tobamoviruses, though predominantly with TMV, have been a favoured subject for examination by electron microscopy of thin sections. Sometimes techniques such as autoradiography and cytochemistry have also been used. The motives for the work have been several: to observe the cytopathology, to discover where in the cell the virus particles are synthesized and assembled, to note features that could serve to distinguish between different Tobamoviruses or between Tobamoviruses and viruses from other groups, and to observe, if possible, what it is about a hypersensitive host that results in local lesion formation instead of systemic spread of the infection.

Large numbers of Tobamovirus particles accumulate in the cytoplasm of systemically infected cells (Figure 9) and often form crystalline arrays[115-119] (as seen in Figures 9 and 10), though with TMV itself organelles such as nuclei, chloroplasts, and mitochondria are not at first greatly affected.

Angled-layer aggregates, in which successive layers of virus particles are oriented at an angle of 60° to the previous layer, are found particularly with the aucuba strain of TMV,[120-122] but also occur with U_5TMV[123] (Figure 11) and the tentative Tobamoviruses PCV and BNYVV.[38,40,45]

At the same time as virus particles first appear in type TMV-infected cells, X-bodies also develop in the cytoplasm. They are regions containing membranous vesicles, ribosomes, X-tubules, and sometimes virus particles (Figure 12). The X-tubules have about the same external diameter as normal microtubules (25 to 30 nm), but are thicker-walled and more densely staining. At early stages of infection they generally occur in threes or pairs (Figure 12B), but in long-standing infections they or tubules of similar aspect may also be found in extensive parallel arrays.

The nature and function (if any) of the X-tubules is obscure. They have been seen in TMV-infected tomato,[115] tobacco,[116,117,124] *Nicotiana clevelandii* (Table 3), potato,[125] and *Chenopodium amaranticolor*,[126] but have not been found despite a deliberate search in plants infected with some other Tobamoviruses (Table 3). Nor were they found in TMV-infected *Nicotiana glutinosa*,[127] and would therefore appear not to be on the direct pathway of viral synthesis. X-bodies containing X-tubules do occur in Samsun tobacco plants held at 35° while infected with the Ni118 mutant of TMV,[128] although at this temperature the coat protein and RNA of this mutant are not assembled into virus particles. X-tubules do not appear to react with antibodies directed against TMV coat protein.[129]

Particles resembling those of TMV were found in the nucleus and chloroplasts as well as the cytoplasm of tobacco cells infected with a Tobamovirus that was thought to have been U_1TMV,[124] but was probably the U_5 strain or one behaving similarly[123,130] (Figures 11 and 13). Shalla et al.[131] obtained evidence that the virus-like particles found in the chloroplasts (though not the cytoplasm) of U_5-infected cells were shorter than 300 nm and not infectious. Virus (or virus-like) particles have also been found to occur, though less frequently, in the chloroplasts of plants infected with U_1TMV, ORSV, and FMV (Table 3). Table 3 lists some other abnormalities found in Tobamovirus-infected cells.

CGMMV[132-134] and CV4 (Table 3) seem unique among Tobamoviruses in causing severe vesiculation and distortion of the mitochondria (Figure 14) like those induced by some viruses with small polyhedral particles (see Chapter 12, Volume 1), whereas most Tobamoviruses examined induce tonoplast vesicles in infected cells similar to those observed in cells infected with cucumber mosaic virus.[135]

One aim of most workers studying the cytopathology of TMV-infected cells has been to obtain evidence on the sites of synthesis and assembly of the virus particles. The evidence is still ambiguous, though it is clear that the particles accumulate in the cytoplasm and that

Table 3
COMPARISON OF CYTOPATHOLOGICAL EFFECTS INDUCED BY SOME TOBAMOVIRUSES[a]

		Presence of				
Virus	Host used	X-body	Tonoplast vesicles	Virus-like particles in chloroplasts	Vesicles in mitochondria	Abnormalities in nucleus
Tobacco mosaic virus (TMV)						
Strain U1[b]	*N. clevelandii*	+	+	+	—[c]	—
Strain U5[b]	*N. clevelandii*	—	+	+ +	—	Virus-like particles in one case
Strain U6[b]	*N. clevelandii*	+	+	—	—	Electron-lucent material
Ribgrass mosaic virus (RMV)	*N. clevelandii*	+	+	—	—	—
Odontoglossum ringspot virus (ORSV)	*N. clevelandii*	+	+	+	—	—
Sunn-hemp mosaic virus (SHMV) = cowpea strain of TMV	*Phaseolus vulgaris*	—	+	—	—	A few virus-like particles
Frangipani mosaic virus (FMV)	*Datura stramonium*	—	+	+	—	A few virus-like particles
Tomato mosaic virus (ToMV)	Tomato	—	(a few)	—	—	Electron-lucent material
Cucumber virus 4 (CV4)	Cucumber	—	—	—	+	—

[a] After T. Hama, unpublished data.
[b] Strains described by Siegel and Wildman.[18] The RNAs of all these strains have been shown to have a very high degree of nucleotide sequence homology.[11]
[c] None observed.

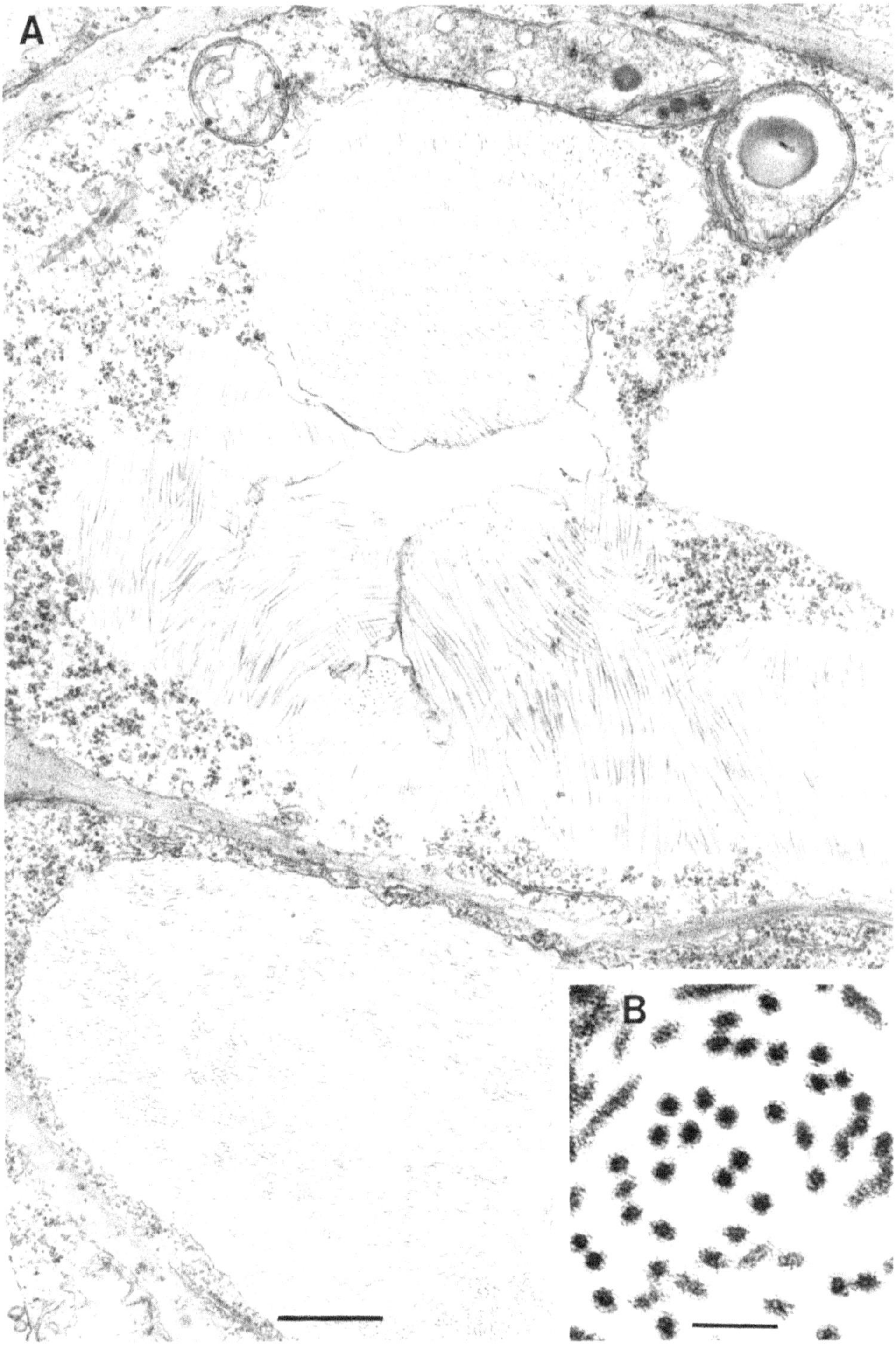

FIGURE 9. A. Thin section of TMV (U_1 strain or type TMV)-infected cells of *Nicotiana clevelandii*, showing masses of virus particles in the cytoplasm; bar = 500 nm. B. Cross-sectioned TMV particles in a Turkish Samsun tobacco leaf cell; bar = 100 nm.

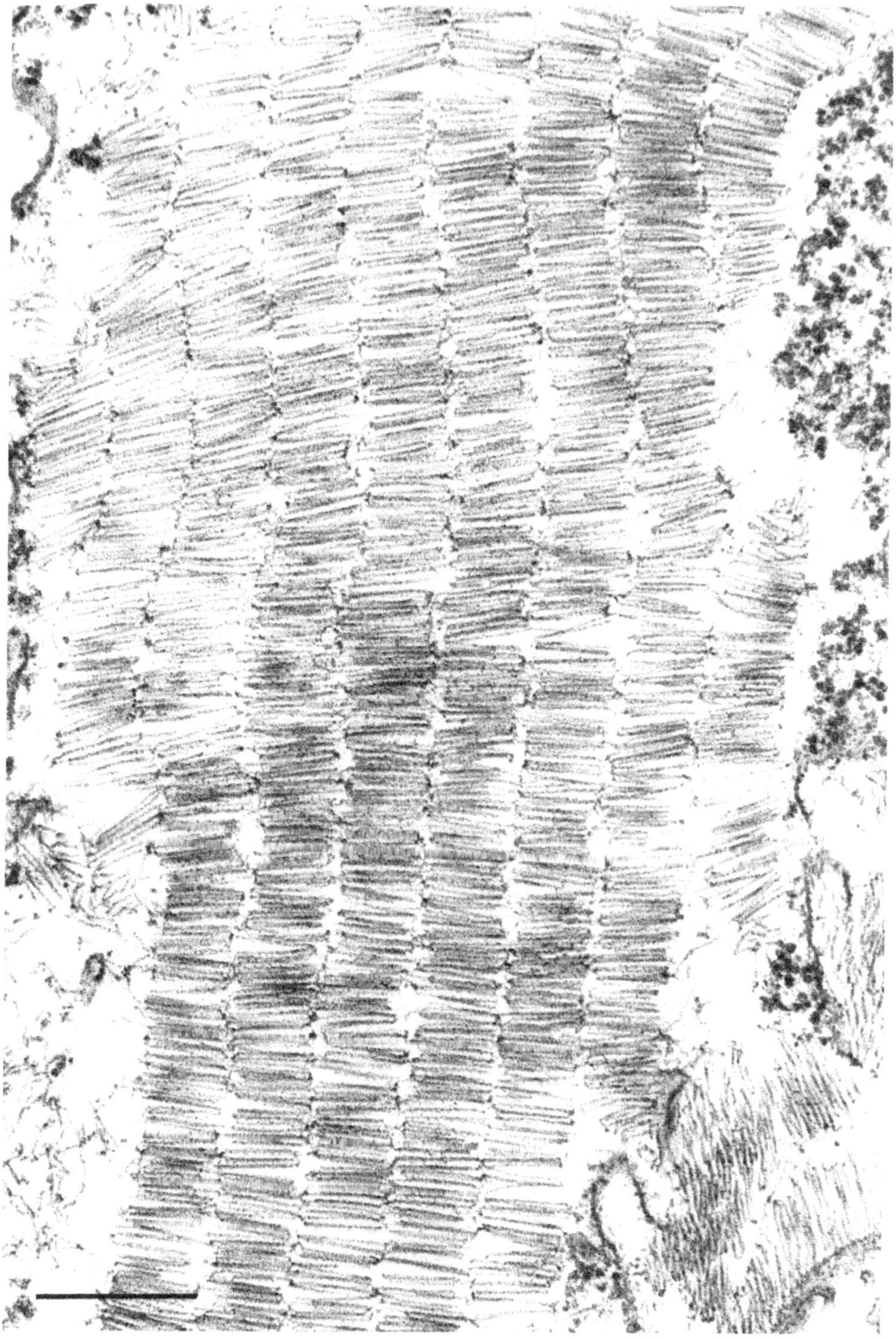

FIGURE 10. Thin section of an array of tomato mosaic virus (ToMV) particles in an infected *Lycopersicon esculentum* (tomato) leaf cell. The ranks of particles have probably been slightly disturbed during fixation. Bar = 500 nm.

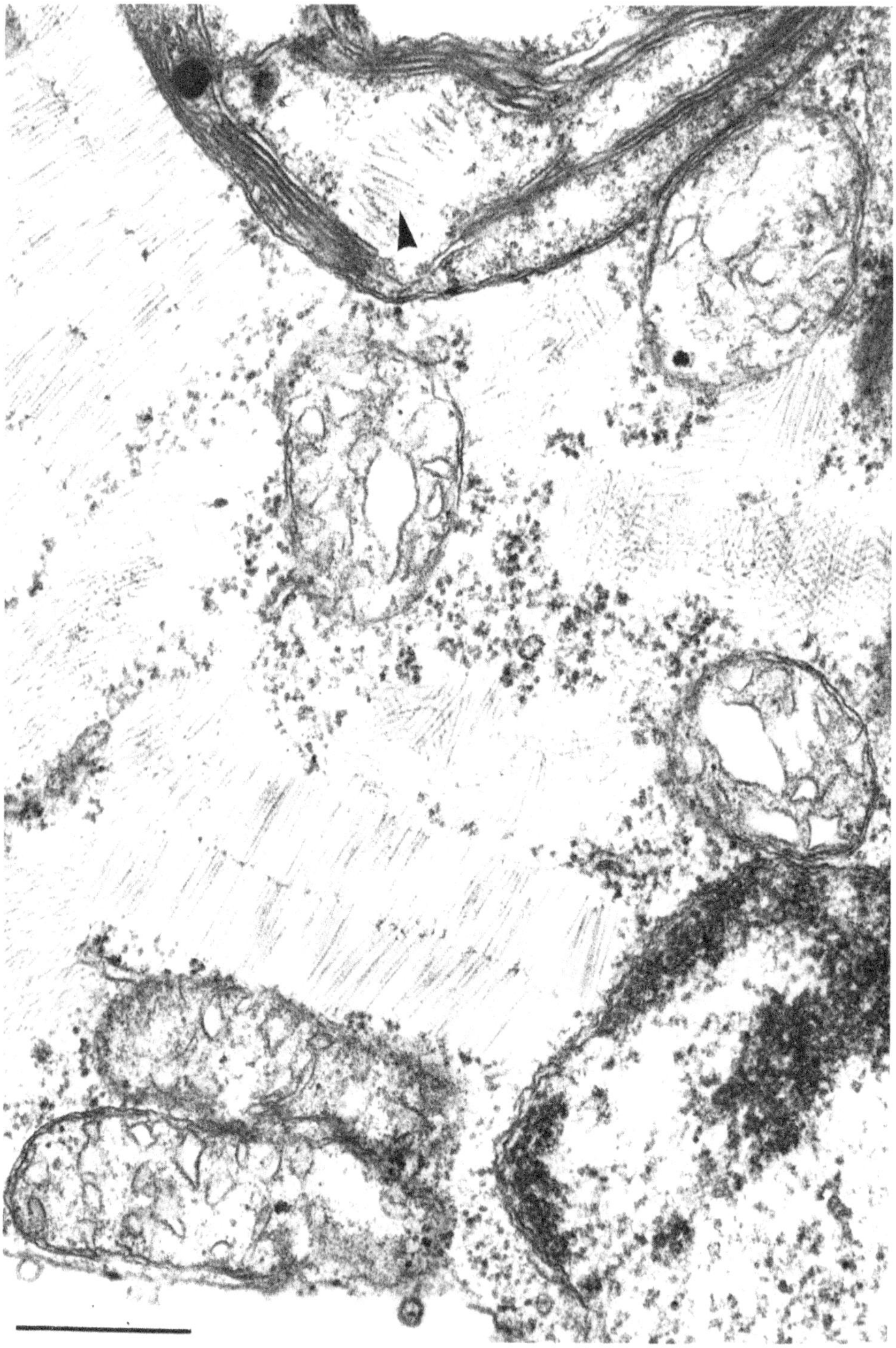

FIGURE 11. Thin section of a *Nicotiana clevelandii* leaf cell infected with U_5TMV, showing a chloroplast containing virus-like particles (arrowhead) and angled layer aggregates of virus particles as well as the more usual perpendicularly stacked virus particles and small vesicles on the tonoplast (bottom). Bar = 500 nm.

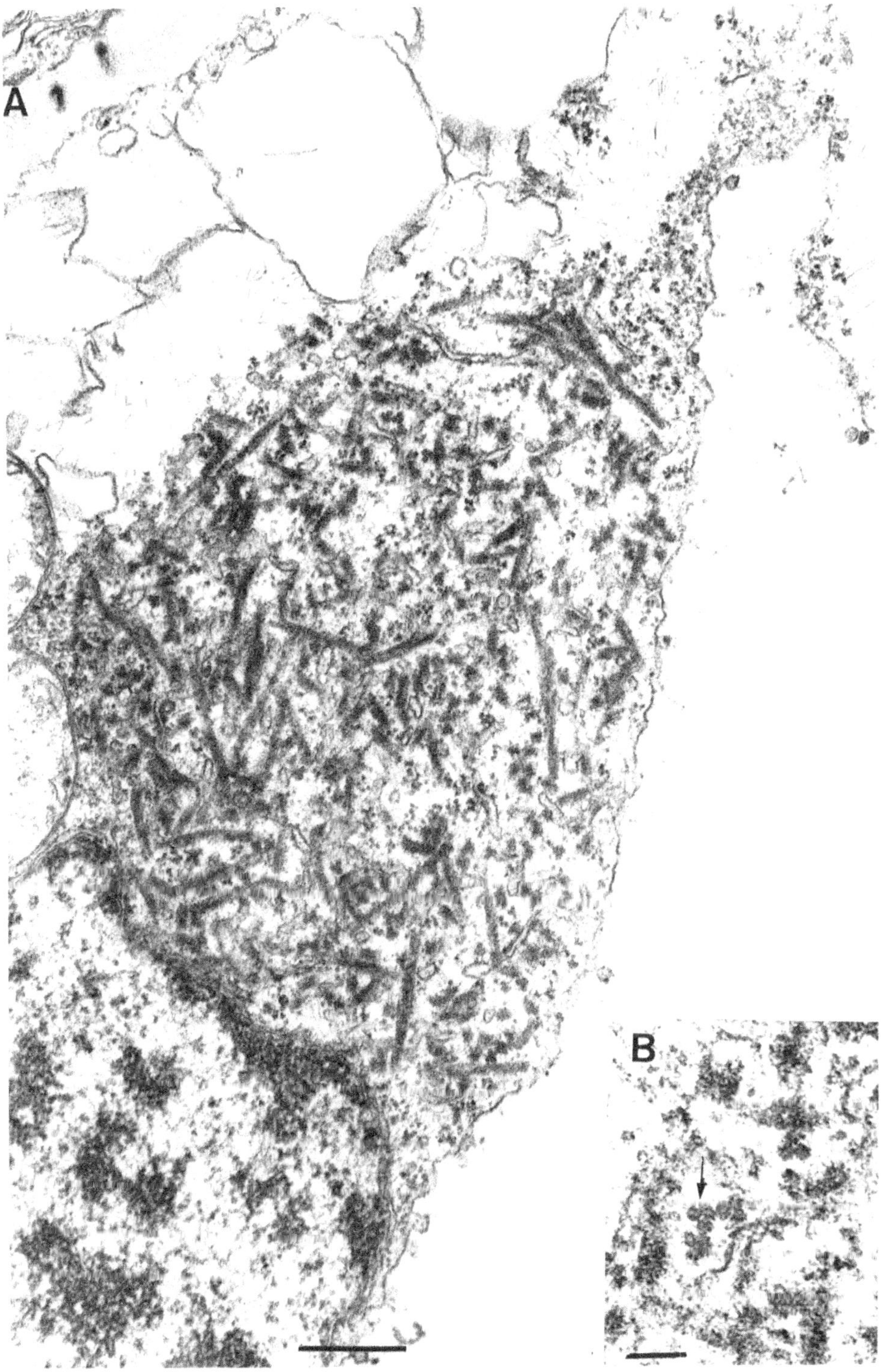

FIGURE 12. A, Thin section of a *Nicotiana clevelandii* leaf cell infected with ribgrass mosaic virus (RMV), showing a typical X-body containing X-tubules and ribosomes; bar = 500 nm. B, Higher magnification of X-tubules seen in cross-section; they are usually grouped in threes (arrow); bar = 100 nm.

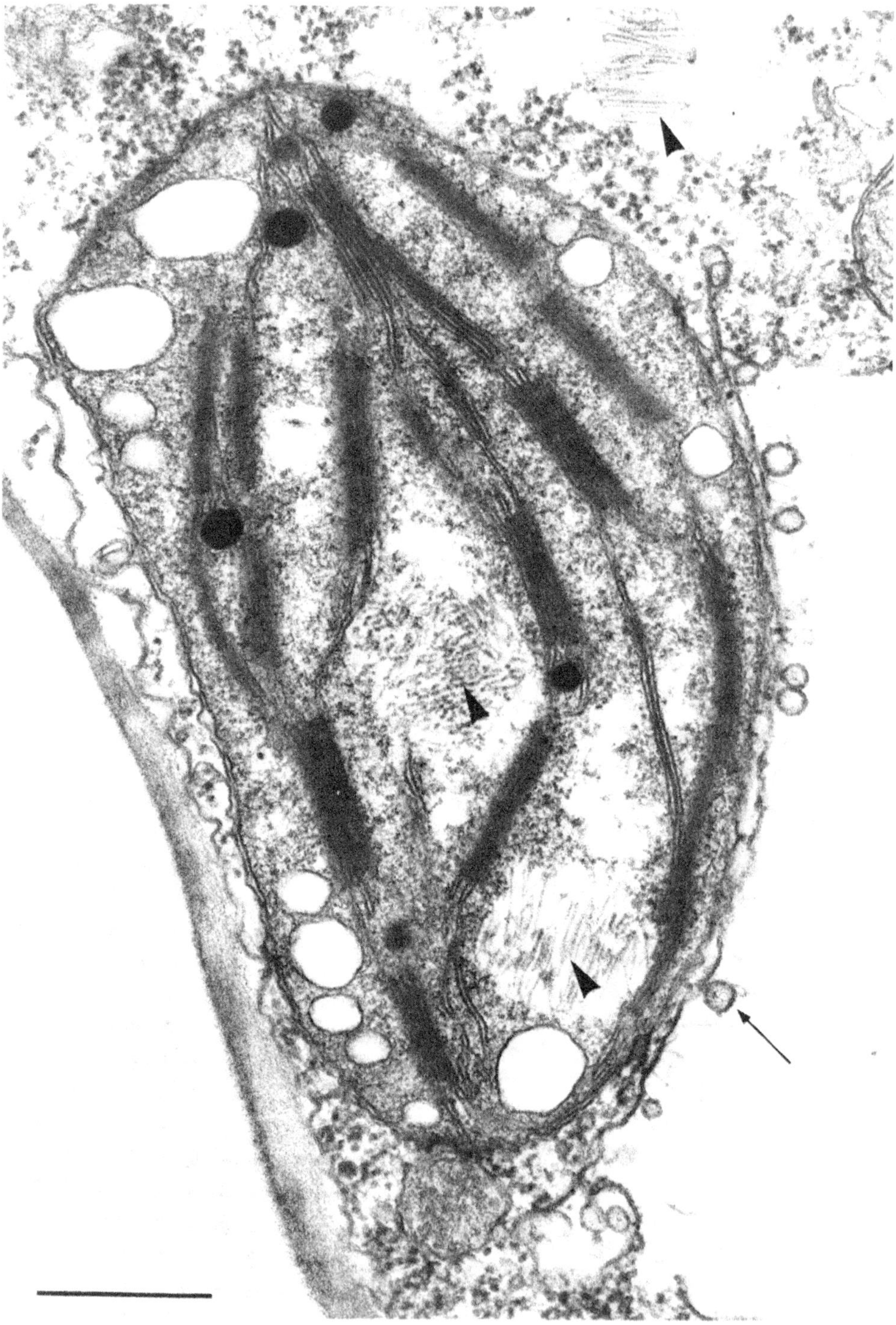

FIGURE 13. Thin section of a *Nicotiana clevelandii* leaf cell infected with U₁TMV, showing virus particles in both chloroplast and cytoplasm (arrowheads) and vesicles associated with the tonoplast (arrow). Some virus particles can also be seen scattered in the vacuole. Bar = 500 nm.

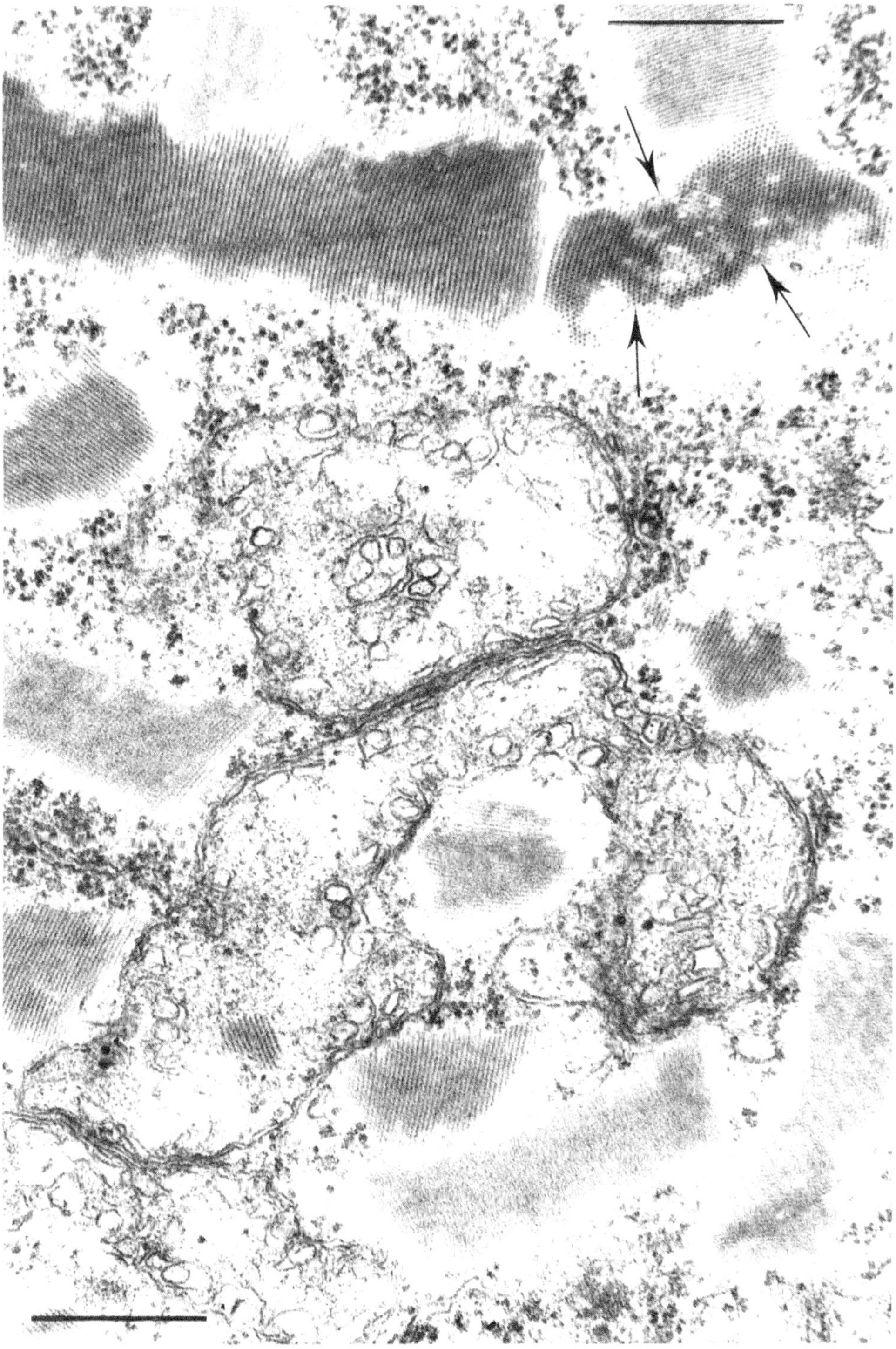

FIGURE 14. Thin section of a *Cucumis sativus* (cucumber) leaf cell infected with cucumber virus 4 (CV4), showing vesiculated mitochondria and characteristic stacking of virus particles (arrows indicate stacks in transverse section). Bar = 500 nm.

the chloroplasts and nuclei are not altered significantly except as a secondary result of infection.

Very early work on the location of TMV replication suggested the chloroplasts as the active sites of protein synthesis and perhaps also RNA replication.[136,137] Later some support was found for the proposition that TMV RNA was synthesized in the nucleus,[138] but there is now stronger cytopathological and biochemical evidence, based on the location of the double-stranded RNA replicative form, that TMV RNA is synthesized in association with cytoplasmic membranes, possibly those of the tonoplast vesicles (Figure 13), and not in nuclei, chloroplasts, or mitochondria.[139-144]

With U_5TMV, Betto et al.[130] studied the incorporation of ^{3}H-uridine using autoradiography of thin sections of infected tobacco leaves. They found that the only above-background label occurred at the interface between the chloroplasts and the cytoplasm. The chloroplast envelope was found, as a consequence of infection, to be invaginated, forming a series of vesicles reminiscent of those observed in Tymovirus-infected cells (see Chapter 8, Volume 1), and it was suggested that with U_5TMV these vesicles may be the site of viral RNA synthesis. With CGMMV and CV4 the mitochondrial vesicles mentioned above[131-134] could similarly be the sites of RNA synthesis for these viruses. In conclusion, it seems plausible that different Tobamoviruses may have different sites for synthesizing RNA, though each site would be associated with cellular or organelle membranes.

In contrast to the situation regarding viral RNA synthesis, the evidence on the site of synthesis of the viral proteins seems reasonably clear. Zaitlin[145] and colleagues have shown that cycloheximide but not chloramphenicol inhibits both U_1 and U_5TMV viral protein and complete virus particle synthesis, indicating that only the cytoplasmic ribosomes are involved in viral protein synthesis.

Shalla and Amici,[139] using ferritin-labeled antibodies to TMV coat protein, achieved labeling of TMV particles, and obtained convincing evidence of the presence of free coat protein in the cytoplasm. There was no labeling of chloroplasts; nuclei were labeled, but only if they were first extracted from a leaf homogenate — a process that might have led to displacement of the antigen from its original site. These results are compatible with those outlined above in indicating that TMV protein is synthesized in the cytoplasm. Virus particle assembly also appears to occur in the cytoplasm, possibly with the exception of the cases where virus (or virus-like) particles have been found in chloroplasts or nuclei as well as in the cytoplasm.

Very many papers have appeared on the cytopathology of local lesions induced by TMV in hypersensitive hosts. The conclusions of all these workers are difficult to synthesize into a general picture except to say that under local lesion conditions fewer virus particles are found in each cell, and in many cells no particles are detected at all. The main interest of the work has been to determine what stops the virus from spreading, and it must be concluded that no clear single answer has emerged. It is possible that several of the mechanisms proposed operate at the same time, or that different mechanisms operate in different hosts or under different environmental conditions. The question has recently been reviewed by Martelli,[146] Sela,[147] and van Loon.[148]

B. Possible Tobamoviruses

Examination of thin sections of plants infected with possible Tobamoviruses has revealed virus particles in the cytoplasm only (Table 2). The viruses do not seem to cause great cytological disturbance or induce specific inclusions, except that HyMV causes formation of crystalline inclusions of an unknown material in the cytoplasm of infected *Nicotiana clevelandii* cells.[43]

With both BNYVV and PCV, angled layer aggregates of virus particles have been observed, similar to those produced by the aucuba strain of TMV and by U_5TMV (Figure 11).

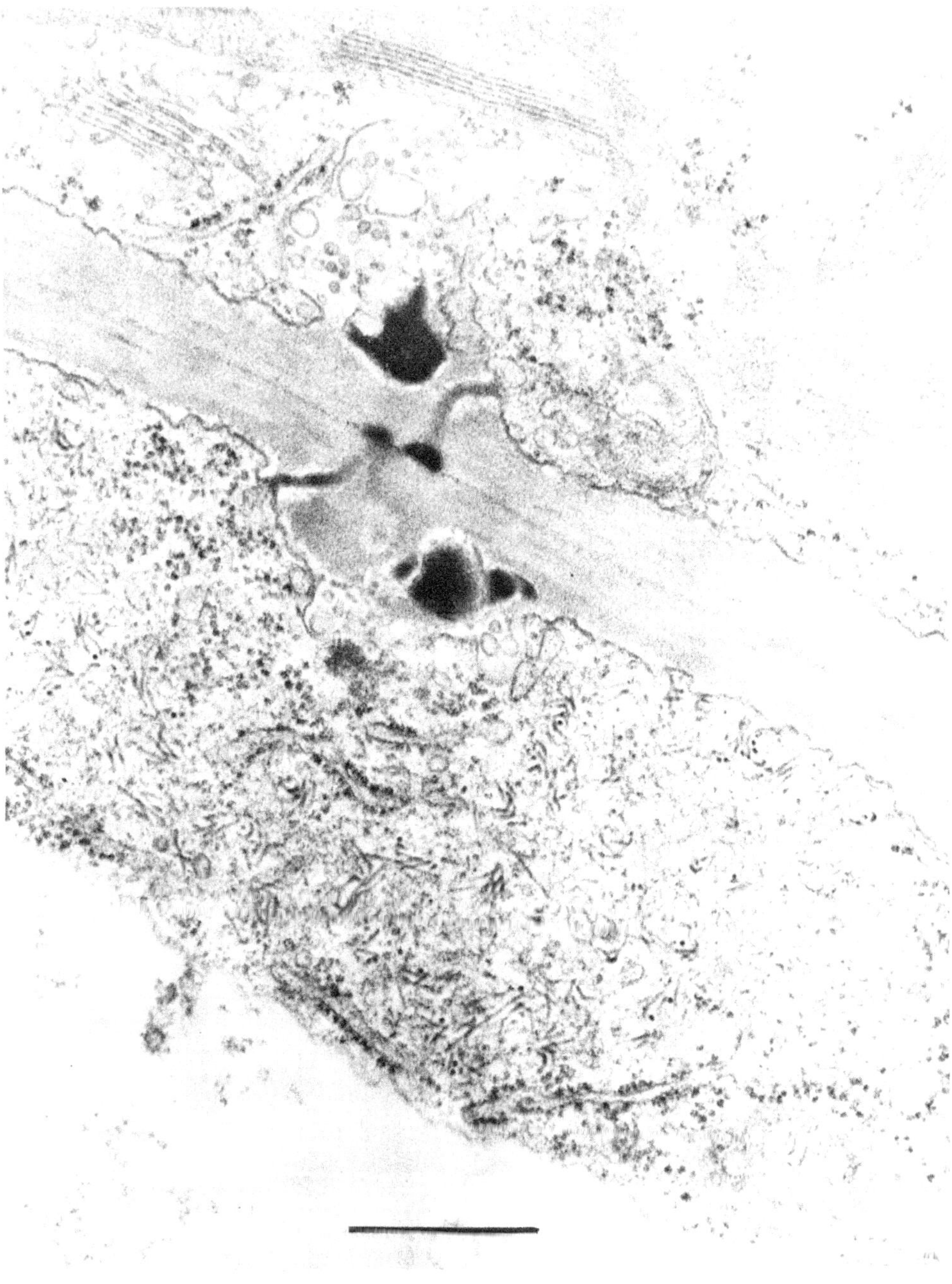

FIGURE 15. Thin section of a *Chenopodium amaranticolor* leaf cell infected with beet necrotic yellow vein virus (BNYVV), showing virus particles scattered in the cytoplasm. Bar = 500 nm.

Only with PMTV have crystalline arrays of virus particles been seen. With possible Tobamoviruses generally, there are fewer particles to be found than in cells infected with true Tobamoviruses and they are often simply scattered in the cytoplasm (Figure 15).

Accumulations of membranous vesicles have been observed in the cytoplasm of BBNV-infected plants and in plants infected with an Italian isolate of BNYVV,[39] but were not reported for French[38] or Japanese[37] isolates or for other possible Tobamoviruses.

Langenberg and Giunchedi[52] investigated by thin sectioning the relationship between SBWMV and BNYVV and their fungal vectors *Polymyxa graminis* and *P. betae*, respectively. There is some evidence that the viruses are carried internally in the vectors,[52] but no virus particles were seen in thin sections of the fungal thalli, although the immediately adjacent host-plant cytoplasm was seen to contain particles. Nevertheless, parts of the fungal cytoplasm were so dense that virus particles might have escaped notice had they been present.

In conclusion, examination of the cytopathology of the possible Tobamoviruses has not so far yielded strongly diagnostic data, except perhaps regarding the inclusions induced by HyMV. The effects observed in infected cells are neither so similar as to suggest strong affinities among the viruses nor so different as to be useful in differentiating between them.

REFERENCES

1. **Gibbs, A. J.,** Tobamovirus group, *CMI/AAB Descriptions of Plant Viruses*, No. 184, 1977.
2. **Shikata, E.,** Tobamovirus (tobacco mosaic virus) group. in *The Atlas of Insect and Plant Viruses*, Maramorosch, K., Ed., Academic Press, New York, 1977, 237.
3. **Van Regenmortel, M. H. V.,** Tobamoviruses, in *Handbook of Plant Virus Infections and Comparative Diagnosis*, Kurstak, E., Ed., Elsevier, Amsterdam, 1981, 541.
4. **Hollings, M., Komuro, Y., and Tochihara, H.,** Cucumber green mottle mosaic virus, *CMI/AAB Description of Plant Viruses*, No. 154, 1975.
5. **Kassanis, B. and Varma, A.,** Sunn-hemp mosaic virus, *CMI/AAB Descriptions of Plant Viruses*, No. 153, 1975.
6. **Hollings, M. and Huttinga, H.,** Tomato mosaic virus, *CMI/AAB Descriptions of Plant Viruses*, No. 156, 1976.
7. **Hennig, B. and Wittmann, H. G.,** Tobacco mosaic virus: mutants and strains, in *Principles and Techniques in Plant Virology*, Kado, C. I. and Agrawal, H. O., Eds., Van Nostrand Reinhold, New York, 1972, 546.
8. **Tóbiás, I., Rast, A. Th.B., and Maat, D. Z.,** Tobamoviruses of pepper, eggplant and tobacco: comparative host reactions and serological relationships, *Neth. J. Plant Pathol.*, 88, 257, 1982.
9. **Matthews, R. E. F.,** Classification and nomenclature of viruses. Fourth report of the International Committee on Taxonomy of Viruses, *Intervirology*, 17, 1, 1982.
10. **VanDeWalle, M. J. and Siegel, A.,** A study of nucleotide sequence homology between strains of tobacco mosaic virus, *Virology*, 73, 413, 1976.
11. **Palukaitis, P. and Symons, R. H.,** Nucleotide sequence homology of thirteen Tobamovirus RNAs as determined by hybridization analysis with complementary DNA, *Virology*, 107, 354, 1980.
12. **Palukaitis, P., Randles, J. W., Tian, Y., Kang, L., and Tien, P.,** Taxonomy of several Tobamoviruses from China as determined by molecular hybridization analysis with complementary DNA, *Intervirology*, 16, 136, 1981.
13. **VanDeWalle, M. J. and Siegel, A.,** Relationships between strains of tobacco mosaic virus and other selected plant viruses, *Phytopathology*, 72, 390, 1982.
14. **Van Regenmortel, M. H. V.,** Serological studies on naturally occurring strains and chemically induced mutants of tobacco mosaic virus, *Virology*, 31, 467, 1967.
15. **Van Regenmortel, M. H. V.,** Antigenic relationships between strains of tobacco mosaic virus, *Virology*, 64, 415, 1975.
16. **Gibbs, A.,** How ancient are the Tobamoviruses?, *Intervirology*, 14, 101, 1980.
17. **Gibbs, A., Tien Po, Kang, L., Tian Y., and Randles, J. W.,** Classification of several tobamoviruses isolated in China on the basis of the amino acid composition of their virion proteins, *Intervirology*, 18, 160, 1982.

18. **Siegel, A. and Wildman, S. G.,** Some natural relationships among strains of tobacco mosaic virus, *Phytopathology*, 44, 277, 1954.
19. **Zaitlin, M. and Israel, H. W.,** Tobacco mosaic virus (type strain). *CMI/AAB Descriptions of Plant Viruses*, No. 151, 1975.
20. **Varma, A. and Gibbs, A. J.,** Frangipani Mosaic virus, *CMI/AAB Descriptions of Plant Viruses*, No. 196, 1978.
21. **Paul, H. L.,** Odontoglossum ringspot virus, *CMI/AAB Descriptions of Plant Viruses*, No. 155, 1975.
22. **Oshima, N. and Harrison, B. D.,** Ribgrass mosaic virus, *CMI/AAB Descriptions of Plant Viruses*, No. 152, 1975.
23. **Sammons, I. M. and Chessin, M.,** Cactus virus in the United States, *Nature (London)*, 191, 517, 1961.
24. **Brandes, J. and Chessin, M.,** An electron microscope study on the size of Sammons' opuntia virus, *Virology*, 25, 673, 1965.
25. **Chang, Y. S., Pei, M. Y., Jen, M. H., Chien, Y. J., Tsao, T. C., and Cheo, C. C.,** The size, shape and structure of Chinese rape (*Brassica campestris* L.-*Youcai*) mosaic virus, YMV_{15}, and some chemical characteristics of its protein subunits, *Scientia Sin.*, 13, 1421, 1964.
26. **Holmes, F. O.,** A distinctive strain of tobacco mosaic virus from *Plantago*. *Phytopathology*, 31, 1089, 1941.
27. **Meshi, T., Ohno, T., and Okada, Y.,** Nucleotide sequence and its character of cistron coding for the 30K protein of TMV (OM strain), *J. Biochem.*, 91, 1441, 1982.
28. **Meshi, T., Ohno, T., Ira, H., and Okada, Y.,** Nucleotide sequence of a cloned cDNA copy of TMV (cowpea strain) RNA, including the assembly origin, the coat protein cistron, and the 3′ noncoding region, *Mol. Gen. Genet.*, 184, 20, 1981.
29. **Knight, C. A., Silva, D. M., Dahl, D., and Tsugita, A.,** Two distinctive strains of tobacco mosaic virus, *Virology*, 16, 236, 1962.
30. **Van Regenmortel, M. H. V.,** Serological methods in the identification and characterization of viruses, *Comp. Virol.*, 17, 183, 1981.
31. **Van Regenmortel, M. H. V.,** *Serology and Immunochemistry of Plant Viruses*, Academic Press, New York, 1982.
32. **Wetter, C. and Bernard, M.,** Identifizierung, Reinigung und serologischer Nachweis von Tabakmosaikvirus und Para-Tabakmosaikvirus aus Zigaretten, *Phytopathol. Z.*, 90, 257, 1977.
33. **Geolet, P., Lomonossoff, G. P., Butler, P. J. G., Okam, M. E., Gait, M. J., and Karn, J.,** Nucleotide sequence of tobacco mosaic virus RNA, *Proc. Natl. Acad. Sci. U.S.A.*, 79, 5818, 1982.
34. **Gibbs, A., Skotnicki, A. H., Gardiner, J. E., Walker, E. S., and Hollings, M.,** A tobamovirus of a green alga, *Virology*, 64, 571, 1975.
35. **Skotnicki, A., Gibbs, A., and Wrigley, N. G.,** Further studies on *Chara corrallina* virus, *Virology*, 75, 457, 1976.
36. **Putz, C.,** Composition and structure of beet necrotic yellow vein virus, *J. Gen. Virol.*, 35, 397, 1977.
37. **Tamada, T.,** Beet necrotic yellow vein virus, *CMI/AAB Descriptions of Plant Viruses*, No. 144, 1975.
38. **Putz, C. and Vuittenez, A.,** The intracellular location of beet necrotic yellow vein virus, *J. Gen. Virol.*, 50, 201, 1980.
39. **Russo, M., Martelli, G. P., and Di Franco, A.,** The fine structure of local lesions of beet necrotic yellow vein virus in *Chenopodium amaranticolor*, *Physiol. Plant Pathol.*, 19, 237, 1981.
40. **Giunchedi, L., Langenberg, W. G., and Marani, F.,** Appearance of beet necrotic yellow vein virus (BNYVV) in host cells, *Phytopathol. Mediterr.*, 20, 112, 1981.
41. **Inouye, T. and Nakasone, W.,** Broad bean necrosis virus, *CMI/AAB Descriptions of Plant Viruses*, No. 223, 1980.
42. **Brunt, A. A. and Stace-Smith, R.,** Some hosts, properties and possible affinities of a labile virus from *Hypochoeris radicata* (Compositae), *Ann. Appl. Biol.*, 90, 205, 1978.
43. **Brunt. A. A. and Stace-Smith, R.,** Electron microscopy of *Nicotiana clevelandii* leaf cells infected with hypochoeris mosaic virus, *J. Gen. Virol.*, 41, 207, 1978.
44. **Randles, J. W.,** Nicotiana velutina mosaic virus, *CMI/AAB Descriptions of Plant Viruses* No. 189, 1978.
45. **Thouvenel, J. C. and Fauquet, C.,** Peanut clump virus, *CMI/AAB Descriptions of Plant Viruses*, No. 235, 1981.
46. **Reddy, D. V. R., Rajeshwari, R., Iizuka, N., Lesemann, D. E., Nolt, B. L., and Goto, T.,** The occurrence of indian peanut clump, a soil-borne virus disease of groundnuts (*Arachis hypogaea*) in India, *Ann. Appl. Biol.*, 102, 305, 1983.
47. **Harrison, B. D.,** Potato mop-top virus, *CMI/AAB Descriptions of Plant Viruses*, No. 138, 1974.
48. **White, R. F., Kassanis, B., and James, M.,** Potato mop-top virus in infected cells, *J. Gen. Virol.*, 15, 175, 1972.
49. **Kassanis, B., Woods, R. D., and White, R. F.,** Some properties of potato mop-top virus and its serological relationship to tobacco mosaic virus, *J. Gen. Virol.*, 14, 123, 1972.

50. **Brakke, M. K.,** Soil-borne wheat mosaic virus, *CMI/AAB Descriptions of Plant Viruses*, No. 77, 1971.
51. **Powell, C. A.,** The relationship between soil-borne wheat mosaic virus and tobacco mosaic virus, *Virology*, 71, 453, 1976.
52. **Langenberg, W. G. and Giunchedi, L.,** Ultrastructure of fungal plant virus vectors *Polymyxa graminis* in soilborne wheat mosaic virus-infected wheat and *P. betae* in beet necrotic yellow vein virus-infected sugar beet, *Phytopathology*, 72, 1152, 1982.
53. **Shirako, Y. and Brakke, M. K.,** Two purified RNAs of soil-borne wheat mosaic virus are needed for infection, *J. Gen. Virol.*, 65, 119, 1984.
54. **Milne, R. G.,** unpublished results.
55. **Randles, J. W., Harrison, B. D., and Roberts, I. M.,** *Nicotiana velutina* mosaic virus: purification, properties and affinities with other rod-shaped viruses, *Ann. Appl. Biol.*, 84, 193, 1976.
56. **Izadpanah, K., Thompson, M. R., and Thornberry, H. H.,** Orchid virus: characterization of a type of tobacco-mosaic virus isolated from *Cattleya* plants with infectious blossom necrosis, *Phytopathol. Z.*, 63, 272, 1968.
57. **Corbett, M. V.,** Some distinguishing characteristics of the orchid strain of tobacco mosaic virus, *Phytopathology*, 57, 164, 1967.
58. **Higgins, T. J. V., Goodwin, P. B., and Whitfeld, P. R.,** Occurence of short particles in beans infected with the cowpea strain of TMV. II. Evidence that short particles contain the cistron for coat-protein, *Virology*, 71, 486, 1976.
59. **Bruening, G., Beachy, R. N., Scalla, R., and Zaitlin, M.,** *In vitro* and *in vivo* translation of the ribonucleic acids of a cowpea strain of tobacco mosaic virus, *Virology*, 71, 498, 1976.
60. **Williams, R. C. and Fisher, H. W.,** Electron microscopy of tobacco mosaic virus under conditions of minimal beam exposure, *J. Mol. Biol.*, 52, 121, 1970.
61. **Markham, R., Hitchborn, J. H., Hills, G. J., and Frey, S.,** The anatomy of the tobacco mosaic virus, *Virology*, 22, 342, 1964.
62. **Hart, R. G. and Yoshiyama, J. M.,** Electron microscopy with reduced beam damage to the specimen: a retractable image intensifier, *J. Ultrastruct. Res.*, 51, 40, 1975.
63. **Fraenkel-Conrat, H.,** Portraits of viruses: tobacco mosaic virus, *Intervirology*, 15, 177, 1981.
64. **Finch, J. T.,** The pitch of tobacco mosaic virus, *Virology*, 38, 182, 1969.
65. **Horne, R. W.,** The formation of virus crystalline and paracrystalline arrays for electron microscopy and image analysis, *Adv. Virus Res.*, 24, 173, 1979.
66. **Franklin, R. E., Caspar, D. L. D., and Klug, A.,** *Plant Pathology: Problems and Progress 1908—1958*, Holton C. S. et al., Eds., 1959, 447.
67. **Finch, J. T.,** The hand of the helix of tobacco mosaic virus, *J. Mol. Biol.*, 66, 291, 1972.
68. **Tsugita, A., Gish, D. T., Young, J., Fraenkel–Conrat, H., Knight, C. A., and Stanley, W. M.,** The complete amino acid sequence of the protein of tobacco mosaic virus, *Proc. Natl. Acad. Sci. U. S. A.*, 46, 1463, 1960.
69. **Anderer, F. A., Uhlig, H., Weber, E., and Schramm, G.,** Primary structure of the protein of tobacco mosaic virus, *Nature (London)*, 186, 922, 1960.
70. **Meshi, T., Kiyama, R., Ohno, T., and Okada, Y.,** Nucleotide sequence of the coat protein cistron and the 3′ noncoding region of cucumber green mottle mosaic virus (watermelon strain) RNA, *Virology*, 127, 54, 1983.
71. **Nonomura, Y. and Ohno, T.,** Direct visualization of the RNA location in cucumber green mottle mosaic virus and tobacco mosaic virus protein disk, *J. Mol. Biol.*, 90, 523, 1974.
72. **Wilson, T. M. A., Perham, R. N., Finch, J. T., and Butler, P. J. G.,** Polarity of the RNA in the tobacco mosaic virus particle and the direction of protein stripping in sodium dodecyl sulphate, *FEBS Lett.*, 64, 285, 1976.
73. **Butler, P. J. G., Finch, J. T., and Zimmern, D.,** Configuration of tobacco mosaic virus RNA during virus assembly, *Nature (London)*, 265, 217, 1977.
74. **Fraenkel–Conrat, H.,** Degradation of tobacco mosaic virus with acetic acid, *Virology*, 4, 1, 1957.
75. **Durham, A. C. H.,** Structures and roles of the polymorphic forms of tobacco mosaic virus protein, *J. Mol. Biol.*, 67, 289, 1972.
76. **Schramm, G., Schumacher, G., and Zillig, W.,** Über die Struktur des Tobakmosaikvirus. III. Der Zerfall in alkalischer Lösung, *Z. Naturforsch.*, 10B, 481, 1955.
77. **Hirth, L. and Richards, K. E.,** Tobacco mosaic virus: model for structure and function of a simple virus, *Adv. Virus Res.*, 26, 145, 1981.
78. **Bloomer, A. C., Graham, J., Hovmöller, S., Butler, P. J. G., and Klug, A.,** Tobacco mosaic virus: interaction of the protein disk with nucleotides and its implications for virus structure and assembly, in *Structural Aspects of Recognition and Assembly in Biological Macromolecules*, Balaban, M., Sussman, J. L., Traub, W., and Yonath, A., Eds., Balaban ISS, Rehovot, 1981, 851.
79. **Shire, S. J., Steckert, J. J., and Schuster, T. M.,** Mechanism of tobacco mosaic virus assembly: incorporation of 4S and 20S protein at pH 7.0 and 20°C, *Proc. Natl. Acad. Sci. U. S. A.*, 78, 256, 1981.

80. **Steckert, J. J. and Schuster, T. M.,** Sequence specificity of trinucleoside diphosphate binding to polymerized tobacco mosaic virus protein, *Nature (London)*, 299, 32, 1982.
81. **Lebeurier, G., Nicolaieff, A., and Richards, K. E.,** Inside-out model for self-assembly of tobacco mosaic virus, *Proc. Natl. Acad. Sci. U. S. A.*, 74, 149, 1977.
82. **McLachlan, A. D., Bloomer, A. C., and Butler, P. J. G.,** Structural repeats and evolution of tobacco mosaic virus coat protein and RNA, *J. Mol. Biol.*, 136, 203, 1980.
83. **Kaplan, I. B., Kozlov, Y. V., Pshennikova, E. S., Taliansky, M. E., and Atabekov, J. G.,** A study of TMV ts mutant Ni2519. III. Location of the reconstitution initiation sites on Ni2519 RNA, *Virology*, 118, 317, 1982.
84. **Fukuda, M. and Okada, Y.,** Mechanism of tobacco mosaic virus assembly: role of subunit and larger aggregate protein, *Proc. Natl. Acad. Sci. U. S. A.*, 79, 5833, 1982.
85. **Lomonossoff, G. P. and Butler, P. J. G.,** Location and encapsidation of the coat protein cistron of Tobacco mosaic virus, *Eur. J. Biochem.*, 93, 157, 1979.
86. **Okada, Y., Fukuda, M., Takebe, I., and Otsuki, Y.,** Initiation site for assembly of several strains of TMV and its relation to occurrence of the short particles in infected plants, *Biosystems*, 12, 257, 1980.
87. **Fukuda, M., Meshi, T., Okada, Y., Otsuki, Y., and Takebe, I.,** Correlation between particle multiplicity and location on virion RNA of the assembly initiation site for viruses of the tobacco mosaic virus group, *Proc. Natl. Acad. Sci. U. S. A.*, 78, 4231, 1981.
88. **Steven, A. C., Trus, B. L., Putz, C., and Wurtz, M.,** The molecular organization of beet necrotic yellow vein virus, *Virology*, 113, 428, 1981.
89. **Veerisetty, V.,** Relationships among structural parameters of virions of helical symmetry, *Virology*, 84, 523, 1978.
90. **Keith, J. and Fraenkel-Conrat, H.,** Tobacco mosaic virus RNA carries 5′ terminal triphosphorylated guanosine blocked by 5′-linked 7-methylguanosine, *FEBS Lett.*, 57, 31, 1975.
91. **Zimmern, D.,** The 5′ end group of tobacco mosaic virus RNA is $m^7G^5ppp^5Gp$, *Nucleic Acids Res.*, 2, 1189, 1975.
92. **Carriquiry, E. and Litvak, S.,** Further studies on the enzymatic aminoacylation of TMV-RNA by histidine, *FEBS Lett.*, 38, 287, 1974.
93. **Beachy, R. N., Zaitlin, M., Bruening, G., and Israel, H. W.,** A genetic map of the cowpea strain of TMV, *Virology*, 73, 498, 1976.
94. **Pelham, H. R. B. and Jackson, R. J.,** An efficient mRNA-dependent translation system from reticulocyte lysates, *Eur. J. Biochem.*, 67, 247, 1976.
95. **Knowland, J.,** Protein systhesis directed by the RNA from a plant virus in a normal animal cell, *Genetics*, 78, 383, 1974.
96. **Hunter, T. R., Hunt, T., Knowland, J., and Zimmern, D.,** Messenger RNA for the coat protein of tobacco mosaic virus, *Nature (London)*, 260, 759, 1976.
97. **Siegel, A., Hari, V., Montgomery, I., and Kolacz, K.,** A messenger RNA for capsid protein isolated from tobacco mosaic virus-infected tissue, *Virology*, 73, 363, 1976.
98. **Jonard, G., Richards, K., Mohier, E., and Gerlinger, P.,** Nucleotide sequence at the 5′ extremity of tobacco-mosaic-virus RNA. II. The coding region (nucleotides 69-236), *Eur. J. Biochem.*, 84, 521, 1978.
99. **Zaitlin, M. and Palukaitis, P.,** The tobacco mosaic virus genome: organisation and gene function, in *Replication of Viral and Cellular Genomes*, Becker, Y., Ed., Martinus Nijhoff, Boston, 1983, 329.
100. **Pelham, H. R. B.,** Leaky UAG termination codon in tobacco mosaic virus RNA, *Nature (London)*, 272, 469, 1978.
101. **Beachy, R. N. and Zaitlin, M.,** Characterization and *in vitro* translation of the RNAs from less-than-full-length, virus related, nucleoprotein rods present in tobacco mosaic virus preparations, *Virology*, 81, 160, 1977.
102. **Meshi, T., Ohno, T., and Okada, Y.,** Nucleotide sequence of the 30K protein cistron of cowpea strain of tobacco mosaic virus, *Nucleic Acids Res.*, 10, 6111, 1982.
103. **Leonard, D. A. and Zaitlin, M.,** A temperature-sensitive strain of tobacco mosaic virus defective in cell-to-cell movement generates an altered viral coded protein, *Virology*, 117, 416, 1982.
104. **Scalla, R., Romaine, P., Asselin, A., Rigaud, J., and Zaitlin, M.,** An *in vitro* study of nonstructural polypeptide synthesized upon TMV infection and its identification with a polypeptide synthesized *in vitro* from TMV RNA, *Virology*, 91, 182, 1978.
105. **Sakai, F. and Takebe, I.,** Protein synthesis in tobacco mesophyll protoplasts induced by tobacco mosaic virus infection, *Virology*, 62, 426, 1974.
106. **Paterson, R. and Knight, C. A.,** Protein synthesis in tobacco protoplasts infected with tobacco mosaic virus, *Virology*, 64, 10, 1975.
107. **Siegel, A., Hari, V., and Kolacz, K.,** The effect of tobacco mosaic virus infection on host and virus-specific protein synthesis in protoplasts, *Virology*, 85, 494, 1978.
108. **Beier, H., Mundry, K. W., and Issinger, O. G.,** *In vivo* and *in vitro* translation of the RNAs, of four tobamoviruses, *Intervirology*, 14, 292, 1980.

109. **Joshi, S., Pleij, C. W. A., Haenni, A. L., Chapeville, F., and Bosch, L.,** Properties of the tobacco mosaic virus intermediate length RNA-2 and its translation, *Virology*, 127, 100, 1983.
110. **Hunter, T., Jackson, R., and Zimmern, D.,** Multiple proteins and subgenomic mRNAs may be derived from a single open reading frame on tobacco mosaic virus RNA, *Nucleic Acids Res.*, 11, 801, 1983.
111. **Goelet, P., and Karn, J.,** Tobacco mosaic virus induces the synthesis of a family of 3′ coterminal messenger RNAs and their complements, *J. Mol. Biol.*, 154, 541, 1982.
112. **Zelcer, A., Weaber, K. F., Balzás, E., and Zaitlin, M.,** The detection and characterization of viral-related double-stranded RNAs in tobacco mosaic virus-infected plants, *Virology*, 113, 417, 1981.
113. **Tsuchizaki, T., Hibino, H., and Saito, Y.,** The biological functions of short and long particles of soil-borne wheat mosaic virus, *Phytopathology*, 65, 523, 1975.
114. **Putz, C., Pinck, M., Fritsch, C., and Pinck, L.,** Identification of the 3′- and 5′-ends of beet necrotic yellow vein virus RNAs, *FEBS Lett.*, 156, 41, 1983.
115. **Shalla, T. A.,** Assembly and aggregation of tobacco mosaic virus in tomato leaflets, *J. Cell Biol.*, 21, 253, 1964.
116. **Kolehmainen, L., Zech, H., and von Wettstein, D.,** The structure of cells during tobacco mosaic virus reproduction. Mesophyll cells containing virus crystals, *J. Cell. Biol.*, 25, 77, 1965.
117. **Milne, R. G.,** Multiplication of tobacco mosaic virus in tobacco leaf palisade cells, *Virology*, 28, 79, 1966.
118. **Warmke, H. E. and Christie, R. G.,** Use of dilute osmium tetroxide for preservation of three-dimensional crystals of tobacco mosaic virus, *Virology*, 32, 534, 1967.
119. **Esau, K.,** *Viruses in Plant Hosts*, University of Wisconsin Press, Madison, 1968, 225 pp.
120. **Warmke, H. E.,** Aucuba strain of tobacco mosaic virus: an unusual aggregate, *Science*, 156, 262, 1967.
121. **Warmke, H. E.,** Fine structure of inclusions formed by the aucuba strain of tobacco mosaic virus, *Virology*, 34, 149, 1968.
122. **Herold, F. and Munz, K.,** An unusual kind of crystalline inclusion of tobacco mosaic virus, *J. Gen. Virol.* 1, 375, 1967.
123. **Shalla, T. A.,** Virus particles in chloroplasts of plants infected with the U5 strain of tobacco mosaic virus, *Virology*, 35, 194, 1968.
124. **Esau, K. and Cronshaw, J.,** Relation of tobacco mosaic virus to the host cells, *J. Cell. Biol.*, 33, 665, 1967.
125. **Liu, K. C. and Boyle, J. S.,** Intracellular morphology of two tobacco mosaic virus strains in, and cytological response of, systemically susceptible potato plants, *Phytopathology*, 62, 1303, 1972.
126. **Milne, R. G.,** Electron microscopy of tobacco mosaic virus in leaves of *Chenopodium amaranticolor*, *Virology*, 28, 520, 1966.
127. **Milne, R. G.,** Electron microscopy of tobacco mosaic virus in leaves of *Nicotiana glutinosa*, *Virology*, 28, 527, 1966.
128. **Kassanis, B. and Milne, R. G.,** An unusual inclusion in plants infected with a tobacco mosaic virus mutant, *J. Gen. Virol.*, 11, 193, 1971.
129. **Granett, A. L. and Shalla, T. A.,** The relation of tobacco mosaic virus X-protein to amorphous cellular inclusions (X-bodies), *Phytopathology*, 60, 426, 1970.
130. **Betto, E., Bassi, M., Favali, M. A., and Conti, G. G.,** An electron microscopic and autoradiographic study of tobacco leaves infected with the U5 strain of tobacco mosaic virus,, *Phytopathol, Z.*, 75, 193, 1972.
131. **Shalla, T. A., Petersen, L. J., and Giunchedi, L.,** Partial characterization of virus-like particles in chloroplasts of plants infected with the U5 strain of TMV, *Virology*, 66, 94, 1975.
132. **Hatta, T., Nakamoto, T., Takagi, Y., and Ushiyama, R.,** Cytological abnormalities of mitochondria induced by infection with cucumber green mottle mosaic virus, *Virology*, 45, 292, 1971.
133. **Hatta, T. and Ushiyama, R.,** Mitochondrial vesiculation associated with cucumber green mottle mosaic virus-infected plants, *J. Gen. Virol.*, 21, 9, 1973.
134. **Sugimura Y. and Ushiyama, R.,** Cucumber green mottle mosaic virus infection and its bearing on cytological alterations in tobacco mesophyll protoplasts, *J. Gen. Virol.*, 29, 93, 1975.
135. **Hatta, T. and Francki, R. I. B.,** Cytopathic structures associated with tonoplasts of plant cells infected with cucumber mosaic and tomato aspermy viruses, *J. Gen. Virol.*, 53, 343, 1981.
136. **Zaitlin, M. and Boardman, N. K.,** The association of tobacco mosaic virus with plastids. I. Isolation of virus from the chloroplast fraction of diseased-leaf homogenates, *Virology*, 6, 743, 1958.
137. **Boardman, N. K. and Zaitlin, M.,** The association of tobacco mosaic virus with plastids. II. Studies on the biological significance of virus as isolated from a chloroplast fraction, *Virology*, 6, 758, 1958.
138. **Reddi, K. K.,** Tobacco mosaic virus with emphasis on the events within the host cell following infection, *Adv. Virus Res.*, 17, 51, 1972.
139. **Shalla, T. A. and Amici, A.,** The distribution of viral antigen in cells infected with tobacco mosaic virus as revealed by electron microscopy, *Virology*, 31, 78, 1967.
140. **Langenberg, W. G. and Schlegel, D. E.,** Localization of tobacco mosaic virus protein in tobacco leaf cells during the early stages of infection, *Virology*, 37, 86, 1969.

141. **Langenberg, W. G. and Schlegel, D. E.,** Autoradiography with ^{125}I-labeled antibodies as a means of localizing TMV antigen in plant cells, *Virology*, 32, 167, 1967.
142. **Ralph, R. K., Bullivant, S., and Wojcik, S. J.,** Cytoplasmic membranes, a possible site of tobacco mosaic virus RNA replication, *Virology*, 43, 713, 1971.
143. **De Zoeten, G. A.,** Cytology of virus infection and virus transport, in *Physiological Plant Pathology, Encyclopedia of Plant Physiology, New Series*, Vol. 4, Heitefuss, R. and Williams, P. H., Eds., Springer Verlag, Basel, 1976, 129.
144. **Zaitlin, M.,** The RNAs of monopartite plant viruses, in *Nucleic Acids in Plants*, Vol. 2, Hall, T. C. and Davies, J. W., Eds., CRC Press, Boca Raton, Fla., 1979, 31.
145. **Zaitlin, M., Spencer, D., and Whitfeld, P. R.,** Studies on the intracellular site of tobacco mosaic virus assembly, in *Biochemical Regulation in Diseased Plants or Injury*, Hirai, T., Ed., The Phytopathological Society of Japan, Tokyo, 1968, 91.
146. **Martelli, G. P.,** Ultrastructural aspects of possible defence reactions in virus-infected plant cells, *Microbiologica*, 3, 369, 1980.
147. **Sela, I.,** Plant-virus interactions related to resistance and localization of viral infections, *Adv. Vir. Res.*, 26, 201, 1981.
148. **van Loon, L. C.,** Mechanisms of resistance in virus-infected plants, in *The Dynamics of Host Defence*, Bailey, J. A. and Deverall, B. J., Eds., Academic Press, New York, 1983, 123.

Chapter 10

HORDEIVIRUS GROUP

The word "hordei" is Latin for "of barley", and refers to the type member of this group, barley stripe mosaic virus (BSMV).[1]

Hordeivirus particles are rigid rods about 20 nm in diameter with a pitch of about 2.5 nm and of two or more length classes in the range 150 and 100 nm. They contain about 4% positive-sense, single-stranded RNA of M_r 1.5 to 1.0 $\times$ 10^6, as a divided genome. There is one coat protein of M_r about 25 $\times$ 10^3. The S values of the two largest particle types are estimated as 192S and 182S, for the type strain of BSMV.[2]

Hordeiviruses have narrow individual host ranges and possess rather stable particles that are transmitted by contact and, with BSMV and lychnis ringspot virus (LRSV), through both gametes to the seed. Rates of seed transmission with these two viruses can be very high,[3-6] but seed transmission has not yet been demonstrated for the other viruses and no specific vectors are known,[2] Hordeivirus particles occur in the cytoplasm and, less frequently, in the nuclei of infected cells.[2]

The symptoms produced by Hordeiviruses are not remarkable except for the "aberrant ratio" genetic effect, where BSMV infection in maize may distort segregation ratios in a small proportion of the ears.[2,7,8]

I. MEMBERS OF THE GROUP AND THEIR RELATIONSHIPS

The group consists of BSMV, the rather similar poa semilatent virus (PSLV),[9] and LRSV (Table 1). There is one possible member, provisionally named anthoxanthum latent blanching virus (ALBV).[10]

Both PSLV[9,21] and LRSV[22] are distantly related serologically to BSMV (SDIs about 7). There appear to be no data on the serological relationship, if any, between PSLV and LRSV. As PSLV has several hosts as well as other properties in common with BSMV, the suggestion has been made that it be considered a serotype of BSMV[21] though the ICTV lists it as a separate virus.[1] PSLV but not BSMV invades *Agropyron trachycaulum* and *Poa palustris*, whereas BSMV but not PSLV gives local lesions in a range of *Chenopodium* species.[21] PSLV is reported to have a wider spread of particle lengths than BSMV, with a higher value for the average length,[21] though an accurate analysis of PSLV particle lengths has not yet been published.

A number of strains of BSMV[24] have been relatively well characterized, and it is worth naming some of them because they possess different numbers of particle components and, apparently, of RNA species (Table 1). The type strain (American Type Culture Collection PV43, not PV69 as listed in some papers) and the Russian strain appear to have two components (but see the section on the viral genome) as do strains C2 and C3 from Manitoba. The Norwich (NW), North Dakota 18 (ND 18), C1, and C4 strains have three components, and the Argentine mild (AM) and ND161 strains have four components. Considerable cytopathological work has been done on the MI-1 strain, which is seed transmitted in barley, and strain 850, which is not (see Secion III).

Whereas BSMV is common and very widespread and PSLV is also probably distributed widely in Canada,[2,9] LRSV has been isolated only once,[23] in a test plant of *Lychnis divaricata* used in screening for sugarbeet viruses. LRSV has an experimental host range confined to a few dicotyledonous families and has not been found to infect monocotyledonous plants.[2,23]

Further study is required to establish the relationship of ALBV to the Hordeiviruses. Its particle morphology, narrow host range, sap transmissibility and lack of known vector suggest that it is a Hordeivirus; however, Catherall and Chamberlain[10] could find no evidence of

Table 1
SOME PROPERTIES OF THE MEMBERS AND OF A POSSIBLE MEMBER OF THE HORDEIVIRUS GROUP

Virus	Natural host family	Particle modal lengths (nm)	Sizes of RNAs ($M_r \times 10^6$)	Size of coat protein ($M_r \times 10^3$)	Ref.
Members and selected strains					
Barley stripe mosaic virus (BSMV)	Gramineae				11 — 19
Type strain		148 + 125— 128	1.43 + 1.24[a] (+0.28)	25	—
Several Manitoba strains (C1 to C4)		141—149 + 119—125 (+ 108—109 for C1 — C4)	ND[b]	ND	20
Norwich (NW) strain, North Dakota 18 (ND18) strain		Two largest + 110	1.43 + 1.24 + 1.1 (+0.28)	ND	2,18,19
Argentina mild (AM) strain, North Dakota 161 (ND161) strain		Three largest + ~100	1.43 + 1.24 + 1.1 + 1.0	ND	2,18
Poa semilatent virus (PSLV)	Gramineae	Mean about 160; range 447 — 85 (including aggregates and fragments)	1.50 + 1.35 + 1.20 + 1.15	ND	9,14,21
Lychnis ringspot virus (LRSV)	Caryophyllaceae	125 (main peak)	1.50 + 1.30 + 1.15 + 0.90	ND	14,22,23
Possible member					
Anthoxanthum latent blanching virus (ALBV)	Gramineae	135 (main peak)	ND	ND	10

[a] RNA fractions contain two molecular species of similar size but different nucleotide sequences.[18]
[b] ND, not determined.

seed transmission or of serological relationship with BSMV. A possible serological relationship to PSLV (or LRSV) has not been tested. Limited host range comparisons[10] have not detected any host species common to both ALBV and PSLV, though they both infect Gramineae. The problem of where to place ALBV reminds us that although true Hordeiviruses and true Tobamoviruses are distinct, the possible Tobamoviruses are a potential bridge between the two groups (compare Figures 1 through 3 and Table 1 in this Chapter and Table 2 in Chapter 9, and see Chapter 9).

II. VIRUS STRUCTURE AND COMPOSITION

A. Particle Structure

The particles of Hordeiviruses are rigid rods about 20 nm in diameter in negative stain, though there is a tendency for the particles to flatten (Figures 1 and 2) and to appear flexuous if the support film is not rigid.[2] A value of 22 nm for the particle diameter was derived from X-ray diffraction measurements on wet gels.[25] The pitch of the helix is about 2.5 nm. This structure, unlike the helix of TMV but like that of possible Tobamoviruses, is very evident even after considerable irradiation in the electron microscope beam.

Due to a tendency of the particles toward aggregation and breakage, obtaining well-resolved length histograms presents difficulties although the particles are bimodal in length, the longer being about 150 nm and the shorter about 100 nm. Chiko[20] noted that the length class interval chosen had to be not greater than 7 nm and preferably 2.5 or 3.5 nm, with a sample size of 600 or more particles in order to detect the bimodal distribution.

Each of the RNAs is packaged separately, with the amount of RNA determining particle length. There is 9.9×10^3 M_r of RNA per nanometer of particle length (as compared to 6.8 and 12.3×10^3 for TMV and tobacco rattle virus, respectively) wound in a helix of radius 5.5 nm (as determined by X-ray diffraction),[25] or 6.3 nm (as found by calculation).[26] The axial canal has a radius of 2 nm; three nucleotides are associated with each coat protein subunit and there are 23 to 26 subunits per turn of the helix.[2,26] The RNA makes up 3.7 to 4.0% of the particle by weight.[12,27]

BSMV has one coat protein which was originally estimated to be of M_r about 21.5×10^3,[12] but a more recent estimate is 25×10^3.[28,29] Furthermore, it has been shown that the protein is glycosylated; the carbohydrate is attached to the tripeptide –Gly–Asp–Ala– by way of an amide bond involving asparagine.[28,29] Partridge and his colleagues[28] suggested that glycosylation of the coat protein could be a determinant of the high seed transmissibility of BSMV; however, there exist many efficiently seed-transmitted viruses whose coat proteins are not glycosylated.

B. The Viral Genome

As seen in Table 1, different Hordeiviruses may appear to have different numbers of RNA components, and it was formerly thought that the type and Russian strains of BSMV had two genomic RNAs whereas the Norwich and ND18 strains had three.[13-16,30] There is now evidence that the type strain has in fact three genomic RNAs, "RNA 2" being a mixture of two molecules (2a and 2b) of similar M_r but different nucleotide sequences and encapsidated in particles of closely similar length (Table 1).[18,19] Thus the type strain of BSMV is a "pseudo-two component" strain and the so-called RNA 2 is an experimental fraction rather than a genetic unit.

With Hordeiviruses, a fourth RNA is frequently encountered but does not appear to be genomic. With the type, Norwich and ND18 strains, this RNA has been shown to be of M_r about 0.28×10^6 (~800 nucleotides) and to be a subgenomic copy of part of the third genomic RNA (i.e., RNA 2b of the type strain and RNA 3 of the Norwich and ND18 strains).[18,19] In the case of the AM strain of BSMV, the fourth or smallest RNA is of M_r

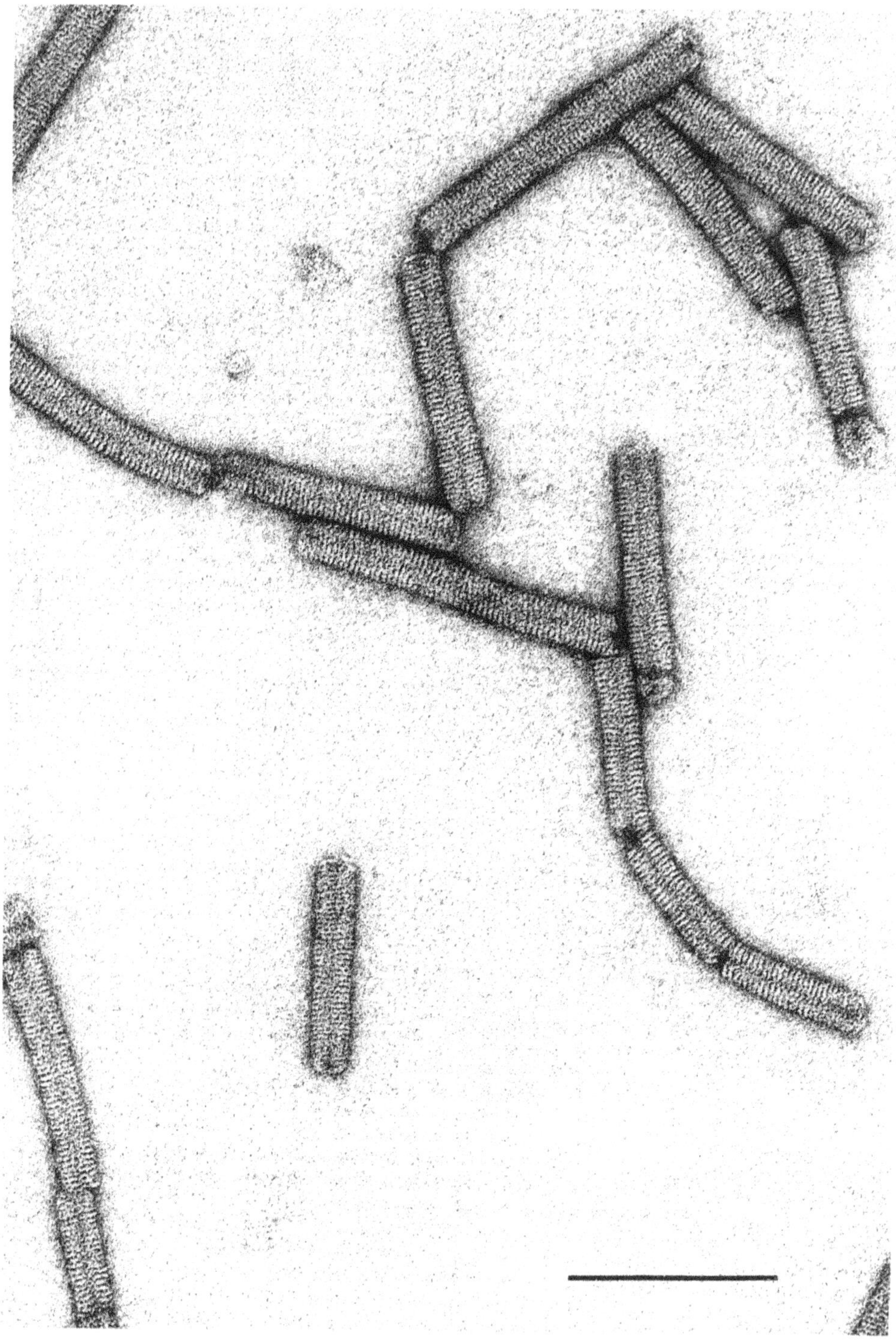

FIGURE 1. Purified particles of barley stripe mosaic virus (BSMV) stained in a thin layer of uranyl formate. (Compare with Figure 3B.) Bar = 100 nm.

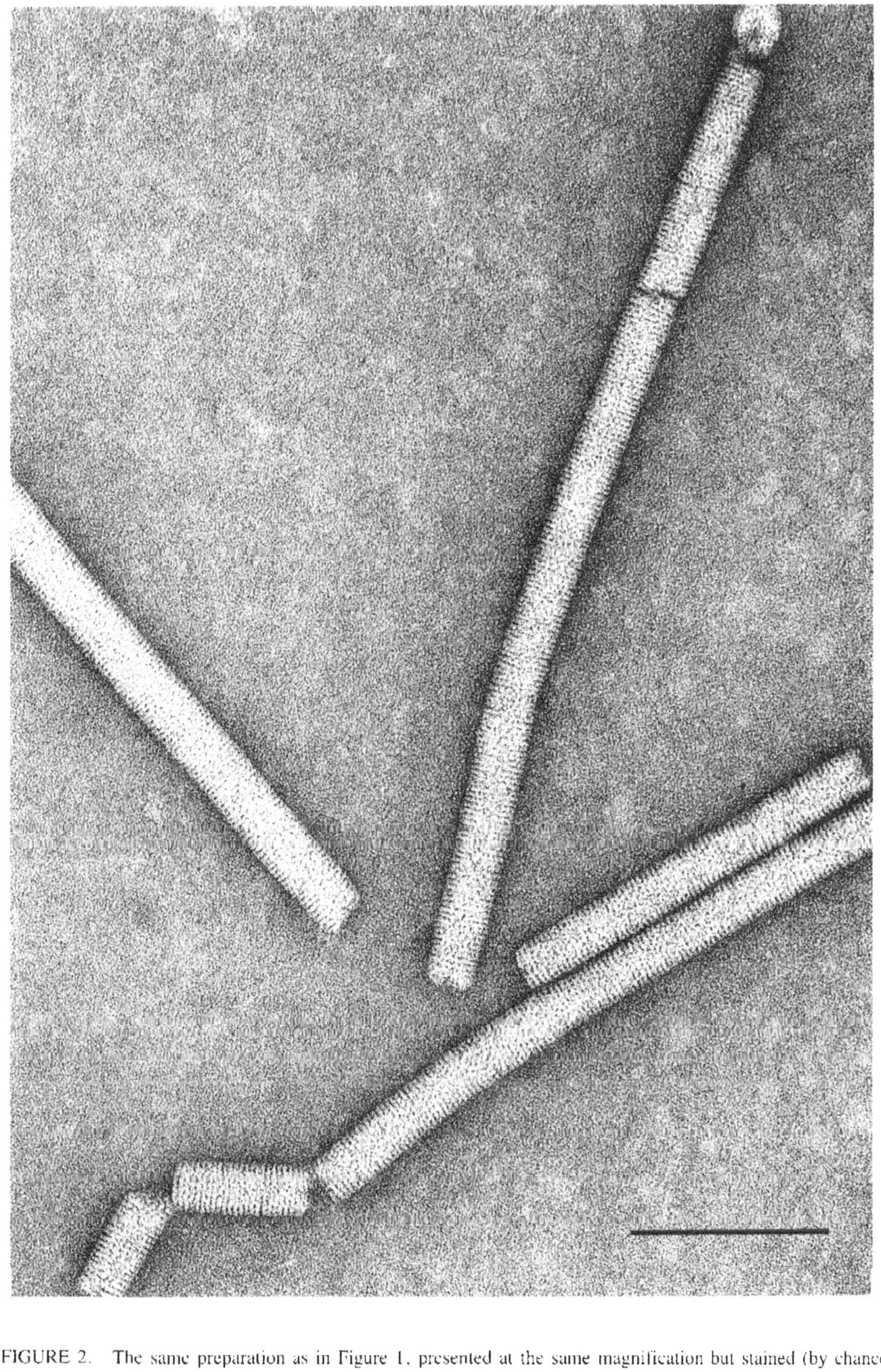

FIGURE 2. The same preparation as in Figure 1, presented at the same magnification but stained (by chance) in a deeper layer of uranyl formate. (The difference in appearance is due to particle flattening and greater exclusion of stain in this preparation.) Bar = 100 nm.

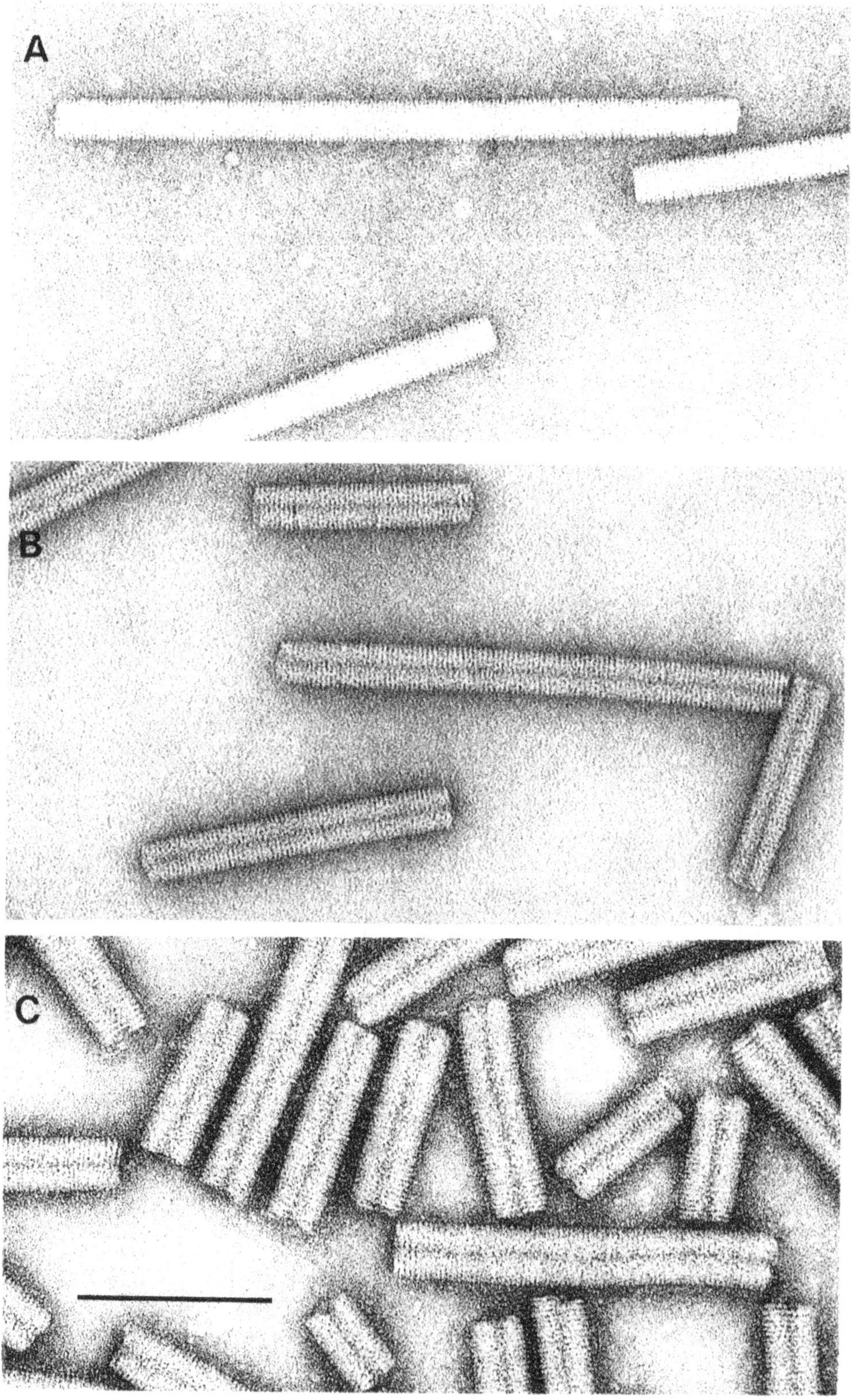

FIGURE 1.

about 1.0 × 10^6 (Table 1). By subculturing this isolate with inocula near the dilution end-point,[16] isolates with two and three RNAs were obtained. It was further shown by analysis with DNA clones specific to the three genomic RNAs of BSMV,[31] that both subcultured AM isolates contained three genomic RNAs, and that no fourth RNA was present as in the original isolate.[31] Thus it appears that all BSMV strains and possibly all Hordeiviruses have tripartite genomes.

Evidence from sequence homology studies[16,18,19,31,32] suggests that RNA 1 of BSMV shares few if any sequences with other viral RNA components. However, there is some disagreement about the extent of sequence homology between RNAs 2 and 3 of the Norwich strain. Atabekov and his colleagues,[32,33] using cDNA:RNA hybridization techniques, concluded that the sequences were highly homologous, whereas Jackson and his colleagues [18,19] using recombinant DNA clones, suggested that probably more than 70% of the sequences of these RNAs were unique.

BSMV RNA appears to have unusual 3′ termini.[33-37] A short internal poly A tract is located about 210 nucleotides from the 3′ end. The intact RNAs can be amino-acylated with tyrosine in vitro and a 210-residue aminoacyl-accepting moiety can be cleaved from the genomic RNAs by RNase H after hybridization of oligo (dT) to the poly A tract.[37] The aminoacylation reaction may be important in replication because encapsidated BSMV RNA is not tyrosylated, whereas some of the viral RNA in cells is.[38] Apparently the BSMV RNA population is heterogeneous in that most molecules have a poly A sequence up to 30 units long, whereas in about 20% of the molecules there are only some eight units in the tract.[33-37] The 5′ terminus of at least RNAs 1 and 3 of the Norwich strain are capped with m^7Gppp, and it is likely that all the RNAs of all strains have this structure.[30,39]

Some translation products of BSMV RNAs have been identified. Jackson and his colleagues[40] found that polysomes from infected barley plants synthesized, in vitro, virus-specific polypeptides of M_r 120, 67, and 25 × 10^3 and that similar products were translated in vivo by BSMV RNA. The 25 × 10^3 M_r polypeptide had characteristics of the coat protein. In wheat germ or reticulocyte lysate translation systems, polypeptides of M_r about 120, 82, 71, and 25 × 10^3 were synthesized. The 120, and 25 × 10^3 M_r products corresponded with those detected in vivo, but it was suggested that the 82, and 71 × 10^3 M_r in vitro products might undergo processing in vivo to give the 67 × 10^3 polypeptide. Atabekov and his colleagues[17] obtained evidence with the Norwich strain that the 120 × 10^3 M_r polypeptide is translated from RNA 1 and the coat protein from RNA 2, and RNA 3 produces a polypeptide of M_r 75 × 10^3. Thus although RNA 2 supposedly included most if not all of the sequences of RNA 3, the two RNAs gave rise to different gene products. It has been proposed[17,30] that different genes may be "open" in the two RNAs. This difficulty resolves itself if, as suggested by Jackson and his colleagues,[19] and discussed above, RNAs 2 and 3 are essentially different molecules.

In the wheat germ system, the subgenomic RNA of both the type and ND18 strains were found to direct the synthesis of two polypeptides of M_r about 20 × 10^3 and 21 × 10^3.[19] How the two peptides are produced is not clear. One possibility is that both are synthesized from a single messenger, with the larger resulting from read-through of a leaky stop signal. It was also not ruled out that the "subgenomic RNA" may contain two RNAs, each of which directs the synthesis of a distinct polypeptide. The peptides have not been detected in vivo, but there is preliminary evidence that the subgenomic RNA is bound to polyribosomes in infected plants, and therefore participates in protein synthesis.[19]

FIGURE 3. Purified particles. A, tobacco mosaic virus (TMV, Tobamovirus); B, soil-borne wheat mosaic virus (SBWMV), possible Tobamovirus; C, tobacco rattle virus (TRV, Tobravirus). All are in uranyl formate stain and at the same nominal magnification. (Compare with Figures 1 and 2, which are of the Hordeivirus, BSMV.) Bar = 100 nm.

III. CYTOPATHOLOGY

The following summary refers to work on BSMV, as the cytopathology of PSLV, LRSV, and ALBV does not appear to have been investigated. Thin sections of BSMV-infected cells have been extensively studied in regard to general cytopathology,[41-43] plastid alterations,[44,45] cell wall thickening and paramural body formation,[46] attachment of virus particles to microtubules,[47-49] and seed transmission.[43,47-53] The cytopathology of BSMV-infected cells and related problems have been extensively reviewed by Jackson and Lane.[2]

Studies of Pring and Timian[54,55] form a useful background to the cytopathology of BSMV. They found, using the ND18 strain in Hulless barley, that in the inoculated mature leaf (leaf 1), viral RNA could be detected 24 hrs after inoculation, whereas virus particles were only detected a day later. Host RNA synthesis, including ribosomal RNA synthesis, was considerably and irreversibly depressed. Viral RNA accumulated in excess of the coat protein, whose synthesis may have been hampered by shortage of functional host ribosomes. Within 24 hrs of inoculation, the "infection principle" had passed to leaf 2 (the first still expanding systemically infected leaf). Here, viral RNA was again at first synthesized in excess of coat protein, but production of the latter and subsequent encapsidation later caught up. Leaf 1 developed mild symptoms only on the 5th day after inoculation, but was killed within 2 weeks. Leaf 2 showed symptoms on the 3rd day after inoculation, became more severely chlorotic and supported higher levels of virus synthesis, but survived a week longer than leaf 1. Symptom severity was not always, however, correlated with virus content.

BSMV particles are easily distinguished in thin sections by their shape and electron density (Figures 4 through 6). They generally appear in the cytoplasm in moderate to large numbers but are not associated with any major organelle or inclusion. Small packets of virus particles are often also found in nuclei (Figure 5A).

It has been noted that the chloroplasts may be variously altered in infected cells and may sometimes bear small flask-shaped invaginations of the membranes[44,45] reminiscent of those found in Tymovirus-infected cells (see Volume 1, Chapter 8). It appears that the outer chloroplast membrane forms the "flask", but the inner membrane may also become invaginated. McMullen and his colleagues[45] observed that some infected cells contained massive amounts of virus while their plastids were not seen to bear any peripheral vesicles, and further work will be necessary to show whether the vesicles are involved in virus synthesis. Another phenomenon of unknown significance is the severe alteration of microbodies caused by an Italian isolate of BSMV (Figure 6).[56]

Infection of barley with strain ND18 of BSMV can, in the primary, acute phase of infection, induce local thickening of the cell wall in association with somewhat distorted, bulbous, and extended plasmodesmata. Paramural bodies, extracytoplasmic sacs, and osmiophilic aggregates often occur adjacent to these regions on either side of the cell wall.[46] Another curious but perhaps incidental phenomenon is that BSMV particles have been found attached at right angles by one end to cytoplasmic or spindle microtubules.[47-49]

Gold and his colleagues[57] first demonstrated the presence of BSMV particles in extracts of embryos, endosperm, and pollen of infected barley, and Gardner[43] first observed BSMV particles in thin sections of infected pollen. Carroll and his colleagues,[47-53] using Atlas barley, described how the seed-transmitted MI-1 strain of BSMV invades both the male and female parts of the floral meristem, and can be found in thin sections of pollen mother cells, male gametes, egg cells, endosperm, embryo sacs, and, eventually, embryos. Virus particles were seen to accumulate largely in the cytoplasm but also in the nuclei. In these studies, they also compared behavior of MI-1 strain of BSMV with that of strain 850, a non-seed-passaged strain that is not transmitted to the offspring by seed or pollen. Virus particles of this strain were found to be excluded from all reproductive tissues, and could only be detected sequestered in the maturing ovary and anther wall cells. The behavior of the MI1

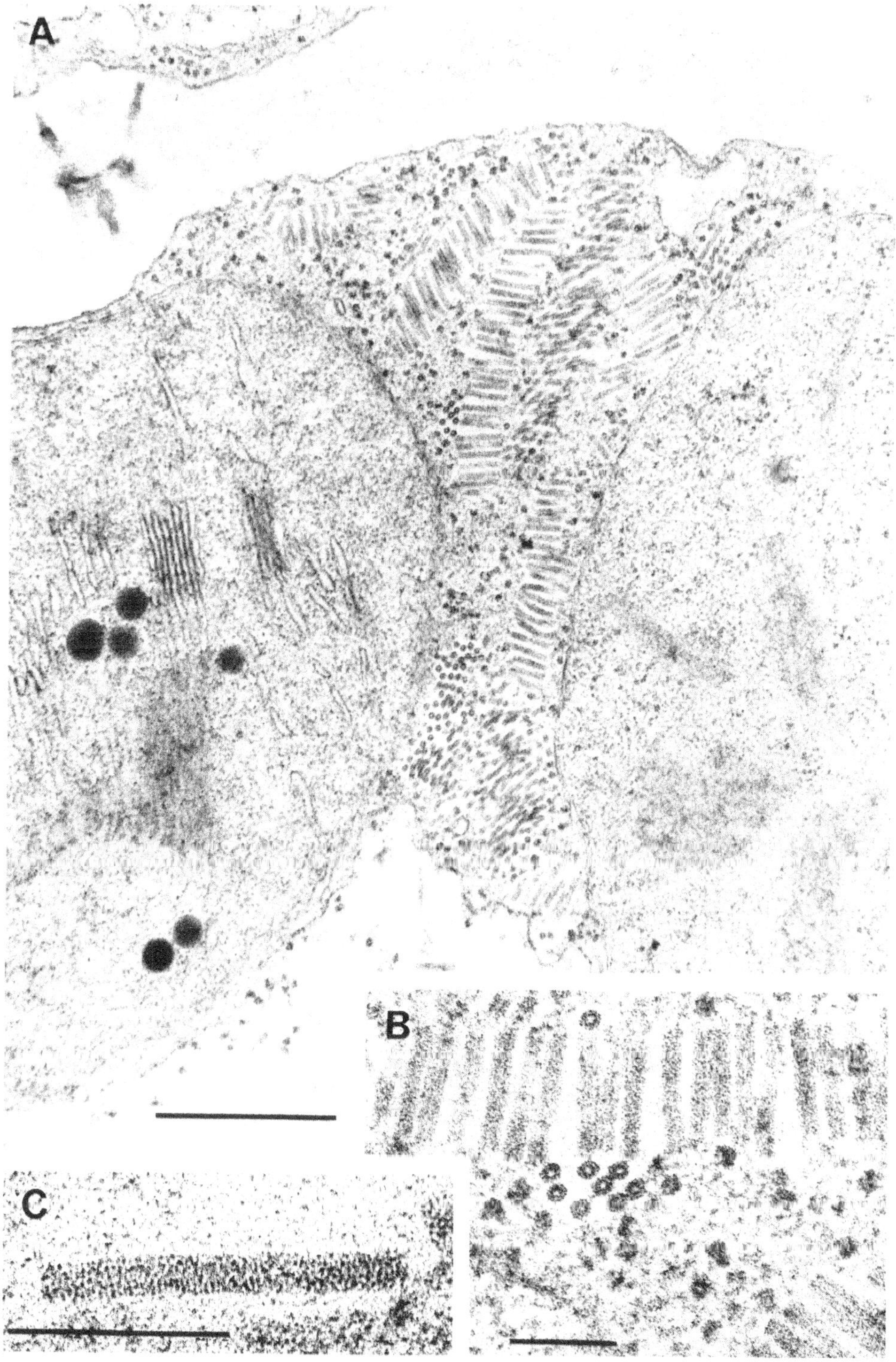

FIGURE 4. Thin sections of a *Hordeum vulgare* (barley) leaf infected with BSMV. A, Note virus particles in the cytoplasm; bar = 500 nm. B, Shows the appearance of the particles in profile and cross-section; bar = 100 nm. C, shows a particle in profile at the same nominal magnification as that of Figures 1 and 2 (note the considerable difference in the size of negatively stained and epoxy-embedded particles and the fine structure in the embedded particle). Bar = 100 nm.

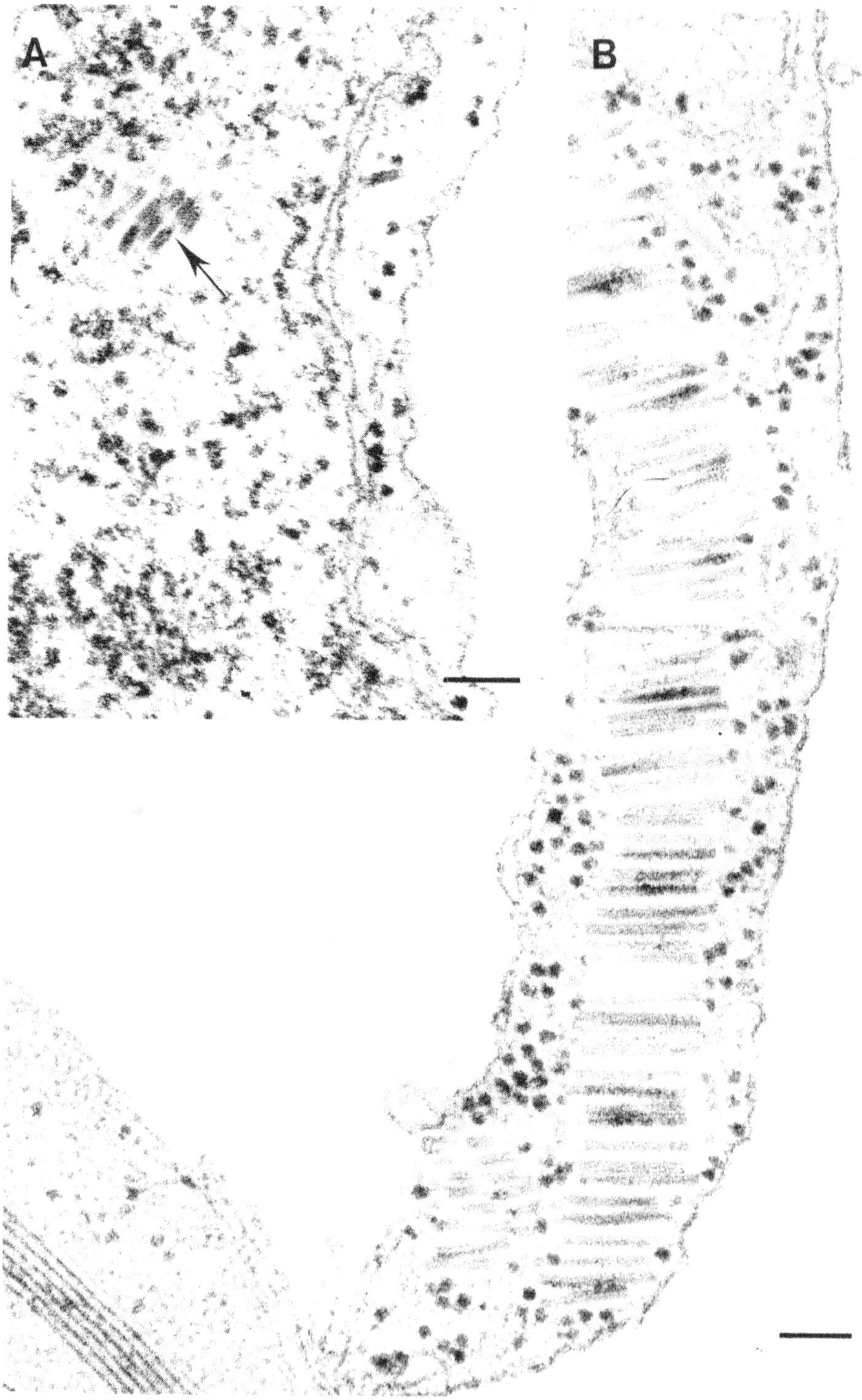

FIGURE 5. Material as in Figure 4. A, a group of virus particles in the nucleus (arrow); B, virus particles in the cytoplasm. The dense circular profiles are ribosomes. Bars = 100 nm.

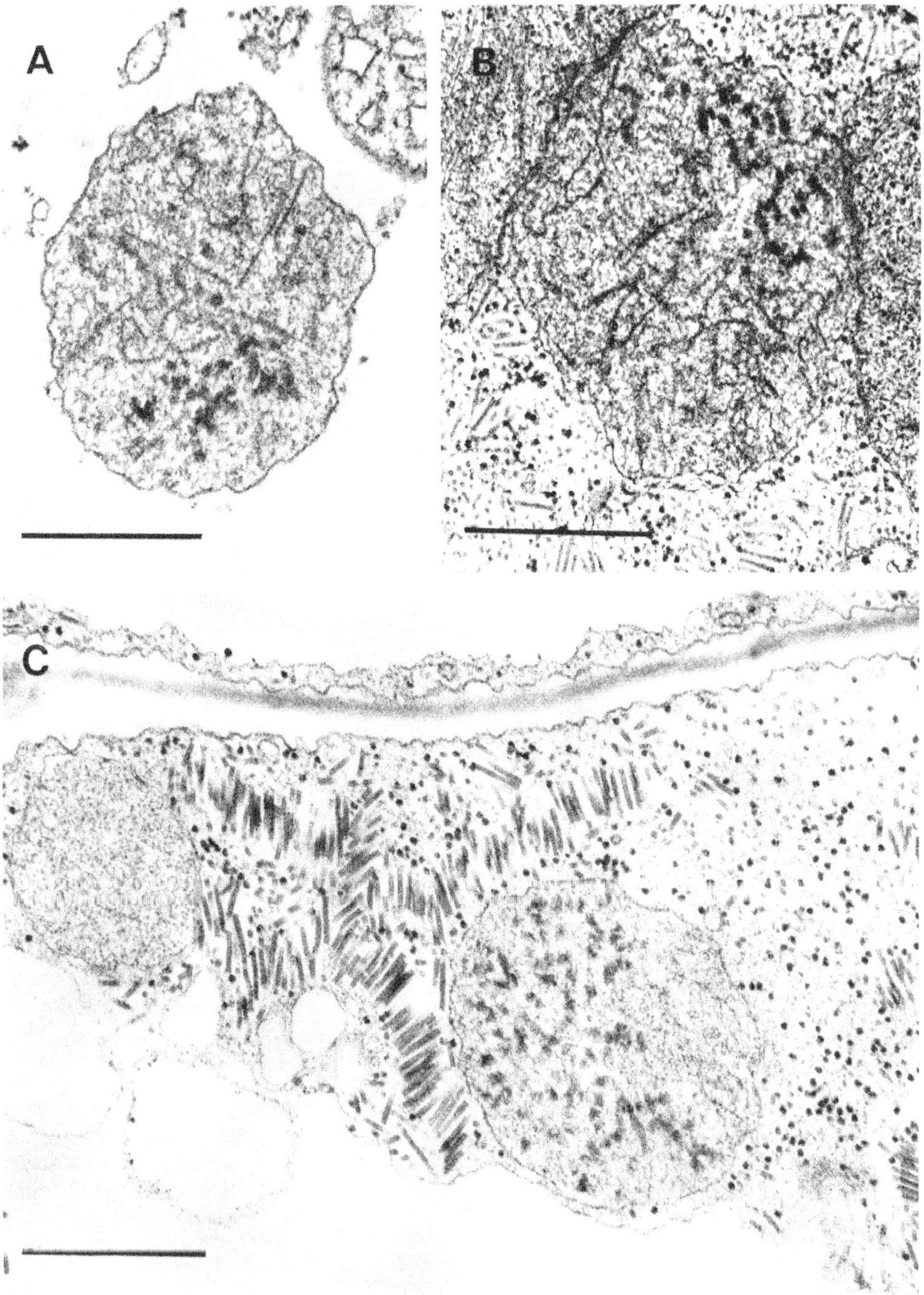

FIGURE 6. Material as in Figures 4 and 5. A, B, and C, microbodies containing abnormal dense patches and rod or plate-like structures of unknown significance, not found in cells of uninfected plants. Virus particles in the cytoplasm can also be seen in (B) and (C). Bars = 500 nm.

and 850 strains in Atlas barley represent, of course, only isolated examples from a matrix of possible interactions between virus strain, host variety, time of infection, and environmental conditions. Seed transmission and its cytopathology are more fully discussed by Jackson and Lane.[2]

REFERENCES

1. **Matthews, R. E. F.,** Classification and nomenclature of viruses. Fourth report of the International Committee on Taxonomy of Viruses, *Intervirology*, 17, 1, 1982.
2. **Jackson, A. O. and Lane, L. C.,** Hordeiviruses, in *Handbook of Plant Virus Infections and Comparative Diagnosis*, Kurstak, E., Ed., Elsevier, Amsterdam, 1981, 565.
3. **Bennett, C. W.,** Seed transmission of plant viruses, *Adv. Virus Res.*, 14, 221, 1969.
4. **Timian, R. G.,** The range of symbiosis of barley and barley stripe mosaic virus, *Phytopathology*, 64, 342, 1974.
5. **Carroll, T. W., Gossel, P. L., and Hockett, E. A.,** Inheritance of resistance to seed transmission of barley stripe mosaic virus in barley, *Phytopathology*, 69, 431, 1979.
6. **Carroll, T. W.,** Barley stripe mosaic virus: its economic importance and control in Montana, *Plant Dis.*, 64, 136, 1980.
7. **Sprague, G. F., McKinney, H. H., and Greeley, L.,** Virus as a mutagenic agent in maize, *Science*, 141, 1052, 1963.
8. **Pring, D. R.,** Barley stripe mosaic virus infection of corn and the "aberrant ratio" genetic effect, *Phytopathology*, 64, 64, 1974.
9. **Slykhuis, J. T.,** Poa semilatent virus from native grasses, *Phytopathology*, 62, 508, 1972.
10. **Catherall, P. L. and Chamberlain, J. A.,** A new virus disease of sweet vernal grass (*Anthoxanthum odoratum* L.), Proc. 3rd. Conf. Virus Diseases of Gramineae in Europe, Rothamsted, 1980, Plumb, R. T., Ed., 1980, 83.
11. **Harrison, B. D., Nixon, H. L., and Woods, R. D.,** Lengths and structure of particles of barley stripe mosaic virus, *Virology*, 26, 284, 1965.
12. **Atabekov, J. G. and Novikov, V. K.,** Barley stripe mosaic virus, *CMI/AAB Descriptions of Plant Viruses*, No. 68, 1971.
13. **Jackson, A. O. and Brakke, M. K.,** Multicomponent properties of barley stripe mosaic virus ribonucleic acid, *Virology*, 55, 483, 1973.
14. **Lane, L. C.,** The components of barley stripe mosaic and related viruses, *Virology*, 58, 323, 1974.
15. **Brakke, M. K. and Palomar, M. K.,** Separation of components of barley stripe mosaic virions by density-gradient centrifugation, *Virology*, 71, 255, 1976.
16. **Palomar, M. K., Brakke, M. K., and Jackson, A. O.,** Base sequence homology in the RNAs of barley stripe mosaic virus, *Virology*, 77, 471, 1977.
17. **Dolja, V. V., Sokolova, N. A., Tjulkina, L. G., and Atabekov, J. G.,** A study of barley stripe mosaic virus (BSMV) genome. II. Translation of individual RNA species of two BSMV strains in a homologous cell-free system, *Mol. Gen. Genet.*, 175, 93, 1979.
18. **Gustafson, G. D., Milner, J. J., McFarland, J. E., Pedersen, K., Larkins, B. A., and Jackson, A. O.,** Investigation of the complexity of barley stripe mosaic virus RNAs with recombinant DNA clones, *Virology*, 120, 182, 1982.
19. **Jackson, A. O., Dawson, J. R. O., Covey, S. N., Hull, R., Davies, J. W., McFarland, J. E., and Gustafson, G. D.,** Sequence relations and coding properties of a subgenomic RNA isolated from barley stripe mosaic virus, *Virology*, 127, 37, 1983.
20. **Chiko, A. W.,** Evidence of multiple virion components in leaf-dip preparations of barley stripe mosaic virus, *Virology*, 63, 115, 1975.
21. **Polák, Z. and Slykhuis, J. T.,** Comparisons of poa semilatent and barley stripe mosaic viruses, *Can. J. Bot.*, 50, 263, 1972.
22. **Gibbs, A. J., Kassanis, B., Nixon, H. L., and Woods, R. D.,** The relationship between barley stripe mosaic and lychnis ringspot viruses, *Virology*, 20, 194, 1963.
23. **Bennett, C. W.,** Lychnis ringspot, *Phytopathology*, 49, 706, 1959.

24. **McKinney, H. H. and Greeley, L. W.,** Biological characteristics of barley stripe-mosaic virus strains and their evolution, *U.S. Dep. Agric. Tech. Bull.*, No. 1324, 1965.
25. **Finch, J. T.,** Preliminary X-ray diffraction studies on tobacco rattle and barley stripe mosaic viruses, *J. Mol. Biol.*, 12, 612, 1965.
26. **Veerisetty, V.,** Relationships among structural parameters of virions of helical symmetry, *Virology*, 84, 523, 1978.
27. **Brakke, M. K.,** Ultraviolet absorption spectra and difference spectra of barley stripe mosaic and tobacco mosaic viruses in buffer and sodium dodecyl sulfate, *Virology*, 98, 76, 1979.
28. **Partridge, J. E., Shannon, L. M., Gumpf, D. J., and Colbaugh, P.,** Glycoprotein in the capsid of plant viruses as a possible determinant of seed transmissibility, *Nature (London)*, 247, 391, 1974.
29. **Gumpf, D. J., Cunningham, D. S., Heick, J. A., and Shannon. L. M.,** Amino acid sequence in the proteolytic glycopeptide of barley stripe mosaic virus, *Virology*, 78, 328, 1977.
30. **Atabekov, J. G. and Morozov, S. Yu.,** Translation of plant virus messenger RNAs, *Adv. Virus Res.*, 25, 1, 1979.
31. **McFarland, J. E., Brakke, M. K., and Jackson, A. O.,** Complexity of the Argentina mild strain of barley stripe mosaic virus, *Virology*, 130, 397, 1983.
32. **Taliansky, M. E., Boykov, S. V., Malyshenko, S. I., Kavsan, V. M., and Atabekov, J . G.,** A study of barley stripe mosaic virus (BSMV) genome. I. Determination of sequence homology between BSMV RNA species, *Mol. Gen. Genet.*, 175, 89, 1979.
33. **Boykov, S. V., Taliansky, M. E., Malyshenko, S. I., Kozlov, Yu. V., and Atabekov, J. G.,** Homology in the 3′ -terminal regions of different barley stripe mosaic virus RNA species, *Virology*, 113, 168, 1981.
34. **Agranovsky, A. A., Dolja, V. V., Kavsan, V. M., and Atabekov, J. G.,** Detection of polyadenylate sequences in RNA components of barley stripe mosaic virus, *Virology*, 91, 95, 1978.
35. **Negruk, V. I., Agranovsky, A. A., Skryabin, K. G., and Atabekov, J. G.,** Size determination of poly(A) sequences in barley stripe mosaic virus RNA, *Anal. Biochem.*, 99, 450, 1979.
36. **Agranovsky, A. A., Dolja, V. V., Gorbulev, V. G., Kozlov, Yu. V., and Atabekov, J. G.,** Aminoacylation of barley stripe mosaic virus RNA: polyadenylate-containing RNA has a 3′ -terminal tyrosine-accepting structure, *Virology*, 113, 174, 1981.
37. **Agranovsky, A. A., Dolja, V. V., and Atabekov, J. G.,** Structure of the 3′ extremity of barley stripe mosaic virus RNA: evidence for internal poly(A) and a 3′ -terminal tRNA-like structure, *Virology*, 119, 51, 1982.
38. **Loesch-Fries, L. S. and Hall, T. C.,** In vivo aminoacylation of brome mosaic and barley stripe mosaic virus RNAs, *Nature (London)*, 298, 771, 1982.
39. **Agranovsky, A. A., Dolja, V. V., Kagramanova, V. K., and Atabekov, J. G.,** The presence of a cap structure at the 5′-end of barley stripe mosaic virus RNA, *Virology*, 95, 208, 1979.
40. **Gustafson, G. D., Larkins, B. A., and Jackson, A. O.,** Comparative analysis of polypeptides synthesized *in vivo* and *in vitro* by two strains of barley stripe mosaic virus, *Virology*, 111, 579, 1981.
41. **Shalla, T. A.,** Relations of tobacco mosaic virus and barley stripe mosaic virus to their host cells as revealed by ultrathin tissue-sectioning for the electron microscope, *Virology*, 7, 193, 1959.
42. **Shalla, T. A.,** Electron microscopy of cells infected with barley stripe mosaic virus as a result of mechanical and seed transmission, in *Viruses of Plants*, Beemster, A. B. R. and Dijkstra, J., Eds., North-Holland, Amsterdam, 1966, 94.
43. **Gardner, W. S.,** Electron microscopy of barley stripe mosaic virus: comparative cytology of tissues infected during different stages of maturity, *Phytopathology*, 57, 1315, 1967.
44. **Carroll, T. W.,** Relation of barley stripe mosaic virus to plastids, *Virology*, 42, 1015, 1970.
45. **McMullen, C. R., Gardner, W. S., and Myers, G. A.,** Aberrant plastids in barley leaf tissue infected with barley stripe mosaic virus, *Phytopathology*, 68, 317, 1978.
46. **McMullen, C. R., Gardner, W. S., and Myers, G. A.,** Ultrastructure of cell-wall thickenings and paramural bodies induced by barley stripe mosaic virus, *Phytopathology*, 67, 462, 1977.
47. **Mayhew, D. E. and Carroll, T. W.,** Barley stripe mosaic virions associated with spindle microtubules, *Science*, 185, 957, 1974.
48. **Carroll, T. W. and Mayhew, D. E.,** Anther and pollen infection in relation to the pollen and seed transmissibility of two strains of barley stripe mosaic virus in barley, *Can. J. Bot.*, 54, 1604, 1976.
49. **Carroll, T. W. and Mayhew, D. E.,** Occurrence of virions in developing ovules and embryo sacs of barley in relation to the seed transmissibility of barley stripe mosaic virus, *Can. J. Bot.*, 54, 2497, 1976.
50. **Carroll T. W.,** Electron microscopic evidence for the presence of barley stripe mosaic virus in cells of barley embryos, *Virology*, 37, 649, 1969.
51. **Carroll, T. W.,** Seed transmissibility of two strains of barley stripe mosaic virus, *Virology*, 48, 323, 1972.
52. **Carroll, T. W.,** Barley stripe mosaic virus in sperm and vegetative cells of barley pollen, *Virology*, 60, 21, 1974.

53. **Mayhew, D. E. and Carroll, T. W.,** Barley stripe mosaic virus in the egg cell and egg sac of infected barley, *Virology*, 58, 561, 1974.
54. **Pring, D. R. and Timian, R. G.,** Physiological effects of barley stripe mosaic virus infection, *Phytopathology*, 59, 1381, 1969.
55. **Pring, D. R.,** Viral and host RNA synthesis in BSMV-infected barley, *Virology*, 44, 54, 1971.
56. **Milne, R. G.,** unpublished results, 1983.
57. **Gold, A. H., Suneson, C. A., Houston, B. R., and Oswald, J. W.,** Electron microscopy and seed and pollen transmission of rod-shaped particles associated with the false stripe virus disease of barley, *Phytopathology*, 44, 115, 1954.

Chapter 11

TOBRAVIRUS GROUP

The Tobravirus group is named after the type member, tobacco rattle virus (TRV).[1,2] Viruses in the group have divided genomes consisting of two positive-sense, single-stranded RNA segments. Each segment is encapsidated separately, giving rise to long (L) and short (S) particles which are straight, rigid rods 21 to 23 nm in diameter, containing about 5% nucleic acid.[3] One coat protein, coded by the smaller RNA, forms the capsid of both the L and S particles.

The particles are rather stable and, generally, Tobraviruses are easily sap transmissible. However, the stable isolates can give rise to unstable variants that owing to the loss of the S particles containing RNA 2, produce neither coat protein nor virus particles. Such strains are sensitive to ribonuclease and to freeze-thawing, and are difficult to propagate mechanically.[4-10] In at least one case, unstable variants can cause important natural infections in crops, and require special methods of diagnosis.[11]

The viruses are soil-borne and may sometimes invade only the roots of plants. The vectors are nematodes of the genera *Trichodorus* and *Paratrichodorus* (Trichodoridae).[12] The ability to transmit Tobraviruses can persist in vectors for over a year but is lost when the nematode molts. The viruses do not multiply in the vectors but are adsorbed specifically to the lining of the pharynx of vector species[13] in a manner similar to that found for the Nepoviruses (Chapter 2). Seed transmission occurs infrequently with TRV[14] but is more important with pea early browning virus (PEBV), where 1 to 10% transmission can occur.[15,16]

Tobraviruses multiply in the cytoplasm of their hosts; with one strain, inclusion bodies involving the mitochondria are formed.[16]

I. MEMBERS OF THE GROUP AND THEIR RELATIONSHIPS

The group consists only of TRV and PEBV and their strains (Table 1). PEBV strains can in principle be distinguished from those of TRV as they infect pea and *Vicia faba* systemically instead of only locally, and possess L particles of modal length 215 nm instead of 190 nm. Generally, the stable isolates of PEBV produce large (2- to 3-mm diameter) local lesions on inoculated primary leaves of *Phaseolus vulgaris* whereas those of TRV produce minute lesions. PEBV, as originally found in Holland, differed serologically from selected TRV isolates by SDIs between 6 and 9.[17]

Unfortunately, however, for taxonomy and diagnostics, the strains of TRV and PEBV together constitute a complex of entities more or less serologically interrelated. As an example of this complexity, Cooper and Mayo[20] cited two Tobravirus isolates, RQ and IT6, that gave TRV-like symptoms on pea and *P. vulgaris* but had the particle lengths and serological properties of PEBV. TRV isolates were earlier separated into three serotypes[21] but more recently it has been suggested [16,17]that, as serotypes I and II intergrade, they should be known collectively as serotype I-II. Serotype III would continue to comprise the Brazilian isolates of TRV; these are serologically rather homogeneous but are only distantly related to members of serotype I-II.

Dutch isolates of PEBV are distantly related serologically to those of TRV serotype I-II. British PEBV isolates are in turn distantly related to Dutch isolates but are apparently unrelated to isolates of TRV serotype I-II.[17] Seven Tobravirus isolates from northern Italy were found to fall into two groups, represented by isolates 5 and 6. Serologically, isolate 5 was similar to TRV-Lisse, whereas isolate 6 resembled the English PEBV strain SP5.[22] A Tobravirus from Morocco had the host reactions and particle morphology of PEBV but was serologically very close to the PRN strain of TRV.[23] TRV strains reported from Japan

Table 1
MEMBERS OF THE TOBRAVIRUS GROUP AND SOME OF THEIR PROPERTIES[a]

Virus	S_{20w} of particles			Modal lengths of particles (nm)			M_r of RNAs (× 10^6)			M_r of coat protein (× 10^3)	Ref.
	L	S	VS[b]	L	S	VS	RNA 1	RNA 2	Other		
Tobacco rattle virus (TRV)											2
Serotype I-II strains											
PRN	304	205	ND[c]	190	75	ND	2.4	1.0	ND	~22	
Lisse	—[d]	—	ND	185	67.6	ND	2.4	0.8	ND	~22	
California B	300	245	200	190	110	45	—	—	—	~22	
California C	300	230	200	190	85—90	45	—	—	—	~22	
Oregon yellow (ORY)	286	212	—	195	90	48	—	—	—	~22	
Oregon severe	287	223	—	195	100	48	—	—	—	~22	
Oregon mild	286	204	—	195	81	48	—	—	—	~22	
Spinach yellow mottle (SYM)	300	230	~150	188	101	57 + 48	2.4	1.5	0.7 + 0.6	~25	
Serotype III Strain											
Campinas (CAM)	300	159	ND	197	52	—	2.4	0.7	—	—	
Pea early-browning virus (PEBV)											15
Dutch isolates	—	—	—	210	105	—	—	—	—	—	
British isolates (SP5)	286	210	—	212	103	—	2.4	1.3	—	~22	

[a] Data from Harrison and Robinson[16,18] and Kurppa and colleagues.[19]
[b] Very short class of particles.
[c] None detected
[d] No data available.

(strains HSN, A, and MD-1) are serologically related either to the PRN strain or to each other, and would all therefore fall within serotype I-II[24,25]

The ability of the different genome segments of a virus to complement one another in artificial mixtures (pseudorecombination) is an indicator of whether or not the components come from closely related virus strains. This test is easily applied with Tobraviruses by seeing whether unstable strains, carrying only RNA 1, can be converted to stable strains by the addition of S particles or of purified RNA 2. It has been found that strains CAM and PRN of TRV cannot complement each other in this way,[26] nor are strains PRN or the related WAG compatible with PEBV.[8,27] However, PRN and the California strains B and C of TRV (all serologically closely related) can all complement each other.[27] Ghabrial and Lister[28] succeeded in obtaining a poorly productive pseudorecombinant between strain ORY of TRV (S particles) and strain TRV-Z, probably closely similar to CAM (L particles) but the reciprocal cross was not viable. In summary, viable pseudorecombinants have been constructed from isolates within the same serotype, but this has only rarely been achieved when one parent comes from serotype I-II and the other from serotype III.[8,16,18,26-32]

Strain ORY of TRV is of some interest. It originally produced a vivid yellow mosaic in *Nicotiana clevelandii* but, over a period of two years of monthly transfer in *N. clevelandii*, gradually lost this characteristic, though remaining fully infective. It was then designated as strain OR. The behavior was observed on two separate occasions, starting with a stock of the virus held at $-20°C$. Strain ORY was found to contain two variants of RNA 2, the larger of these carrying the information for the expression of the yellow symptom. On passage, the proportion of this RNA gradually decreased, and the smaller, alternative RNA 2 dominated the culture.[33]

It has been observed that peanut clump virus (PCV) has properties intermediate between those of a Hordeivirus and a Tobamovirus[34] but is listed by Matthews[1] as a possible member of both the Tobamovirus and Tobravirus groups. In fact PCV has properties that could make it an anomalous member of any of these groups. The virus is soil-borne, and probably transmitted by a plasmodiophoromycete fungus,[34] though the transmitting soils also contained numerous *Trichodorus* nematodes.[35] The virus particles are about 21 nm wide and of two modal lengths, 245 and 190 nm, and the genome is composed of two RNA species of M_r about 2.1 and 1.7×10^6. The virus does not appear to be serologically related to any other, and while the mode of transmission remains in doubt, PCV must for the present rest in taxonomic limbo. We have provisionally designated it as a possible Tobamovirus (see Chapter 9).

II. VIRUS STRUCTURE AND COMPOSITION

A. Particle Structure

Electron micrographs of negatively stained Tobravirus particles[18,36] (Figure 1) show rigid rods 21 to 23 nm in diameter, with one end slightly convex and the other concave. The pitch of the clearly visible helix is 2.5 nm.[37,38] The particles are generally of two modal lengths, long (L) and short (S), though minor third and fourth length classes are sometimes detected (VS particles, Table 1). The L particles of all TRV strains are 185 to 195 nm long, whereas the S particles have lengths that are characteristic of particular isolates. When TRV strains are compared, the lengths of the S particles correlate imperfectly with serological groupings.[18] PEBV L particles appear to be about 10% longer and 6% narrower than those of TRV; the S particles of PEBV, though usually about 105 nm long, have been reported as being shorter in some Dutch isolates.[15,22] A point of technical interest is that Tobravirus particle length histograms may have greater resolution and sharper peaks if the virus preparations are first fixed in glutaraldehyde to minimize particle breakage.[39]

The TRV coat protein has been prepared from the virus particles by disruption in 67% acetic acid,[40] urea followed by acetic acid,[24] or 1.5 *M* calcium chloride.[41] There is a single

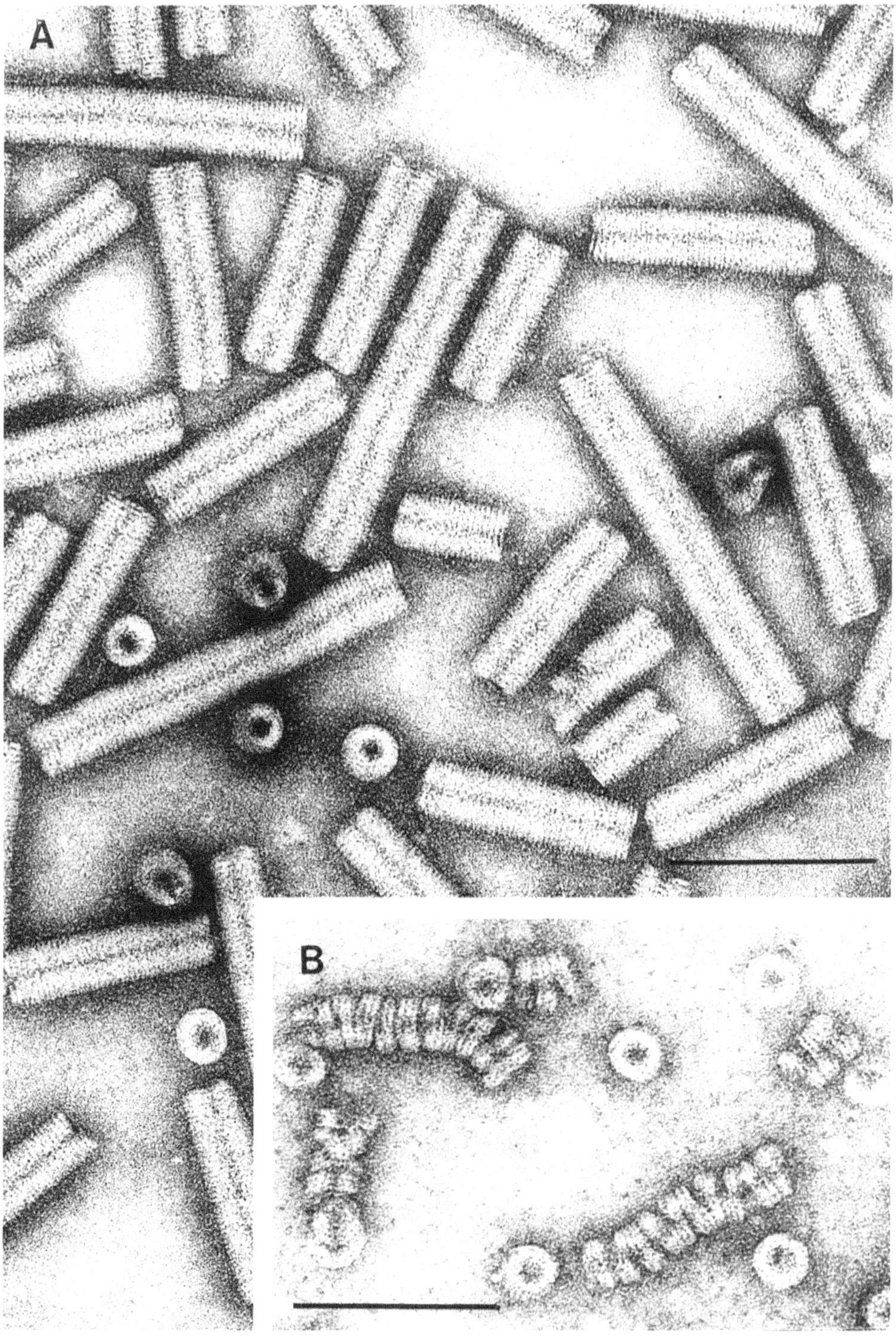

FIGURE 1. A, Purified particles of an Italian isolate of tobacco rattle virus (TRV) in uranyl formate stain; some of the shorter fragments are viewed on end. B, Preparation of TRV (CAM strain) coat protein in disc form; some discs are aggregated and some standing on edge. Uranyl acetate stain. Bars = 100 nm. (Courtesy of Dr. I. M. Roberts, Scottish Crop Research Institute, Dundee, Scotland.)

protein common to both L and S particles[24,42] A value of about 21.7 × 10^3 was obtained by sedimentation equilibrium measurements for the M_r of the coat protein of the PRN strain of TRV[43] but values obtained by SDS polyacrylamide-gel electrophoresis (SDS-PAGE) have ranged fairly widely. This is partly because the coat proteins of Tobraviruses behave anomalously in this system.[41,43] After a review of available data, Mayo and Robinson[43] suggested a most probable value of 22 × 10^3 M_r for the coat protein of both TRV and PEBV. They noted that although TRV and PEBV coat proteins migrated at different rates in gel electrophoresis, they did not differ significantly in size, as had been reported earlier.[42]

The coat protein from TRV preparations stored under nonsterile conditions may migrate more rapidly in SDS polyacrylamide gels than protein from fresh preparations or those stored in the absence of bacteria.[44] The effect can be reproduced using a protease derived from a pseudomonad bacterium found in stored TRV preparations[45] and is similar to that of incubation with papain.[44] The pseudomonad protease-mediated modification had no apprent effect on infectivity, resistance to RNase, electron-microscopic appearance, and several other physical parameters,[44] although a new antigenic determinant was detected on the modified virus particles.[46]

The number of protein subunits per turn of the Tobravirus helix is still in doubt, though X-ray crystallography of oriented gels of purified virus particles indicated an integral number of subunits in three turns.[38] Early work indicated a value near 25 subunits per turn.[37,47,48] Tollin and Wilson[38] estimated that if the M_r of the coat protein subunit was 24 × 10^3, there would be 25$^1/_3$ subunits per turn, while an M_r of 19 × 10^3 would indicate 31$^1/_3$ subunits per turn. An estimate of 32 subunits per turn was arrived at by calculations based on the mass and structure of the TRV coat protein disc aggregate.[49,50] If there are four nucleotide residues per protein subunit in the primary helix of the virus particle,[38,51] then a figure near 25 protein subunits per turn of the helix would be likely.[38]

TRV coat protein when isolated can exist as oligomers of the basic subunit, sedimenting at about 4S or depending on concentration, ionic strength, pH, and temperature, can form "discs" (Figure 1B) of 2 to 3 turns of subunits forming a right-hand helix, sedimenting at 36 to 40S.[18,40,49,50,52-54]

The discs may also dislocate and continue growth to form loose spirals.[50,53] However, the narrow rod-shaped particles suggested to be a polymerization form of TRV protein [53] were almost certainly contaminating particles of tobacco mosaic virus.[55,56]

Reconstitution of nucleoprotein particles of the California C strain of TRV (serotype I-II) was studied by Semancik and Reynolds[57] and Morris and Semancik.[53] Dialysis of RNA and protein mixtures against low-ionic-strength buffer, pH 7.5 to 8.0, at low temperature resulted in at least a proportion of particles indistinguishable from native virus. In a different laboratory, with the CAM strain of TRV (serotype III), these conditions did not result in the reconstitution of native particles, though inefficient reconstitution was achieved in high molarity phosphate buffer, pH 4.7, at 1°C.[58,59] At least with the CAM strain, discs are probably a necessary precursor for reconstitution, and under a wide range of conditions the 5′ ends of both RNAs 1 and 2 bind to the discs and are then protected from RNase digestion. The reconstitution is thus probably polar, starting at or very near the 5′ end of the RNAs and proceeding in the 3′ direction. The oligomeric 4S form of the protein, rather than discs, appears necessary for rod elongation, which requires rather particular conditions in order to proceed.[59,60] At physiological temperatures and ionic strengths, in vitro reconstitution of the CAM strain has not yet been demonstrated, and this difference from the California C strain should be reinvestigated.

B. The Viral Genome

Information on Tobravirus genomes available up until 1978 has been very fully reviewed.[18,61-63] As Table 1 shows, there are two major RNAs (RNA 1 and 2) with, in some strains, other RNA molecules also being detected in small amounts.

RNA/RNA molecular hybridization experiments[64,65] using RNAs 1 and 2 of the CAM strain of TRV have indicated that between 20 and 30% of the sequences of RNA 2 are present in RNA 1. There is evidence that the 3′ terminal 100 nucleotide residues of RNA 1 and 2 have similar though not identical sequences, and that there may be regions of greater homology elsewhere.[66] Robinson[33] has confirmed that with the CAM strain of TRV, RNAs 1 and 2 probably contain a common sequence of about 450 nucleotide residues. He also reported, however, that with strains SYM and OR, there seemed to be little homology between RNA 1 and 2. Thus an extensive sequence common to the two parts of the genome appears not to be an essential feature of TRV.

Hybridization methods using cDNA were also used to estimate the amount of homology between the RNAs of the CAM, SYM, and OR strains.[33] The results indicated that the RNA 1s of SYM and OR are almost identical, but quite different from that of CAM. On the other hand, no sequences in common were detected between the RNA 2s of CAM, SYM, and OR. This is interesting, as RNA 2 codes for the coat protein (see below), and SYM and OR are of the same serotype (serotype I-II).

Both RNA 1 and 2 of the CAM strain of TRV possess the 3′ terminal sequence $GpCpCpC_{OH}$,and there appear to be neither phosphorylation nor polyadenylation at the 3′ end.[67,68] Unconfirmed data suggest that RNA 2 but, curiously, not RNA 1 of the CAM strain is capped at the 5′ end by 7-methyl guanosine.[60]

As RNA 1 can multiply in plants in the absence of RNA 2, it may well contain an RNA replicase cistron. RNA 1 also determines some aspects of symptom expression[4,8,18,28-30] but it is not known whether these effects are controlled by the putative replicase or by a different protein or proteins.

When translated in vitro in either the wheat germ or rabbit reticulocyte systems, RNA 1 of TRV-CAM produced two major proteins of M_r 170 × 10³ and 140 × 10³.[69,70] The size of these proteins exceeds the "straight" coding capacity of the RNA, implying some coding overlap. When isolated leaf protoplasts were infected with RNA 1 only, two major polypeptides were detected, not found in uninfected protoplasts.[71] They had estimated M_rs of 187 × 10³ and 142 × 10³ and similar electrophoretic mobilities to the two major products of in vitro translation. It thus appears that RNA 1 codes for two proteins but the precise method of coding and the functions of the proteins are not yet known. Similar results were obtained by translating RNA 1 of the Lisse and PRN strains of TRV in reticulocyte lysates.[72] Two major proteins of M_r about 170 × 10³ and 120 × 10³ were produced. Addition of Mg^{2+} increased yields of the larger protein, apparently by facilitating readthrough of a leaky codon terminating the smaller protein. This codon appeared to be neither UAG nor UAA.

The RNA 2s of different Tobravirus strains differ widely in size (from 1.5 × 10⁶ to 0.7 × 10⁶ M_r) but this difference has not yet been correlated with any function. That RNA 2 codes for the coat protein has been confirmed by the finding that RNA 2 of TRV-CAM can be translated in vitro to give, as the only major product, a single polypeptide which was identified as coat protein.[69,70] In contrast to the results with the CAM strain, RNA 2 of the PRN strain could be translated to give both the coat protein and a second protein of M_r about 31 × 10³.[70] The authors suggested that CAM strain RNA 2 may also carry this second cistron whose expression, however, requires the generation of a subgenomic messenger.

Further understanding of the way in which RNA 2 is translated has come from the work of Bisaro and Siegel.[73] They found that, in CAM-infected tobacco tissue, a third RNA could be detected, besides the genomic RNAs 1 and 2. This RNA 3, of M_r 0.5 × 10⁶, was shown using cDNA techniques to be derived from RNA 2, and it coded for a protein of M_r about 30 × 10³ in the wheat germ system. Pelham[72] obtained similar results with the PRN strain and suggested that his RNA 3 was derived from the 3′ portion of RNA 1. Bisaro and Siegel[73] suggested that perhaps all Tobraviruses generate an RNA 3 during infection, but that different amounts of it are encapsidated by different strains. Some strains (Table 1) produce small amounts of a particle 44 to 48 nm in length, that could represent the encapsidated RNA 3.

It was also suggested[73] that the coat protein coding sequence lies toward the 5′ end of RNA 2, and that the gene for the 30 × 10^3 M_r protein lies near the 3′ terminus; this gene would only, however, be expressed after processing into the monocistronic subgenomic RNA 3.

That a second gene may exist in RNA 2, besides that of the coat protein, was first suggested by Lister and Bracker.[31] This idea was strengthened by experiments on a variant of TRV-CAM (CAM-YS) producing yellow symptoms in several hosts.[74] The yellowing character is controlled by RNA 2 but the variant possesses a coat protein indistinguishable from that of CAM itself when investigated by serology, PAGE, and tryptic peptide mapping. It was concluded that the mutation was outside the coat protein cistron in RNA 2.[74]

Recently the virus-specific single-stranded RNAs present in TRV-CAM infected tissue have been reinvestigated using cDNA and blotting techniques.[75] Five subgenomic RNAs were detected, with estimated M_rs of 0.5, 0.22, 1.1, 0.36, and 0.16 × 10^6 (RNAs 3 to 7, respectively). RNA 3 was present in large amounts and as previously reported, hybridized to cDNA derived from RNA 2. The other RNAs were detected in small amounts. RNA 5 appeared to be generated from RNA 1, and all the others from RNA 2. It was noted that, as with RNA 1, the combined size of the two proteins produced by RNA 2 appears to exceed the "straight" coding capacity of the RNA. The messenger activities of RNAs 4 to 7 have not yet been tested, and their significance is not known.

III. CYTOPATHOLOGY

Tobraviruses appear to multiply in the cytoplasm of infected plants[2,15] and of protoplasts.[76] The characteristic particles of Tobraviruses are very evident in thin sections of infected tissues (Figures 2 and 3), and their structure resembles that of embedded and sectioned purified virus.[38] Most interest has centered on serotype III strains of TRV because of a curious association between the virus particles and mitochondria.[77-81]

As noted first by Kitajima,[77] the L particles of serotype III strains of TRV become attached by their ends to the outer membranes of mitochondria in leaf cells, producing a characteristic "hedgehog" configuration (Figure 3), while the S particles remain oriented at random in the cytoplasm.[77-79] However, it was noted that in pollen of infected tomato plants (but in this tissue only) the S particles, segregated from the L particles, also lined up perpendicularly to the mitochondrial outer membranes.[77,80] The reason for this alignment of L and, in certain conditions, also S particles is not clear; it does not appear to occur with serotype I-II strains of TRV or with PEBV[2,15,82-84] (Figure 2).

The unstable form of the CAM strain of TRV in *Nicotiana clevelandii* leaves produces inclusions containing mitochondria, ribosomes, and darkly staining granules.[81] The mitochondria were found to be abnormal, with many membranous vesicles formed both internally and around their periphery. The inclusions were rich in protein and single-stranded RNA which was, however, protected from RNase digestion by the protein. Normal isolates of the PRN strain also produced similar inclusions, containing in addition some virus particles, but the inclusions were found less frequently and they persisted for only a few days.[81] The mitochondrial membrane may be involved in viral RNA synthesis, but there is no direct evidence of this.[18] As cycloheximide but not chloramphenicol can inhibit TRV accumulation in infected leaf discs,[85] cytoplasmic rather than mitochondrial or chloroplast ribosomes are implicated in viral protein synthesis.

De Zoeten,[83] using a California strain of TRV, noted severe chloroplast breakdown in infected chlorotic cells. On the other hand, a local Italian strain of TRV (serotype I-II) severely damaged the mitochondria (Figure 2C) but left the chloroplasts apparently unaffected.[86] Chen and his colleagues[84] investigated the cytopathology of an unidentified (probably North American) strain of TRV in the roots of mechanically inoculated plants. The virus particles were found in crystalline arrays in the cytoplasm but there was no association

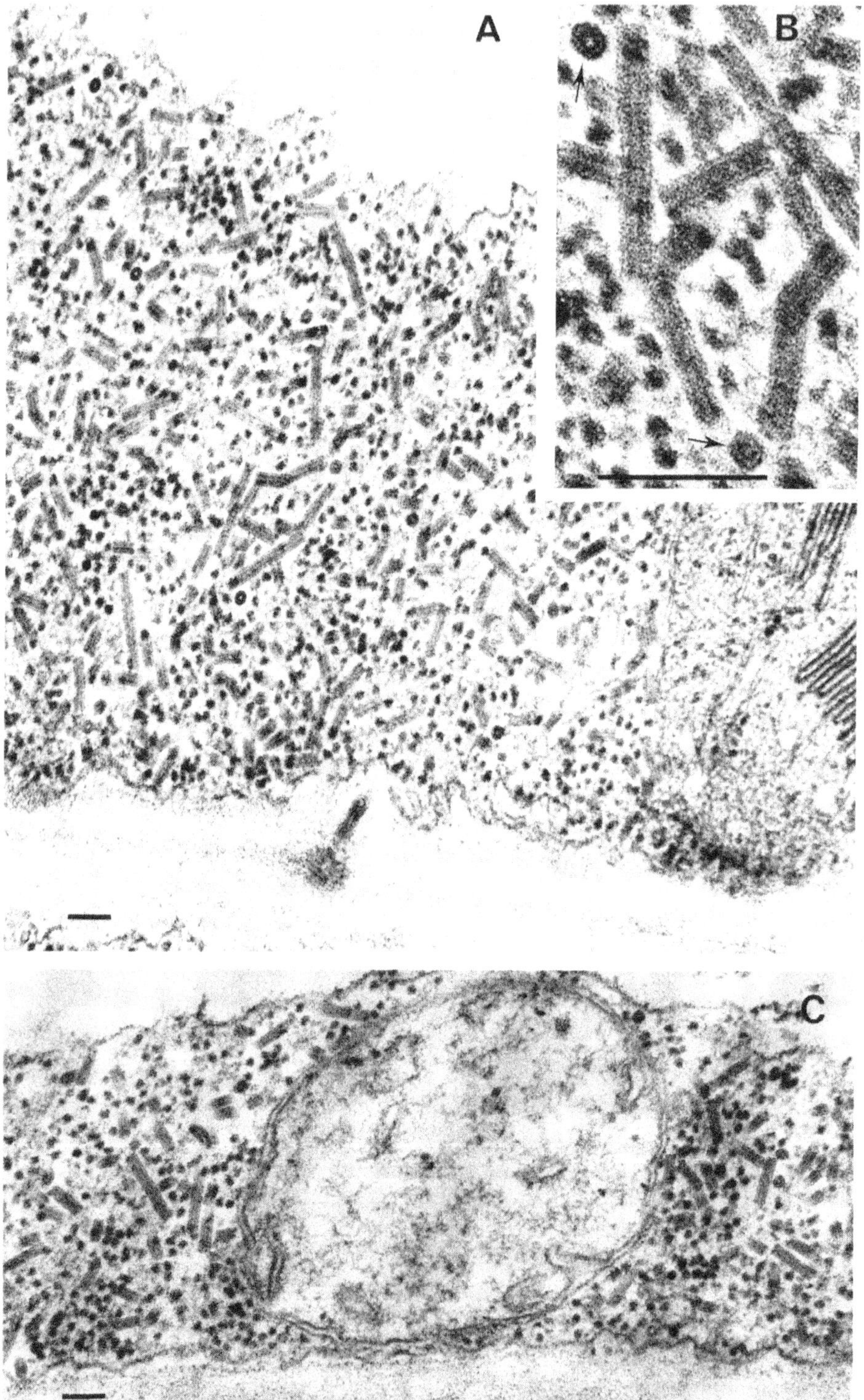

FIGURE 2. Thin section of leaves of *Nicotiana tabacum* (White Burley tobacco) infected with an Italian isolate of TRV (serotype I-II). A, Cytoplasm with many scattered virus particles; B, Particles in (A) enlarged , showing the darker staining of the viral RNA in the particles (arrows point to particles in transverse section); C, A typically damaged mitochondrion surrounded by cytoplasm containing virus particles. Note that the virus particles are not attached to the mitochondrial surface and compare with Figure 3. Bars = 100 nm.

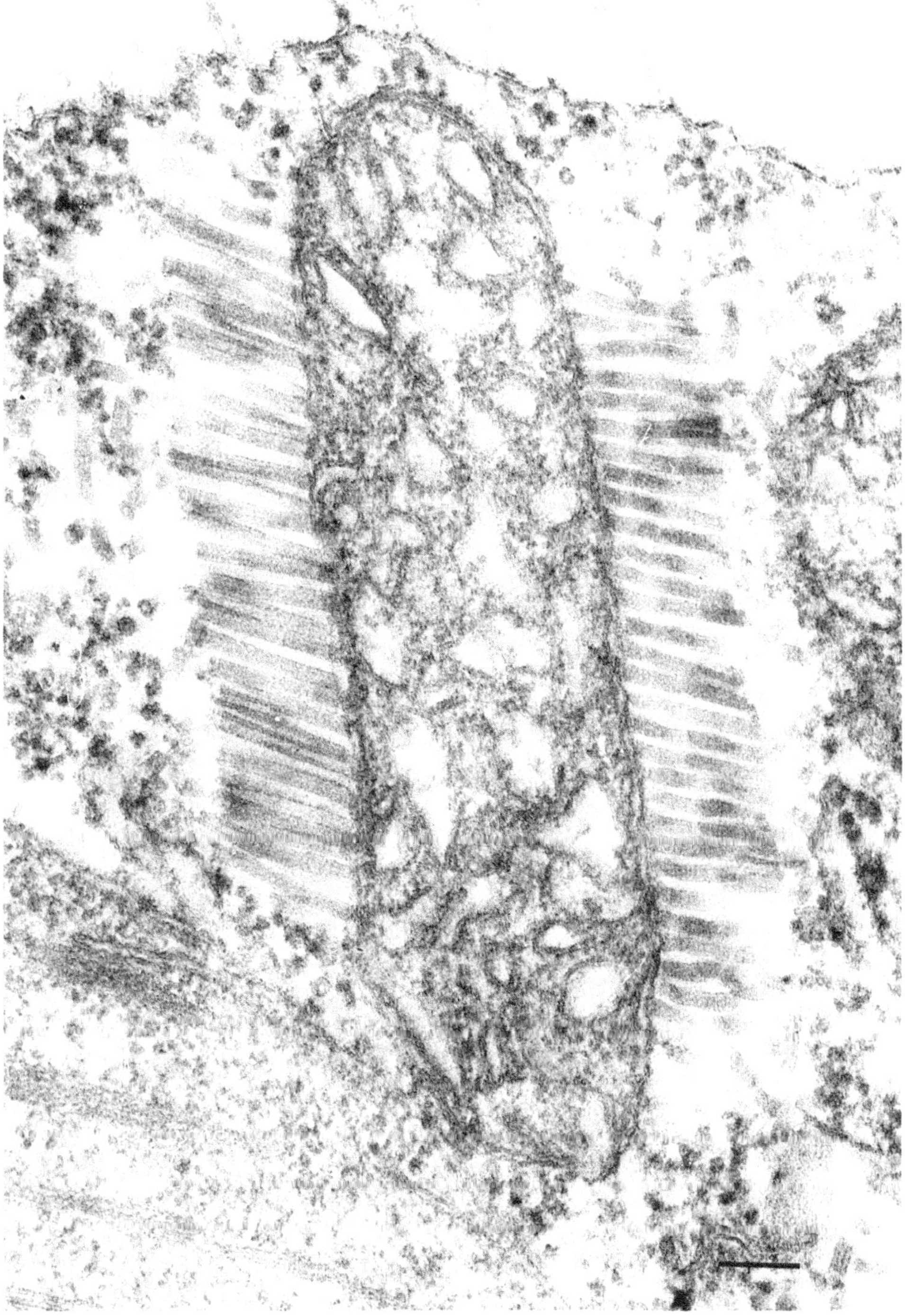

FIGURE 3. Thin section of a leaf of *Nicotiana clevelandii* infected with the CAM strain of TRV (serotype III). Note the long particles associated with the mitochondrial outer membrane. Bar = 100 nm. (Courtesy of Dr. I. M. Roberts, Scottish Crop Research Institute, Dundee, Scotland.)

of particles with any organelle, nor was any apparent cytological disturbance caused by the infection.

With PEBV, only the L particles were observed to be aligned in crystalline arrays; the S particles were disposed at random in the cytoplasm.[87] The L particles appeared to be joined by fine cross-bridges in a manner not observed with TRV.

REFERENCES

1. **Matthews, R. E. F.,** Classification and nomenclature of viruses. Fourth report of the International Committee on Taxonomy of Viruses, *Intervirology*, 17, 1, 1982.
2. **Harrison, B. D.,** Tobacco rattle virus, *CMI/AAB Descriptions of Plant Viruses*, No. 12, 1970.
3. **Harrison, B. D. and Nixon, H. L.,** Separation and properties of particles of tobacco rattle virus with different lengths, *J. Gen. Microbiol.*, 21, 569, 1959.
4. **Cadman, C. H. and Harrison, B. D.,** Studies on the properties of soil-borne viruses of the tobacco-rattle type occurring in Scotland, *Ann. Appl. Biol.*, 47, 542, 1959.
5. **Gibbs, A. J. and Harrison, B. D.,** A form of pea early-browning virus found in Britain, *Ann. Appl. Biol.*, 54, 1, 1964.
6. **Lister, R. M.,** Possible relationships of virus-specific products of tobacco rattle virus infections, *Virology*, 28, 350, 1966.
7. **Lister, R. M.,** A symptomatological difference between some unstable and stable variants of pea early browning virus, *Virology*, 31, 739, 1967.
8. **Lister, R. M.,** Functional relationships between virus-specific products of infection by viruses of the tobacco rattle type, *J. Gen. Virol.*, 2, 43, 1968.
9. **Huttinga, H.,** Interaction between components of pea early-browning virus, *Neth. J. Plant Pathol.*, 75, 338, 1969.
10. **Semancik, J. S. and Odening, L. A.,** Growth curve of unstable form of TRV infection and conversion of unstable to stable form infection, *Virology*, 39, 613, 1969.
11. **Harrison, B. D. and Robinson, D. J.,** Genome reconstitution and nucleic acid hybridization as methods of identifying particle-deficient isolates of tobacco rattle virus in potato plants with stem-mottle disease, *J. Virol. Meth.*, 5, 255, 1982.
12. **Taylor, C. E. and Cadman, C. H.,** Nematode vectors, in *Viruses, Vectors and Vegetation*, Maramorosch, K., Ed., Interscience, New York, 1969, 55.
13. **Taylor, C. E. and Robertson, W. M.,** Location of tobacco rattle virus in the nematode vector, *Trichodorus pachydermus* Seinhorst, *J. Gen. Virol.*, 6, 179, 1970.
14. **Lister, R. M. and Murant, A. F.,** Seed-transmission of nematode-borne viruses, *Ann. Appl. Biol.*, 59, 49, 1967.
15. **Harrison, B. D.,** Pea early-browning virus, *CMI/AAB Descriptions of Plant Viruses*, No. 120, 1973.
16. **Harrison, B. D. and Robinson, D. J.,** Tobraviruses, in *Handbook of Plant Virus Infections and Comparative Diagnosis*, Kurstak, E. Ed., Elsevier, Amsterdam, 1981, 515.
17. **Maat, D. Z.,** Pea early-browning virus and tobacco rattle virus — two different, but serologically related viruses, *Neth. J. Plant Pathol.*, 69, 287, 1963.
18. **Harrison, B. D. and Robinson, D. J.,** The tobraviruses, *Adv. Virus Res.*, 23, 25, 1978.
19. **Kurppa, A., Jones, A. T., Harrison, B. D., and Bailiss, K. W.,** Properties of spinach yellow mottle, a distinctive strain of tobacco rattle virus, *Ann. Appl. Biol.*, 98, 243, 1981.
20. **Cooper, J. I. and Mayo, M. A.,** Comparison of some tobravirus isolates, *Rep. Scott. Hortic. Res. Inst. 1972* , p. 57, 1973.
21. **Harrison, B. D. and Woods, R. D.,** Serotypes and particle dimensions of tobacco rattle viruses from Europe and America, *Virology*, 28, 610, 1966.
22. **van Hoof, H. A., Maat, D. Z., and Seinhorst, J. W.,** Viruses of the tobacco rattle virus group in northern Italy: their vectors and serological relationships, *Neth. J. Plant Pathol.*, 72, 253, 1966.
23. **Lockhart, B. E. L. and Fischer, H. U.,** Some properties of an isolate of pea early-browning virus occurring in Morocco, *Phytopathology*, 66, 1391, 1976.
24. **Miki, T. and Okada, Y.,** Comparative studies on some strains of tobacco rattle virus, *Virology*, 42, 993, 1970.

25. **Tomaru, K., Takanami, Y., and Udagawa, A.,** Strains of tobacco rattle virus isolated from tobacco plants in Japan, *Ann. Phytopathol. Soc. Jpn.*, 36, 275, 1970.
26. **Frost, R. R., Harrison, B. D., and Woods, R. D.,** Apparent symbiotic interaction between particles of tobacco rattle virus, *J. Gen. Virol.*, 1, 57, 1967.
27. **Semancik, J. S. and Kajiyama, M. A.,** Enhancement of tobacco rattle virus stable form infection by heterologous short particles, *Virology* , 34, 170, 1968.
28. **Ghabrial, S. A. and Lister, R. M.,** Coat protein and symptom specification in tobacco rattle virus, *Virology*, 52, 1, 1973.
29. **Sänger, H. L.,** Characteristics of tobacco rattle virus. I. Evidence that its two particles are functionally defective and mutually complementing, *Mol. Gen. Genet.*, 101, 346, 1968.
30. **Sänger, H. L.,** Functions of the two particles of tobacco rattle virus, *J. Virol.*, 3, 304, 1969.
31. **Lister, R.M. and Bracker, C. E.,** Defectiveness and dependence in three related strains of tobacco rattle virus, *Virology*, 37, 262, 1969.
32. **Morris, T. J. and Semancik, J. S.,** Homologous interference between the components of tobacco rattle virus, *Virology*, 52, 314, 1973.
33. **Robinson, D. J.,** RNA species of tobacco rattle virus strains and their nucleotide sequence relationships, *J. Gen. Virol.*, 64, 657, 1983.
34. **Thouvenel, J. C. and Fauquet, C.,** Further properties of peanut clump virus and studies on its natural transmission, *Ann. Appl. Biol.*, 97, 99, 1981.
35. **Thouvenel, J. C., Dollet, M., and Fauquet, C.,** Some properties of peanut clump, a newly discovered virus, *Ann. Appl. Biol.*, 84, 311, 1976.
36. **Williams, R. C. and Fisher, H. W.,** *An Electron Micrographic Atlas of Viruses*, Charles C Thomas, Springfield, Ill., 1974, 128 pp.
37. **Offord, R. E.,** Electron microscopic observations on the substructure of tobacco rattle virus, *J. Mol. Biol.*, 17, 370, 1966.
38. **Tollin, P. and Wilson, H. R.,** Some observations on the structure of the Campinas strain of tobacco rattle virus, *J. Gen. Virol.* 13, 433,1971.
39. **Locatelli, E. A. and Allen, T. C.,** Effects of electron microscope preparatory techniques on lengths of tobacco rattle virus, *Phytopathol. Z.*, 96, 273, 1979.
40. **Semancik, J. S.,** Identity of structural protein from two isolates of TRV with different length of associated short particles, *Virology*, 40, 618, 1970.
41. **Ghabrial, S. A. and Lister, R. M.,** Anomalies in molecular weight determinations of tobacco rattle virus protein by SDS-polyacrylamide gel electrophoresis, *Virology*, 51, 485, 1973.
42. **Cooper, J. I. and Mayo, M. A.,** Some properties of the particles of three tobravirus isolates, *J. Gen. Virol.* 16, 285, 1972.
43. **Mayo, M. A. and Robinson, D. J.,** Revision of estimates of the molecular weights of tobravirus coat proteins, *Intervirology*, 5, 313, 1975.
44. **Mayo, M. A. and Cooper, J. I.,** Partial degradation of the protein in tobacco rattle virus during storage, *J. Gen. Virol.*, 18, 281, 1973.
45. **Mayo, M. A., Robinson, D. J., And Pérombelon, M. C. M.,** Some properties of a bacterial protease with a specific effect on the protein in tobacco rattle virus particles, *J. Gen. Microbiol.* 85, 121, 1974.
46. **Robinson, D. J.,** Proteins of tobraviruses, *Rep. Scott. Hortic. Res. Inst. 1972*, p. 58, 1973.
47. **Nixon, H. L. and Harrison, B. D.,** Electron microscopic evidence on the structure of the particles of tobacco rattle virus, *J. Gen. Microbiol.*, 21, 582, 1959.
48. **Finch, J. T.,** Preliminary X-ray diffraction studies on tobacco rattle and barley stripe mosaic viruses, *J. Mol. Biol.*, 12, 612, 1965.
49. **Mayo, M. A. and Dancy De Marcillac, G.,** Analytical centrifugation of the disk aggregates of tobacco rattle virus protein, *Virology*, 76, 560, 1977.
50. **Roberts, I. M. and Mayo, M. A.,** Electron microscope studies of the structure of the disk aggregate of tobacco rattle virus protein, *J. Ultrastruct. Res.*, 71, 49, 1980.
51. **Veerisetty, V.,** Relationships among structural parameters of virions of helical symmetry, *Virology*, 84, 523, 1978.
52. **Fritsch, C., Witz, J., Abou Haidar, M., and Hirth, L.,** Polymerization of tobacco rattle virus protein, *FEBS Lett.*, 29, 211, 1973.
53. **Morris, T. J. and Semancik, J. S.,** *In vitro* protein polymerization and nucleoprotein reconstitution of tobacco rattle virus, *Virology*, 53, 215, 1973.
54. **Gugerli, P.,** Different states of aggregation of tobacco rattle virus coat protein, *J. Gen. Virol.*, 33, 297, 1976.
55. **Fritsch, C. and Milne, R. G.,** unpublished results, 1980.
56. **Milne, R. G., Masenga, V., and Lovisolo, O.,** Viruses associated with white bryony (*Bryonia cretica* L.) mosaic in Northern Italy, *Phytopathol. Mediterr.*, 19, 115, 1973.

57. **Semancik, J. S. and Reynolds, D. A.,** Assembly of protein and nucleoprotein particles from extracted tobacco rattle virus protein and RNA. *Science,* 164, 559, 1969.
58. **Abou Haidar, M., Pfeiffer, P., Fritsch, C., and Hirth, L.,** Sequential reconstitution of tobacco rattle virus, *J. Gen. Virol.,* 21, 83, 1973.
59. **Abou Haidar, M.,** Tobacco rattle virus RNA-protein interactions, *Philos. Trans. R. Soc. London Ser. B,* 276, 165, 1976.
60. **Abou Haidar M. and Hirth, L.,** 5'-Terminal structure of tobacco rattle virus RNA: evidence for polarity of reconstitution, *Virology,* 76, 173, 1977.
61. **Bruening, G.,** Plant covirus systems: two-component systems, in *Comprehensive Virology,* Vol. 11, Fraenkel-Conrat, H. and Wagner, R. R., Eds., Plenum Press, New York, 1977, 55.
62. **Davies, J. W.,** Translation of plant virus ribonucleic acids in extracts from eukaryotic cells, in *Nucleic Acids in Plants,* Vol. 2, Hall, T. C. and Davies, J. W., Eds., CRC Press, Boca Raton, Fla., 1979, 111.
63. **Lane, L. C.** The nucleic acids of multipartite, defective and satellite plant viruses, in *Nucleic Acids in Plants* , Vol. 2, Hall, T. C. and Davies, J. W., Eds., CRC Press, Boca Raton, Fla., 1979, 65.
64. **Minson, A. C. and Darby, G.,** A study of sequence homology between tobacco rattle virus ribonucleic acids, *J. Gen. Virol.,* 19, 253, 1973.
65. **Darby, G. and Minson, A. C.,** The structure of tobacco rattle virus ribonucleic acids: common nucleotide sequences in the RNA species, *J. Gen. Virol.,* 21, 285, 1973.
66. **Gugerli, P., Darby, G., and Minson, A. C.,** The structure of tobacco rattle virus ribonucleic acids: comparison of large oligonucleotides derived from the 3' ends, *J. Gen. Virol.,* 38, 273, 1978.
67. **Darby, G. and Minson, A. C.,** The structure of tobacco rattle virus ribonucleic acids: nature of the 3' -terminal nucleosides, *J. Gen. Virol.,* 14, 199, 1972.
68. **Minson, T. and Darby, G.,** 3' -Terminal oligonucleotide fragments of tobacco rattle virus ribonucleic acids, *J. Mol. Biol.,* 77, 337, 1973.
69. **Mayo, M. A., Fritsch, C., and Hirth, L.,** Translation of tobacco rattle virus RNA *in vitro* using wheat germ extracts, *Virology* , 69, 408, 1976.
70. **Fritsch, C., Mayo, M. A., and Hirth, L.,** Further studies on the translation products of tobacco rattle virus RNA *in vitro, Virology,* 77, 722, 1977.
71. **Mayo, M. A.,** Polypeptides induced by tobacco rattle virus during multiplication in tobacco protoplasts, *Intervirology* , 17, 240, 1982.
72. **Pelham, H. R. B.,** Translation of tobacco rattle virus RNAs *in vitro:* four proteins from three RNAs, *Virology,* 97, 256, 1980.
73. **Bisaro, D. M. and Siegel, A.,** A new viral RNA species in tobacco rattle virus-infected tissue, *Virology,* 107, 194, 1979.
74. **Robinson, D. J.,** A variant of tobacco rattle virus: evidence for a second gene in RNA-2, *J. Gen. Virol.,* 35, 37, 1977.
75. **Bisaro, D. M. and Siegel, A.,** Subgenomic components of tobacco rattle virus in infected tissue, *Virology,* 118, 411, 1982.
76. **Kubo, S., Harrison, B. D., Robinson, D. J., and Mayo, M. A.,** Tobacco rattle virus in tobacco mesophyll protoplasts: infection and virus multiplication, *J. Gen. Virol.,* 27, 293, 1975.
77. **Kitajima, E. W.,** Morfologia das Particulas de Virus do Anel do Pimentão e Ultraestrutura dos Tecidos Infetidos, Ph.D. thesis, University of Sao Paolo, 1967, 54.
78. **Harrison, B. D. and Roberts, I. M.,** Association of tobacco rattle virus with mitochondria, *J. Gen. Virol.,* 3, 121, 1968.
79. **Kitajima, E. W. and Costa, A. S.,** Association of pepper ringspot virus (Brazilian tobacco rattle virus) and host cell mitochondria, *J. Gen. Virol.,* 4, 177, 1969.
80. **Camargo, I. J. B., Kitajima, E. W., and Costa, A. A.,** Visualization of the pepper ring spot virus in tomato pollen, *Phytopathol. Z.,* 64, 282, 1969.
81. **Harrison, B. D., Stefanac, Z., and Roberts, I. M.,** Role of mitochondria in the formation of X-bodies in cells of *Nicotiana clevelandii* infected by tobacco rattle viruses, *J. Gen. Virol.,* 6, 127, 1970.
82. **de Zoeten, G. A.,** California tobacco rattle virus, its intracellular appearance, and the cytology of the infected cell, *Phytopathology,* 56, 744, 1966.
83. **de Zoeten, G. A.,** Winter-type California tobacco rattle virus in its relation to the infected cell, *Neth. J. Plant Pathol.,* 72, 311, 1966.
84. **Chen, T. A., Hirumi, H., and Maramorosch, K.,** Tobacco rattle virus in roots of mechanically inoculated plants, *Phytopathol. Z.,* 65, 15, 1969.
85. **Harrison, B. D. and Crockatt, A. A.,** Effects of cycloheximide on the accumulation of tobacco rattle virus in leaf discs of *Nicotiana clevelandii, J. Gen. Virol.,* 12, 183, 1971.
86. **Milne, R. G. and Masenga, V.,** unpublished observations, 1983.
87. **Hills, G. J. and Plaskitt, K. A.,** Observations of linked plant virus particles, *J. Gen. Virol.,* 44, 569, 1979.

Chapter 12

POTEXVIRUS GROUP

This rather homogeneous group is named after the type member, potato virus X (PVX).[1] The particles are robust flexuous rods 11 to 13 nm wide and 470 to 580 nm in modal length. Each particle contains a single molecule of positive-sense, single-stranded RNA of M_r about 2×10^6. Each virus has only one capsid protein of M_r between 18 and 28 $\times 10^3$; however, the lower values may be spurious, due to partial proteolysis. Numerous virus particles are found in the cytoplasm of infected cells, often forming large aggregates visible even in the light microscope.

Each of the Potexviruses has a fairly narrow host range causing mosaic or ringspot symptoms. The viruses are naturally transmitted by contact and there are no known vectors. Some possible members are aphid transmitted, at least under some circumstances. Seed transmission is unknown except that both clover yellow mosaic virus (ClYMV) and white clover mosaic virus (WClMV) are seed transmitted to a level of 5 to 10% in *Trifolium pratense*[2,3] and possibly in other legumes; foxtail mosaic virus (FoMV) is seedborne in *Briza maxima* and in oats to a level of 1 to 2%.[4] There are a number of recent reviews of the Potexviruses.[5-7]

I. MEMBERS OF THE GROUP AND THEIR RELATIONSHIPS

Table 1 lists those Potexviruses that are serologically related to at least one other member of the group, as well as possessing other typical characteristics of the group. Table 2 lists other viruses that appear to have a good claim to full membership, but whose serological relationship to any of the viruses in Table 1 has not been demonstrated. Lesemann and Koenig[5] suggest that only serologically interrelated Potexviruses should be considered definitive members, but there is a case, which we endorse, for considering all the viruses in both Tables 1 and 2 as full members.

Although all the individual viruses in Table 1 are distantly interrelated serologically, the degree of most of the relationships has not been determined precisely.[6,7] The particles of Potexviruses do not generally diffuse well in agar, so that tube precipitation or ring interface tests have largely been used for studies with intact virus particles. Virus preparations can also be fragmented by sonication, and then reacted with antiserum prepared to intact virus particles in gel-diffusion tests.[38] An alternative is to prepare degraded protein subunits (D-protein) by treating the particles with pyridine, and to use the preparations as both immunogen and antigen.[39] Generally, whole (or sonicated) virus as opposed to D-protein has been the more successful in detecting distant relationships among Potexviruses.

The values for modal lengths of Potexviruses listed in Tables 1 and 2 show some differences. However, their accuracy and taxonomic significance are not clear. Similarly, though values for the size of the RNA and coat protein are available for several Potexviruses (Tables 1 and 2), it is not yet clear how far the variation is intrinsic and how far it may be due to differing experimental procedures.

Two further comments may be made concerning the viruses in Table 2. First, the genus *Nandina*, host of nandina mosaic virus (NanMV) has the English vernacular name of "sacred bamboo". This plant is not a true bamboo but a dicotyledonous plant of the family Berberidaceae. Second, the possible serological relationship between plantago severe mottle virus (PlSMV) and plantain virus X (PlVX) has not been investigated. However, on present evidence (modal length, host range, and test plant reaction) the viruses seem to be different.[36]

Table 3 lists possible Potexviruses for which data are incomplete or partly at variance

Table 1
SEROLOGICALLY INTERRELATED MEMBERS OF THE POTEXVIRUS GROUP

Virus	Particle modal length (nm)	M_r of RNA (× 10⁶)	M_r of coat protein (× 10³)	Ref.
Potato virus X (PVX)	515	2.1	23—26	8—10
Cactus virus X(CVX)	520	2.1	20	3
= Barrel cactus virus (BaCV)	520	ND[a]	22—23.5	11
? = Zygocactus virus X (ZVX)	519	ND	ND	12
Clover yellow mosaic virus (ClYMV)	539	ND	ND	2
Commelina virus X (ComVX)	550	ND	ND	13
Cymbidium mosaic virus (CybMV)	475—490	ND	27.6	14—16
Foxtail mosaic virus (FoMV)	~500	2.03—2.24	21.2	4,17
Hydrangea ringspot virus (HRSV)	490	2.0—2.1	22	18,19
Lily virus X (LVX)	550			13
Narcissus mosaic virus (NaMV)	550—554	ND	ND	20,21
Nerine virus X (NeVX)	525—550	2.5	25.6	21,22
Papaya mosaic virus (PapMV)	530	2.2	19.4—22	23,24
= Ullucus virus X (UVX)				22
Pepino mosaic virus (PepMV)	508	ND	26.6	25
Tulip virus X (TuVX)	495	2.05	22.5	26,27
Viola mottle virus (VMV)	464—496	2.2—2.6	21—21.5	28
White clover mosaic virus (WClMV)	480	2.4	20	29,30

[a] ND, not determined.

Table 2
OTHER POTEXVIRUSES[a]

Virus	Particle modal length (nm)	M_r of RNA (× 10⁶)	M_r of coat protein (× 10³)	Ref.
Bamboo mosaic virus (BaMV)	490	—	—	31,32
Cassava common mosaic virus (CsCMV)	495	—	—	33
Daphne virus X (DVX)	499	—	23	34
Nandina mosaic virus (NanMV)	438 — 471	—	20.5	35
Plantago severe mottle virus (PlSMV)	536	2.19	23.2	36
Plantain virus X(PlVX)	570 — 580	2.1	26 — 29	37

[a] Viruses with good claims to be considered full members of the group but not bearing a demonstrated serological relationship to viruses in Table 1.

with generally accepted Potexvirus characteristics. Boletus virus (BoV) is a fungal virus with rod-shaped particles of 500 nm modal length, but there are no other data suggesting whether it is or is not a Potexvirus.[41] The taxonomic status of zygocactus virus (Table 3) as a separate virus remains in doubt, but it is listed separately because an antiserum to the virus did not react when tested against preparations of zygocactus virus X (ZVX).[12,42] Impatiens latent virus (ILV) may not be a Potexvirus but a Carlavirus.[43]

Table 3
POSSIBLE MEMBERS OF THE POTEXVIRUS GROUP[a]

Virus	Modal length (nm)
Artichoke curly dwarf virus	582
Barley B-1 virus	512
Boletus virus	500
Centrosema mosaic virus	580
Dioscorea latent virus	395 — 445
Impatiens latent virus	520[b]
Malva veinal necrosis virus	525
Negro coffee mosaic virus	550 — 580
Parsley virus 5	497
Parsnip virus 3	~500
Potato aucuba mosaic virus	580
Rhododendron necrotic ringspot virus	504
Rhubarb virus I and II	478
Wineberry latent virus	512
Zygocactus virus	580

[a] Modified from Koenig and Lesemann.[6]
[b] Lockhart and Betzold.[40]

Some of the viruses in Table 3 are aphid transmitted, though this type of transmission has not been found to occur with definitive Potexviruses. Potato aucuba mosaic virus (PAuMV) is transmitted nonpersistently by *Myzus persicae* if the donor plants already carry a Potyvirus such as potato virus A or Y.[44] PAuMV is also apparently not serologically related to any definitive Potexvirus, though its other properties would place it within the group. Somewhat like PAuMV, parsley virus 5 (PaIV5) was also transmitted (inefficiently, in the persistent manner) by the aphid *Cavariella aegopodii*, but only from plants also containing the carrot motley dwarf virus complex.[45] PaIV5 and parsnip virus 3 (ParV3) were found to be serologically related, and both also appear to be serologically related to PVX.[45] The only other possible Potexvirus transmitted by aphids is centrosema mosaic virus, found to be transmitted by four aphid species.[46,47]

II. VIRUS STRUCTURE AND COMPOSITION

A. Particle Structure

Potexvirus particles are flexuous rods with modal lengths ranging from 470 to 580 nm (Tables 1 and 2). The fine structure of the particles (Figure 1) is characteristic in that both cross-banding and longitudinal files of subunits are visible. This is in contrast to the particles of Carlaviruses (generally only longitudinal files are visible; see Chapter 13, Figure 1) and Potyviruses (neither cross-banding nor files are clearly visible; Chapter 14, Figure 1), as first observed by Varma and his colleagues.[48]

Potexvirus structure has been studied using electron microscopy of negatively stained particles, either randomly arrayed or oriented in parallel.[49] The images have been analyzed by optical diffraction techniques and the results compared with those of X-ray diffraction.[24,48,50-56] The potexviruses studied turn out to possess a rather uniform structure.[56-58] The protein and nucleic acid form a one-start helix, and the pitch of a typical particle is 3.4 to

3.5 nm; the number of protein subunits per turn varies slightly from one virus to another, but is fractionally less than nine. The nucleic acid lies at a radius of 3.3 to 3.6 nm, and there is an axial canal of radius 1.5 nm. As an example of Potexvirus particle construction, the papaya mosaic virus (PapMV) particle, contains about 6800 nucleotides and 1410 protein subunits with, almost certainly, five nucleotides being associated with each protein subunit.[24]

The estimated sizes of some Potexvirus coat proteins are listed in Tables 1 and 2. Most of these values have been obtained by polyacrylamide-gel electrophoresis (PAGE) and hence some errors must be expected. The amino acid composition of PVX coat protein has been determined by many authors, with rather closely concordant results.[9,59-63] However, determinations of the protein size do not agree so well; values derived from the amino acid composition and tryptic peptide maps were 26.8×10^3 (257 amino acid residues) published by Tung and Knight,[61] 29.0×10^3 (272 amino acid residues) by Tremaine and Agrawal,[62] and 23.0×10^3 (213 amino acid residues) by Goodman and his colleagues.[63]

At least some Potexvirus coat proteins behave anomalously during PAGE, as the charge on the denatured protein is not proportional to its molecular size.[64] A further problem, most closely investigated with PVX, is that the protein subunits of virus particles can easily be degraded at both the N- and C-terminal ends by proteases present in plant sap. Degradation is exacerbated by reducing agents commonly incorporated in virus extraction buffers such as ascorbate or sulphite.[9] From amino acid and PAGE data, Koenig and her colleagues[9] concluded that the native PVX subunit contains 249 amino acid residues and is of M_r 25.8 $\times 10^3$, whereas one plant protease-degraded product had lost 12 amino acid residues and was of M_r 24.8 $\times$ 10.[3]

Subunits of PVX coat protein have been prepared by disrupting the virus in lithium chloride;[65] while such subunits differ somewhat from the D-protein produced by pyridine treatment,[39] they are serologically indistinguishable in gel-diffusion tests.[10] Together with viral RNA, the native subunits were used at pH 6.0 to 6.2 and in low-ionic-strength buffer to reconstitute rods apparently identical to native virus.[10,63] It was suggested that hydrophobic sites on the subunit, no longer detected in the reconstituted particle, are involved in assembly.[66] Kaftanova and her colleagues[67] also found that helical rods could be reassembled under low-salt conditions but that stacks of discs were formed if the ionic strength was increased.

ClYMV has also been reconstituted.[68,69] Assembly of the particles takes place near neutral pH at low ionic strength. The initiation complexes contain the cap structure m^7GpppGp found at the 5′ end of the RNA,[70] indicating that initiation starts at or near the 5′ end. This appears to be followed by polar maturation in the direction 5′ to 3′.

Another Potexvirus, PapMV has also been reconstituted using low-ionic-strength conditions at pH 8.[71-73] A rapid initiation phase is followed by a slower elongation phase, and initiation occurs at the 5′ end or within 200 nucleotides of it.

There is, therefore, evidence from studies with several Potexviruses that particle assembly starts at or very near the 5′ end of the RNA and proceeds toward the 3′ end. This is in marked contrast to the internal initiation at a loop structure found with tobacco mosaic virus (see Chapter 9). It has been suggested[74] that the initial step of "melting" the secondary structure of the RNA may be different with Potexviruses and Tobamoviruses because PapMV protein has some "melting" ability but tobacco mosaic virus protein does not.

B. The Viral Genome

The size of the genomic RNAs of several Potexviruses have been estimated by PAGE and occasionally other methods. The reported values range from M_r about 2.0 to 2.6×10^6 (Tables 1 and 2). At least some of this variation is probably due to experimental errors.

The 5′ ends of two viruses, PVX and ClYMV, have been examined and both were shown to be capped with m^7GpppG....[70,75] However, no other structural studies appear to have been attempted.

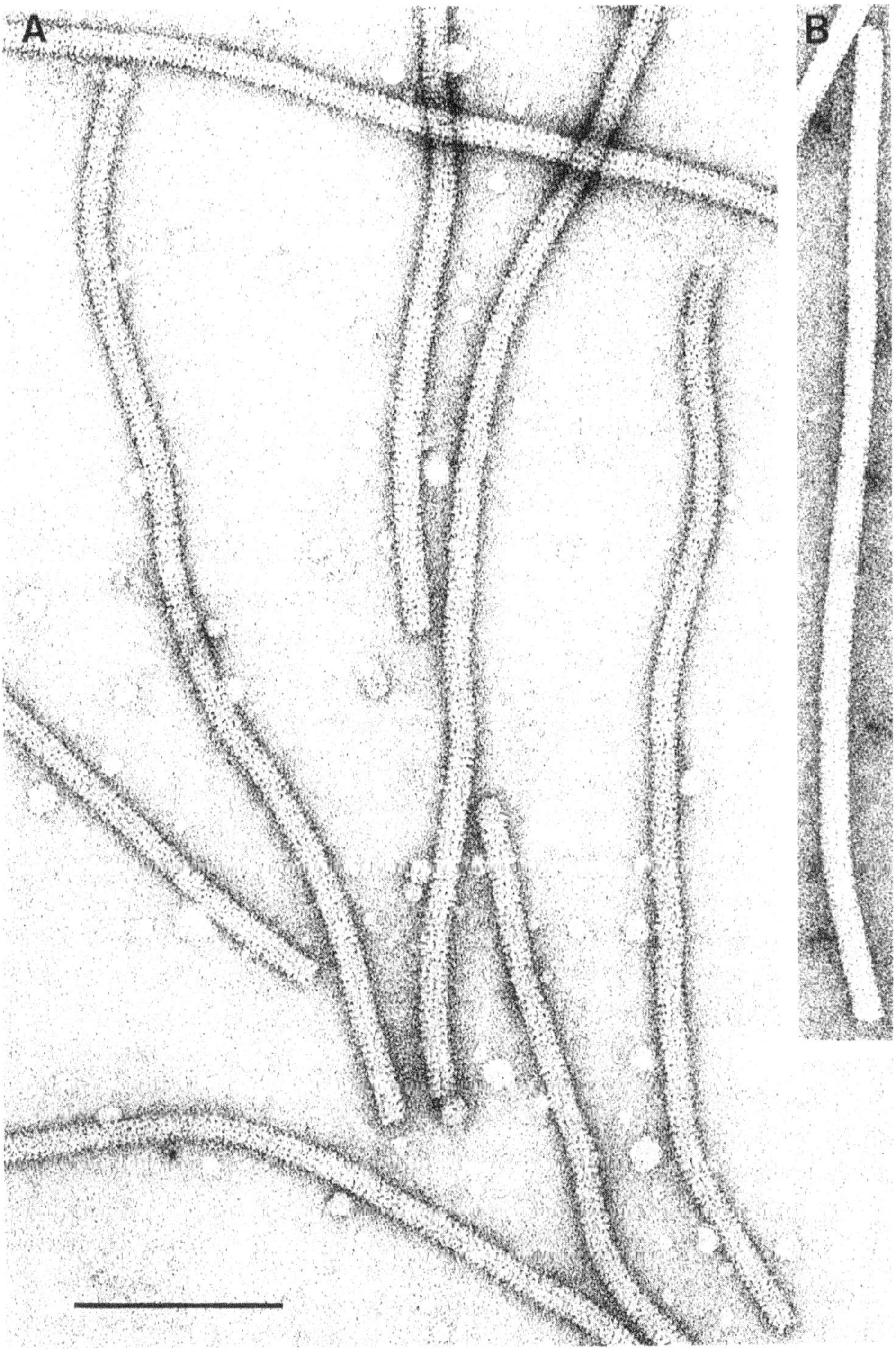

FIGURE 1. A, Purified particles of potato virus X(PVX), and B, those of viola mottle virus (VMX) stained in uranyl formate stain and photographed at the same magnification. In A, and to some extent in B, the characteristic "zipper" structure of Potexviruses is evident, due to both cross-banding and the appearance of longitudinal files of subunits. Bar = 100 nm.

PVX RNA has been translated in vitro in both the wheat germ and rabbit reticulocyte lysate systems.[76,77] The principal products were found to be two large proteins of M_r about 180 and 145 $\times$ 10^3, but no polypeptide corresponding to the coat protein was detected.[76,77] Evidence was reported to support the conclusion that both the large protein products shared amino acid sequences, suggesting that PVX may have a similar translation strategy to that of tobacco mosaic virus[76] (see Chapter 9 and Chapter 9, Figure 8).

III. CYTOPATHOLOGY

Potexviruses (with the one notable exception of PVX itself) seem to induce remarkably uniform cytopathic effects. The virus particles accumulate in the cytoplasm, forming massed aggregates that may exhibit conspicuous banding (Figures 2 and 3). No particular organelle appears to be involved in virus replication, at least as regards morphological evidence, though there is an increase in the number and activity of lysosome-like bodies in ClYMV-infected white clover root nodules and leaves.[78] Also, small vesicles on the tonoplasts of cells infected with some Potexviruses have been observed (Figure 3). They are very similar to those seen in cells infected with Cucumoviruses (see Volume 2, Chapter 5), in which they have been implicated as sites of viral RNA replication.[79] Meristematic dome cells may be full of conspicuous aggregates of Potexvirus particles[80,81] (Figure 2) or may be apparently virus-free.[82]

Local lesions induced by PVX in *Gomphrena globosa* were studied by Allison and Shalla[83] who found abnormal callose deposition at a greater radius than virus particles. They suggested that the callose deposition, together with exclusion of virus particles from the protoplast, might constitute a mechanism for limiting virus spread. Pennazio and his colleages,[84] also using PVX in *Gomphrena globosa*, found on the contrary that the virus infection outran callus formation, so that callose deposition was unlikely to function as an effective barrier. Hanchey and colleagues[85] examining nonlimited local lesions of cymbidium mosaic virus (CybMV) in *Cattleya* flowers, found no cell wall modifications.

PVX and its strains produce a unique type of cellular inclusion (Figures 4 and 5) that has been described as an aggregate of laminated inclusion components or beaded sheets.[83,86-95] These inclusions are usually confined to the cytoplasm (Figure 4) but have also been detected in nuclei (Figure 5). The most informative studies are those of Kozar and Sheludko[88] and Shalla and Shepard.[90] The sheets are about 3 nm thick, and may or may not be studded with beads on both sides; the beads are 11 to 14 nm in diameter. Virus particles usually occur between the sheets. Both sheets and beads can be digested by the proteolytic enzyme subtilisin, and probably consist of protein, but they do not react with antisera to PVX or its D-protein.[90] It is generally agreed that virus particles make their appearance before the beaded sheets are formed, and hence it is unlikely that the latter are involved in virus synthesis. No other Potexvirus has been found to induce beaded sheets in infected cells.

Apart from PVX, the only Potexviruses reported to induce cytoplasmic inclusions not constructed of virus particles are ClYMV[96] and hydrangea ringspot virus (HRSV).[97] The ClYMV inclusions consist of amorphous masses that appear to be aggregated coat protein because they react with ClYMV antiserum. HRSV is reported to induce cylindrical inclusions,[97] but this finding needs confirmation and extension before it can be evaluated.

Amorphous nuclear inclusions are induced by PapMV in papaya and *Chenopodium amaranticolor*,[98] and nuclear inclusions made up of virus particles occur in *Nicotiana clevelandii* infected with narcissus mosaic virus.[99] Nuclear inclusions of unknown nature were also reported in *Opuntia* infected wtih cactus virus X.[100]

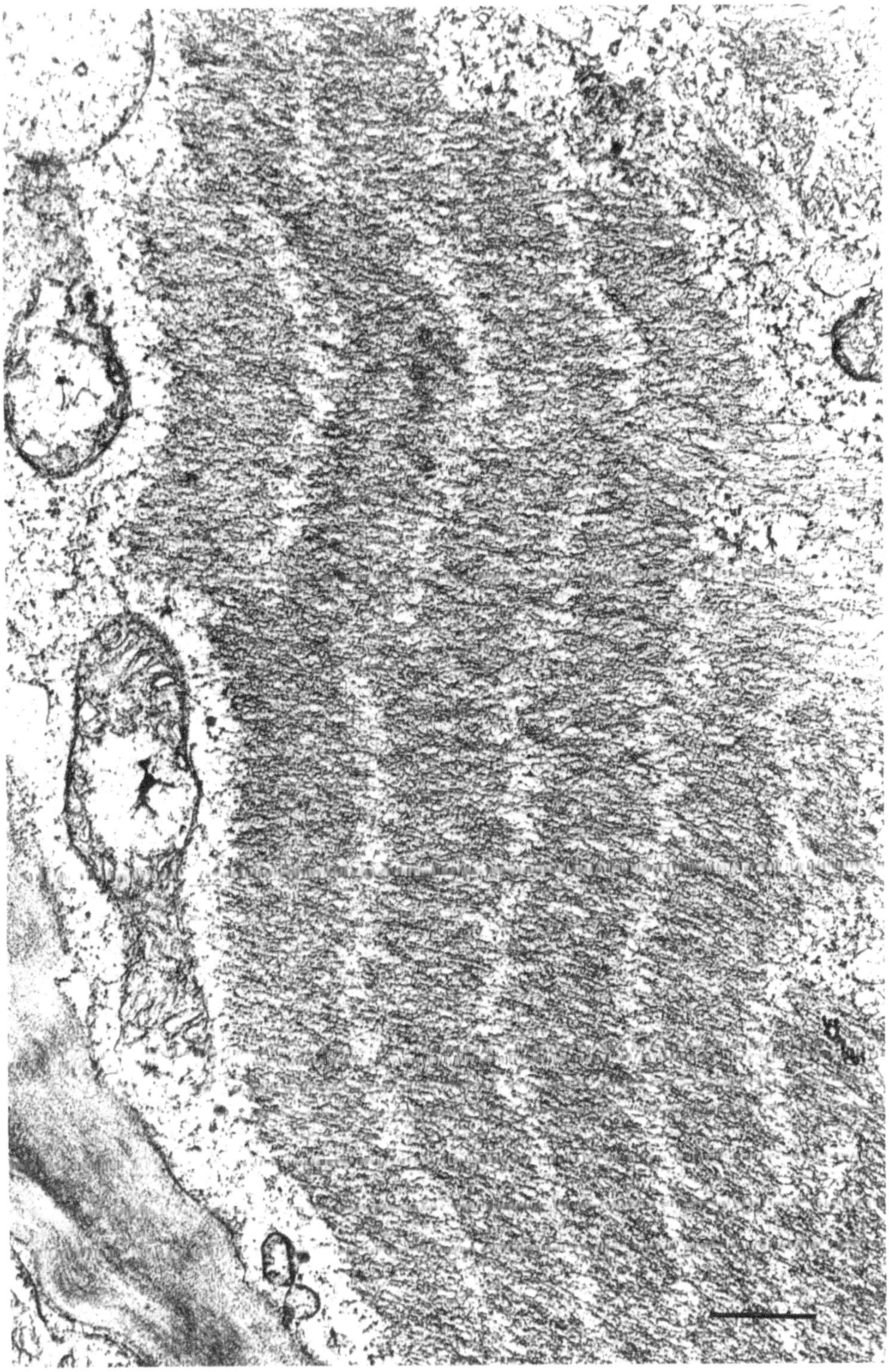

FIGURE 2. Thin section of a meristematic dome cell of a *Datura stramonium* plant infected with PVX. Note the conspicuous banded aggregate composed of virus particles. Bar = 500 nm. (Courtesy of Dr. A. Appiano, Instituto di Fitovirologia Applicata, Torino, Italy).

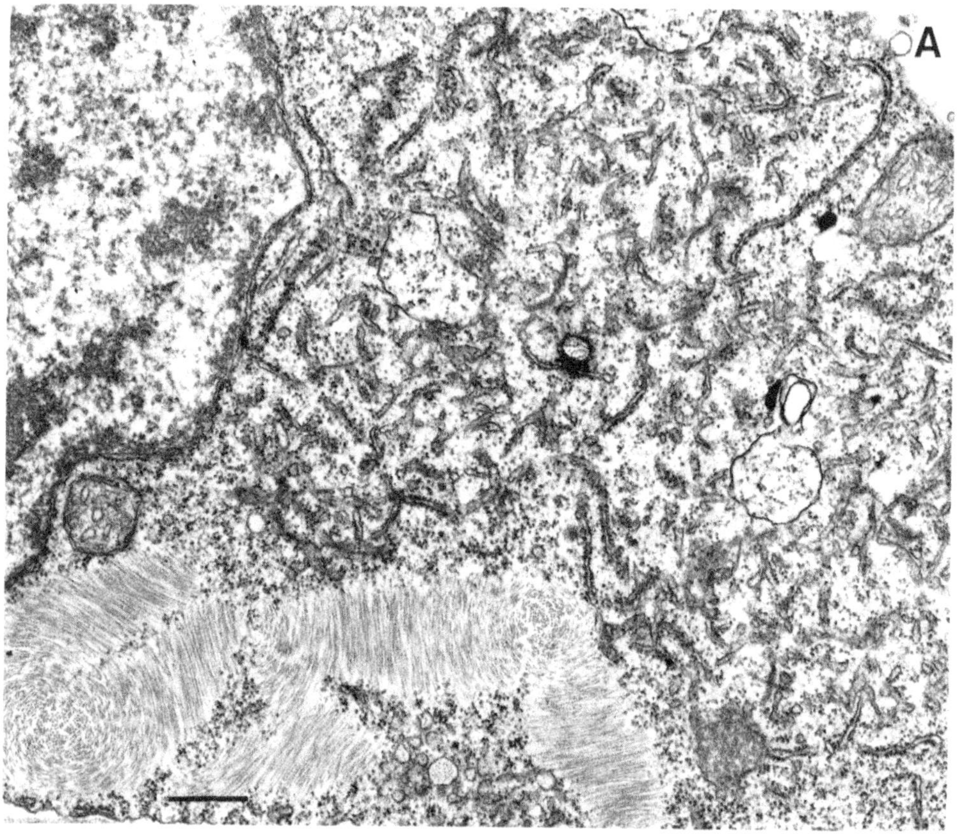

FIGURE 3. Thin sections of *Nicotiana glutinosa* leaves infected wtih pepino mosaic virus (PepMV). A, Virus particle aggregates and membrane proliferation in the cytoplasm; B, banded aggregate of virus particles in the cytoplasm and some vesicles associated with the tonoplast that are characteristic of infection. Bars = 500 nm. (Courtesy of Dr. D. E. Lesemann, Institut für Viruskrankheiten der Pflanzen, Braunschweig, Federal Republic of Germany.)

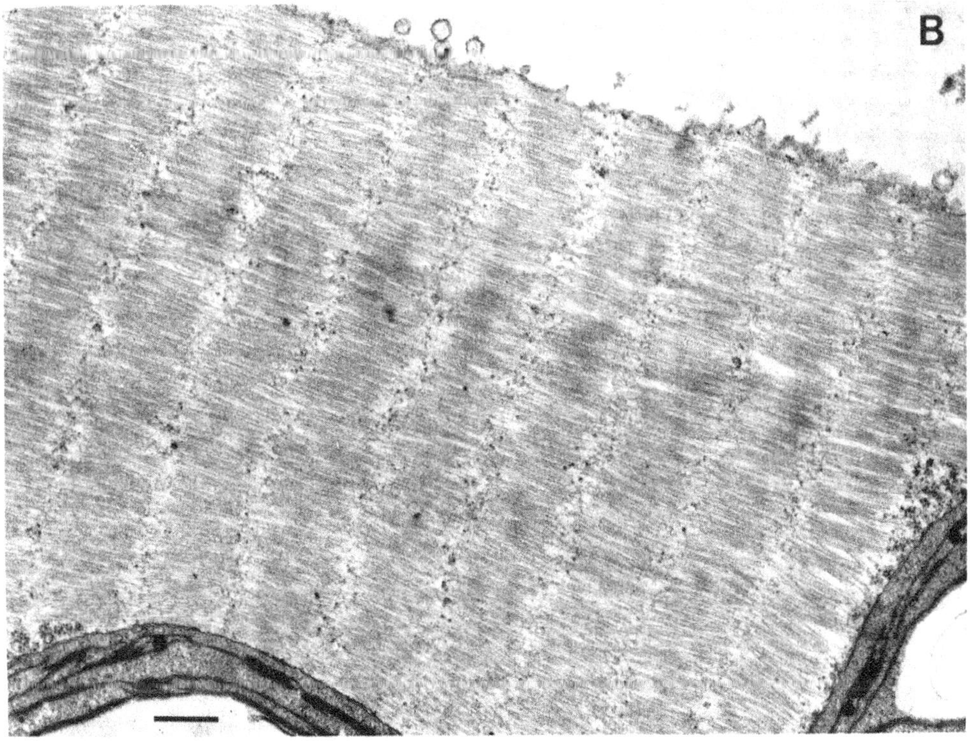

FIGURE 3B.

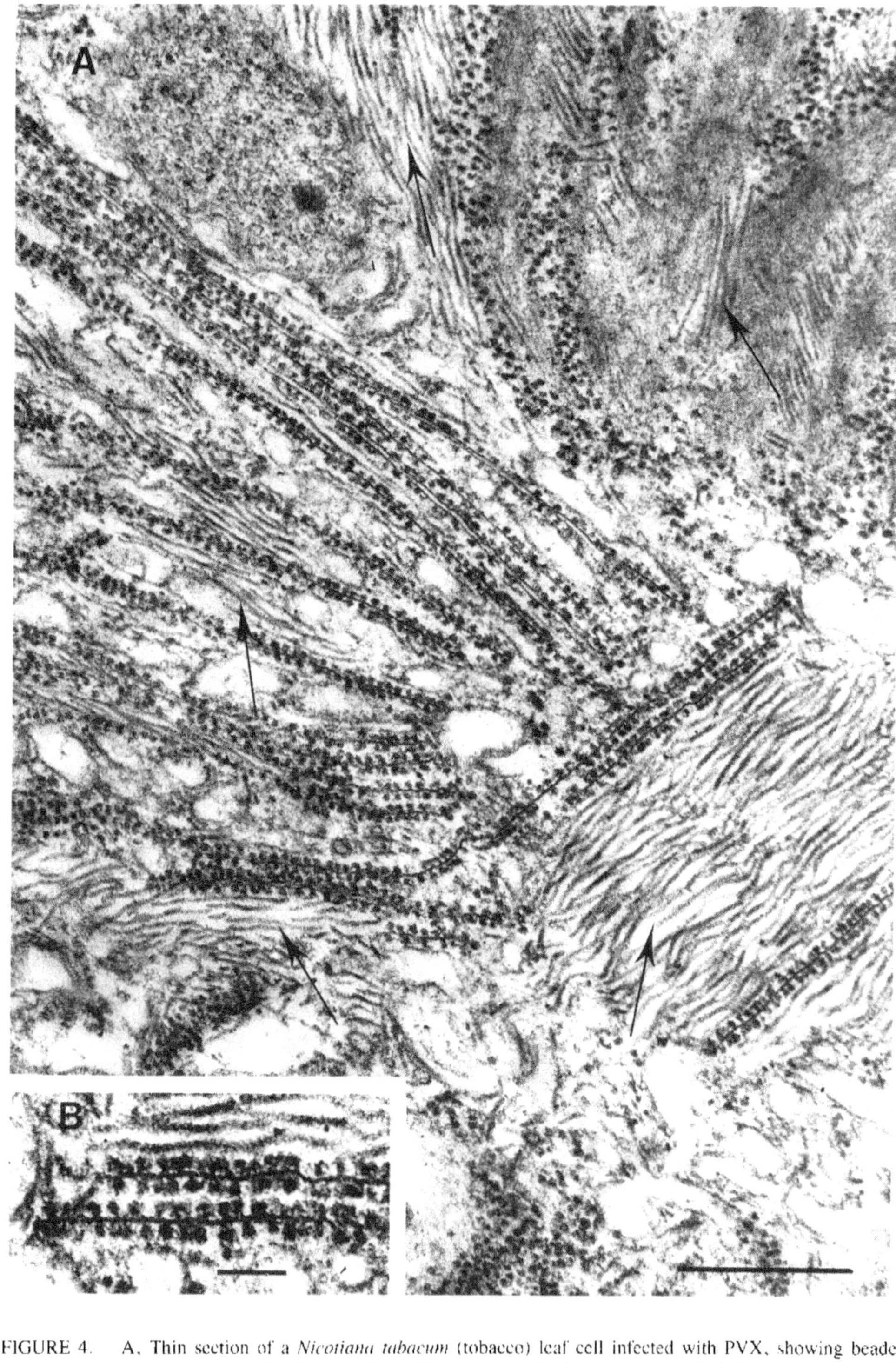

FIGURE 4. A, Thin section of a *Nicotiana tabacum* (tobacco) leaf cell infected with PVX, showing beaded sheets and virus particles (arrows) in the cytoplasm. The center of the figure shows beaded sheets cut transversely whereas at the top right, they are seen cut almost tangentially. Bar = 500 nm. B, Detail of two transversely sectioned beaded sheets and (above them) some virus particles. Bar = 100 nm. (Courtesy of the late Dr. T. A. Shalla, Department of Plant Pathology, University of California, Davis.)

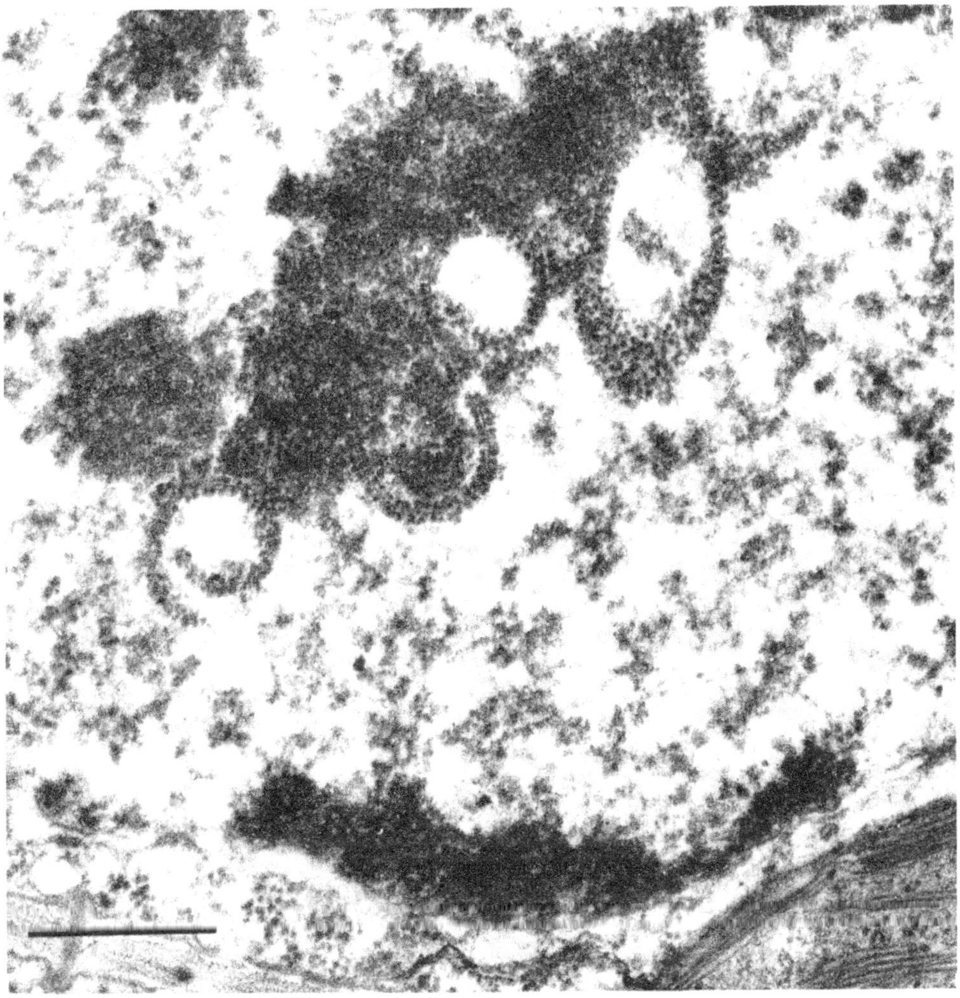

FIGURE 5. Thin section of part of an apical meristem cell of *Datura stramonium* infected with PVX. Note the presence of beaded sheets inside the nucleus. Bar = 500 nm. (Courtesy of Dr. A. Appiano, Instituto di Fitovirologia. Applicata, Torino, Italy.)

REFERENCES

1. **Matthews, R. E. F.,** Classificaton and nomenclature of viruses. Fourth report of the International Committee on Taxonomy of Viruses, *Intervirology*, 17, 1, 1982.
2. **Bos, L.,** Clover yellow mosaic virus, *CMI/AAB Descriptions of Plant Viruses*, No. 111, 1973.
3. **Bercks, R.,** Cactus virus X, *CMI/AAB Descriptions of Plant Viruses*, No. 58, 1971.
4. **Paulsen, A. Q. and Niblett, C. L.,** Purification and properties of foxtail mosaic virus, *Phytopathology*, 67, 1346, 1977.
5. **Lesemann, D. -E. and Koenig R.,** Potexvirus (potato virus X) group, in *The Atlas of Insect and Plant Viruses*, Maramorosch, K., Ed., Academic Press, New York, 1977, 331.
6. **Koenig, R. and Lesemann, D.-E.,** Potexvirus group, *CMI/AAB Descriptions of Plant Viruses*, No. 200, 1978.

7. **Purcifull, D. E. and Edwardson, J. R.,** Potexviruses, in *Handbook of Plant Virus Infections and Comparative Diagnosis*, Kurstak, E., Ed., Elsevier, Amsterdam, 1981, 627.
8. **Bercks, R.,** Potato virus X, *CMI/AAB Descriptions of Plant Viruses*, No. 4, 1970.
9. **Koenig, R., Tremaine, J. H., and Shepard, J. F.,** *In situ* degradation of the protein chain of potato virus X at the N- and C-termini, *J. Gen. Virol.*, 38, 329, 1978.
10. **Goodman, R. M.,** Reconstitution of potato virus X *in vitro*. I. Properties of the dissociated protein structural subunits, *Virology*, 68, 287, 1975.
11. **Attathom, S., Weathers, L. G., and Gumpf, D. J.,** Identification and characterization of a potexvirus from California barrel cactus, *Phytopathology*, 68, 1401, 1978.
12. **Giri, L. and Chessin, M.,** Zygocactus virus X, *Phytopathol. Z.*, 83, 40, 1975.
13. **Stone, O. M.,** Two new potexviruses from monocotyledons, *Acta Hortic.*, 110, 59, 1980.
14. **Francki, R. I. B.,** Cymbidium mosaic virus, *CMI/AAB Descriptions of Plant Viruses*, No. 27, 1970.
15. **Frowd, J. A. and Tremaine, J. H.,** Physical, chemical, and serological properties of Cymbidium mosaic virus, *Phytopathology*, 67, 43, 1977.
16. **Faccioli, G. and Marani, F.,** Cymbidium mosaic virus associated with flower necrosis in Cattleya orchids, *Phytopathol. Mediterr.*, 18, 21, 1979.
17. **Short, M. N.,** Foxtail mosaic virus, *CMI/AAB Descriptions of Plant Viruses*, No. 264, 1983.
18. **Koenig, R.,** Hydrangea ringspot virus, *CMI/AAB Descriptions of Plant Viruses* , No. 114, 1973.
19. **Hill, J. H., Benner, H. I., and Zeyen, R. J.,** Properties of hydrangea ringspot virus ribonucleic acid, *J. Gen. Virol.*, 34, 115, 1977.
20. **Mowat, W. P.,** Narcissus mosaic virus, *CMI/AAB Descriptions of Plant Viruses*, No. 45, 1971.
21. **Maat, D. Z.,** Two potexviruses in Nerine, *Neth. J. Plant Pathol.*, 82, 95, 1976.
22. **Phillips, S. and Brunt, A. A.,** Some hosts and properties of an isolate of nerine virus X from *Agapanthus praecox* subsp. *orientalis*, *Acta Hortic.*, 110, 65, 1980.
23. **Purcifull, D. E. and Hiebert, E.,** Papaya mosaic virus, *CMI/AAB Descriptions of Plant Viruses*, No. 56, 1971.
24. **Tollin, P., Bancroft, J. B., Richardson, J. F., Payne, N. C., and Beveridge, T. J.,** Diffraction studies of papaya mosaic virus, *Virology*, 98, 108, 1979.
25. **Jones, R. A. C., Koenig, R., and Lesemann, D.-E.,** Pepino mosaic virus, a new potexvirus from pepino *(Solanum muricatum)*, *Ann. Appl. Biol.*, 94, 61, 1980.
26. **Mowat, W. P.,** A potexvirus from tulip, *Scottish Hortic. Res. Inst. Ann. Rept.* for 1978, p. 102, 1979.
27. **Radwan, M. M., Wilson, H. R., and Duncan, G. H.,** Diffraction studies of tulip virus X particles, *J. Gen. Virol.*, 56, 297, 1981.
28. **Lisa, V., Boccardo, G., and Milne, R. G.,** Viola mottle virus, *CMI/AAB Descriptions of Plant Viruses*, No. 247, 1982.
29. **Bercks, R.,** White clover mosaic virus, *CMI/AAB Descriptions of Plant Viruses*, No. 41, 1971.
30. **Koenig, R., Stegemann, H., Francksen, H., and Paul, H. L.,** Protein subunits in the potato virus X group. Determination of the molecular weights by polyacrylamide electrophoresis, *Biochim. Biophys. Acta*, 207, 184, 1970.
31. **Kitajima, E. W., Lin, M. T., Cupertino, F. P., and Costa, C. L.,** Electron microscopy of bamboo mosaic virus-infected leaf tissues, *Phytopathol. Z.*, 90, 180, 1977.
32. **Lin, M. T., Kitajima, E. W., Cupertino, F. P., and Costa, C. L.,** Partial purification and some properties of bamboo mosaic virus, *Phytopathology*, 67, 1439, 1977.
33. **Costa, A. S. and Kitajima, E. W.,** Cassava common mosaic virus, *CMI/AAB Descriptions of Plant Viruses*, No. 90, 1972.
34. **Forster, R. L. and Milne, K. S.,** Daphne virus X, *CMI/AAB Descriptions of Plant Viruses* , No. 195, 1978.
35. **Zettler, F. W., Hiebert, E., Maciel-Zambolim, E., Christie, R. G., and Abo El-Nil, M. M.,** A potexvirus infecting nandina domestica' 'Harbor Dwarf', *Acta Hortic.*, 110, 71, 1980.
36. **Rowhani, A. and Peterson, J. F.,** Characterization of a flexuous rod-shaped virus from *Plantago*, *Can. J. Plant Pathol.*, 2, 12, 1980.
37. **Hammond, J. and Hull, R.,** Plantain virus X: a new potexvirus from *Plantago lanceolata*, *J. Gen. Virol.*, 54, 75, 1981.
38. **Tomlinson, J. A. and Walkey, D. G. A.,** Effects of ultrasonic treatment on turnip mosaic virus and potato virus X, *Virology*, 32, 267, 1967.
39. **Shalla, T. A. and Shepard, J. F.,** An antigenic analysis of potato virus X and of its degraded protein. II. Evidence for a conformational change associated with the depolymerization of structural protein, *Virology*, 42, 835, 1970.

40. **Lockhart, B. E. L. and Betzold, J. A.,** Latent infection of Impatiens by a potexvirus, *Acta Hortic.*, 110, 81, 1980.
41. **Huttinga, H., Wichers, H. J., and Dieleman-Van Zaayen, A.,** Filamentous and polyhedral virus-like particles in *Boletus edulis*. *Neth. J. Plant Pathol.*, 81, 102, 1975.
42. **Casper, R. and Brandes, J.,** A new cactus virus, *J. Gen. Virol.*, 5, 155, 1969.
43. **Koenig, R.,** personal communication, 1982.
44. **Kassanis, B. and Govier, D. A.,** Potato aucuba mosaic virus, *CMI/AAB Descriptions of Plant Viruses*, No.98, 1972.
45. **Frowd, J. A. and Tomlinson, J. A.,** The isolation and identification of parsley viruses occurring in Britain, *Ann. Appl. Biol.*, 72, 177, 1972.
46. **van Velsen, R. J. and Crowley, N. C.,** *Centrosema* mosaic: a plant virus disease transmitted by both aphids and plant bugs, *Nature* (London), 189, 858, 1961.
47. **Crowley, N. C. and Francki, R. I. B.,** Purification and some properties of Centrosema mosaic virus, *Aust. J. Biol. Sci.*, 16, 468, 1963.
48. **Varma, A., Gibbs, A. J., Woods, R. D., and Finch, J. T.,** Some observations on the structure of the filamentous particles of several plant viruses, *J. Gen. Virol.*, 2, 107, 1968.
49. **Horne, R. W.,** The formation of virus crystalline and paracrystalline arrays for electron microscopy and image analysis, *Adv. Virus Res.*, 24, 173, 1979.
50. **Robinson, D. J., Hutcheson, A., Tollin, P., and Wilson, H. R.,** A double-helical structure for re-aggregated protein of narcissus mosaic virus, *J. Gen. Virol.*, 29, 325, 1975.
51. **Wilson, H. R., Al-Mukhtar, J., Tollin, P., and Hutcheson, A. M.,** Observations on the structure of particles of white clover mosaic virus, *J. Gen. Virol.*, 39, 361, 1978.
52. **Tollin, P., Wilson, H. R., and Mowat, W. P.,** Optical diffraction from particles of narcissus mosaic virus, *J. Gen. Virol*, 29, 331, 1975.
53. **Tollin, P., Wilson, H. R., and Bancroft, J. B.,** Further observations on the structure of particles of potato virus X, *J. Gen. Virol.*, 49, 407, 1980.
54. **Tollin, P., Wilson, H. R., Bancroft, J. B., Richardson, J. F., Payne, N. C., and Alford, W. P.,** Diffraction studies of clover yellow mosaic virus, *J. Gen. Virol.*, 52, 205, 1981.
55. **Bancroft, J. B., Hills, G. J., and Richardson, J. F.,** A re-evaluation of the structure of narcissus mosaic virus and polymers made from its protein, *J. Gen. Virol.*, 50, 451, 1980.
56. **Richardson, J. F., Tollin, P., and Bancroft, J. B.,** The architecture of the potexviruses, *Virology*, 112, 34, 1981.
57. **Veerisetty, V.,** Relationships among structural parameters of virions of helical symmetry, *Virology*, 84, 523, 1978.
58. **Veerisetty, V.,** Suggestions for the classification and nomenclature of helical plant viruses, *Intervirology* 11, 167, 1979.
59. **Shaw, J. G., Reichmann, M. E., and Hatt, D. L.,** Amino acid analyses of the proteins of two strains of potato virus X, *Virology*, 18, 79, 1962.
60. **Miki, T. and Knight, C. A.,** The protein subunit of potato virus X, *Virology*, 36, 168, 1968.
61. **Tung, J. -S. and Knight, C. A.,** The coat protein subunits of potato virus X and white clover mosaic virus, a comparison of methods for determining their molecular weights and some *in situ* degradation products of potato virus X protein, *Virology*, 49, 214, 1972.
62. **Tremaine, J. H. and Agrawal, H. O.,** Limited proteolysis of potato virus X by trypsin and plant proteases, *Virology*, 49, 735, 1972.
63. **Goodman, R. M., Horne, R. W., and Hobart, J. M.,** Reconstitution of potato virus X *in vitro*. II. Characterization of the reconstituted product, *Virology*, 68, 299, 1975.
64. **Koenig, R.,** Anomalous behaviour of the coat proteins of potato virus X and cactus virus X during electrophoresis in dodecyl sulfate-containing polyacrylamide gels, *Virology*, 50, 263, 1972.
65. **Francki, R. I. B. and McLean, G. D.,** Purification of potato virus X and preparation of infectious ribonucleic acid by degradation with lithium chloride, *Aust. J. Biol. Sci.*, 21, 1311, 1968.
66. **Goodman, R. M.,** Reconstitution of potato virus X *in vitro*. III. Evidence for a role for hydrophobic interactions, *Virology* , 76, 72, 1977.
67. **Kaftanova, A. S., Kiselev, N. A., Novikov, V. K., and Atabekov, J. G.,** Structure of products of protein reassembly and reconstruction of potato virus X, *Virology*, 67, 283, 1975.
68. **Bancroft, J. B., AbouHaidar, M., and Erickson, J. W.,** The assembly of clover yellow mosaic virus and its protein, *Virology*, 98, 121, 1979.
69. **AbouHaidar, M.,** Polar assembly of clover yellow mosaic virus, *J. Gen. Virol.*, 57, 199, 1981.
70. **AbouHaidar, M. and Bancroft, J. B.,** The structure of the 5′ -terminus of papaya mosaic virus RNA, *J. Gen. Virol.*, 39, 559, 1978.
71. **Erickson, J. W. and Bancroft, J. B.,** The self-assembly of papaya mosaic virus, *Virology*, 90, 36, 1978.

72. **Erickson, J. W. and Bancroft, J. B.,** The kinetics of papaya mosaic virus assembly, *Virology*, 90, 47, 1978.
73. **AbouHaidar, M. and Bancroft, J. B.,** The initiation of papaya mosaic virus assembly, *Virology*, 90, 54, 1978.
74. **Erickson, J. W. and Bancroft, J. B.,** Melting of viral RNA by coat protein: assembly strategies for elongated plant viruses, *Virology*, 108, 235, 1981.
75. **Sonenberg, N., Shatkin, A. J., Ricciardi, R. P., Rubin, M., and Goodman, R. M.,** Analysis of terminal structures of RNA from potato virus X, *Nucleic Acids Res.*, 5, 2501, 1978.
76. **Wodnar-Filipowicz, A., Skrzeczkowski, L. J., and Filipowicz, W.,** Translation of potato virus X RNA into high molecular weight proteins, *FEBS Lett.*, 109, 151, 1980.
77. **Ricciardi, R. P., Goodman, R. M., and Gottlieb, D.,** Translation of PVX RNA *in vitro* by wheat germ. I. Characterization of the reaction and product size, *Virology*, 85, 310, 1978.
78. **Tu, J. C.,** Lysosomal distribution and acid phosphatase activity in white clover infected with clover yellow mosaic virus, *Phytopathology*, 66, 588, 1976.
79. **Hatta, T. and Francki, R. I. B.,** Cytopathic structures associated with tonoplasts of plant cells infected with cucumber mosaic and tomato aspermy viruses, *J. Gen. Virol.*, 53, 343, 1981.
80. **Appiano, A. and Pennazio, S.,** Electron microscopy of potato meristem tips infected with potato virus X, *J. Gen. Virol.*, 14, 273, 1972.
81. **Pennazio, S. and Appiano, A.,** Potato virus X aggregates in ultrathin sections of *Datura* meristematic dome, *Phytopathol. Mediterr.*, 14, 12, 1975.
82. **Lawson, R. H. and Hearon, S. S.,** Distribution and staining properties of Cymbidium mosaic virus in buds of *Cattleya* orchids, in *3rd Int. Symp. Virus Ornamental Plants 1972, Acta. Hortic.*, 36, 195, 1974.
83. **Allison, A. V. and Shalla, T. A.,** The ultrastructure of local lesions induced by potato virus X: a sequence of cytological events in the course of infection, *Phytopathology*, 64, 784, 1974.
84. **Pennazio, S., Redolfi, P., and Sapetti, C.,** Callose formation and permeability changes during the partly localized reaction of *Gomphrena globosa* to potato virus X, *Phytopathol. Z.*, 100, 172, 1981.
85. **Hanchey, P., Livingston, C. H., and Reeves, F. B.,** Cytology of flower necrosis in Cattleyas infected by Cymbidium mosaic virus, *Physiol. Plant Pathol.*, 6, 227, 1975.
86. **Borges, L. and Ferreira, J. F. D.,** Comparative study of cell structure in *Datura metel* L. healthy and infected with potato virus X or potato virus Y, *Rev. Biol.*, 6, 421, 1968.
87. **Turner, R. H. and Milne, R. G.,** Electron microscopy of thin sections of potato virus X-induced inclusions (Abstract), 1st Int. Congr. Plant Pathol., London, 1968.
88. **Kozar, F. E. and Sheludko, Y. M.,** Ultrastructure of potato and *Datura stramonium* plant cells infected with potato virus X, *Virology*, 38, 220, 1969.
89. **Stols, A. L. H., Hill-van der Meulen, G. W., and Toen, M. K. I.,** Electron microscopy of *Nicotiana glutinosa* leaf cells infected with potato virus X, *Virology*, 40, 168, 1970.
90. **Shalla, T. A. and Shepard, J. F.,** The structure and antigenic analysis of amorphous inclusion bodies induced by potato virus X, *Virology*, 49, 654, 1972.
91. **Shalla, T. A. and Peterson, L. J.,** Infection of isolated plant protoplasts with potato virus X, *Phytopathology*, 63, 1125, 1973.
92. **Doraiswamy, S. and Lesemann, D.,** Comparison of cytological effects induced by three members of the potexvirus group, *Phytopathol. Z.*, 81, 314, 1974.
93. **Otuski, Y., Takebe, I., Honda, Y., Kajita, S., and Matsui, C.,** Infection of tobacco mesophyll protoplasts by potato virus X, *J. Gen. Virol.*, 22, 375, 1974.
94. **Honda, Y., Kajita, S., Matsui, C., Otsuki, Y., and Takebe, I.,** An ultrastructural study of the infection of tobacco mesophyll protoplasts by potato virus X, *Phytopathol. Z.*, 84, 66, 1975.
95. **Reunov, A. V.,** Ultrastructural features of the parenchyma of young leaves of *Datura stramonium* L. systemically infected with potato virus X, *Virology*, 96, 265, 1979.
96. **Schlegel, D. E. and Delisle, D. E.,** Viral protein in early stages of clover yellow mosaic virus infection of *Vicia faba*, *Virology*, 45, 747, 1971.
97. **Zeyen, R. J.,** Intercellular inclusions in hydrangea ringspot virus-infected hydrangea, *2nd Int. Congr. Plant Pathol. Minneapolis 1973, Abstracts, No.* 0275, 1973.
98. **Christie, R. G. and Edwardson, J. R.,** Light and electron microscopy of plant virus inclusions, *Fla. Exp. Stn. Monograph.*, No. 9, 1977, 150 pp.
99. **Turner, R. H.,** Electron microscopy of cells infected with narcissus mosaic virus, *J. Gen. Virol.*, 13, 177, 1971.
100. **Milicić, D.,** Viruskorper und Zellteilungsanomalien in Opuntia brasiliensis, *Protoplasma*, 43, 228, 1954.

Chapter 13

CARLAVIRUS GROUP

The Carlaviruses form a rather homogeneous group. Some of them cause serious disease but a number are symptomless at least in certain hosts; this is true for carnation latent virus (CLV), which gives its name to the group.[1,2]

Carlavirus particles are fairly stiff but flexible rods 12 to 13 nm in diameter and between 610 and 700 nm in length (though Potyviruses have an overlapping modal length range of 680 to 900 nm; see Chapter 14). Where the composition of the virus particles is known, there is one coat protein subunit of M_r about 33 × 10^3 and one molecule of positive-sense single-stranded RNA of M_r between 2.3 and 2.6 × 10^6.[1,2]

The host ranges of the individual viruses are rather narrow. Where natural vectors are known, they are aphids, and tranmission is nonpersistent. Seed transmission has occasionally been reported but seems unimportant. The viruses are mechanically transmissible. There is no tissue specificity, and virus particles are found in the cytoplasm without, however, producing very distinctive inclusion bodies or cytopathic effects.[2]

I. MEMBERS OF THE GROUP AND THEIR RELATIONSHIPS

Many viruses have been isolated with some characteristics of Carlaviruses; however, decisions as to what constitute separate viruses are difficult because many members are distinguished according to their host ranges while being related serologically. The serological relationships of the members have been quite extensively tested but not all combinations have been tried, and certainly not in the same laboratory with the same virus isolates and sets of antisera. As an example of the difficulty, Veerisetty and Brakke[3] found that alfalfa latent virus (ALV) was not serologically related to pea streak virus (PeSV) whereas Hampton[4] found the two viruses to be serologically indistinguishable.

Viruses which we consider qualify for inclusion in the Carlavirus group as distinct entities are listed in Table 1. However, we realize that as more data become available, modification to this list will almost certainly have to be made. Those viruses about whose taxonomic status we have real doubts are listed in Table 2 as possible members of the group.

The better known Carlaviruses, CLV, chrysanthemum virus B (CVB), cactus virus 2 (CV2), passiflora latent virus (PaLV), PeSV, red clover vein mosaic virus (RCVMV), potato virus M (PVM), and potato virus S (PVS) are all serologically interrelated, and for some pairs of viruses sufficient data are available to express their relationships as serological differentiation indices (SDIs). The following SDIs were derived from the literature by Wetter and Milne.[2] PVS-CVB, 3.5; CLV-CVB, 3.7; CLV-PVM, 5.0; PVM-CVB, 5.2; CLV-PVS, 5.9; RCVMV-PeSV, 6.2; PVM-PVS, 6.4. It is interesting to note that the two most closely related pairs of viruses in the above list (PVS-CVB and CLV-CVB) have very different natural hosts, whereas the two most distantly related pairs (RCVMV-PeSV and PVM-PVS) share a number of natural hosts.

Poplar mosaic virus (PopMV), until recently in a rather isolated position serologically, is now known to be serologically related to lonicera latent virus (LLV), which in turn is related to shallot latent virus (SLV) and RCVMV.[43] Hop latent virus (HoLV) and American hop latent virus (AHLV) are serologically related to hop mosaic virus (HoMV) but are at present considered to be distinct viruses; HoMV is related to CLV and PVM.[11-14,44]

There may be some differences in the modal lengths of certain Carlaviruses, and this is a character that might be useful taxonomically. Nevertheless, a difference of 30 nm, representing about 5%, can easily arise due to experimental errors, especially if the viruses are

Table 1
MEMBERS OF THE CARLAVIRUS GROUP

Virus	Modal length (nm)	Ref.
Carnation latent virus (CLV)	652	5
Cactus virus 2 (CV2)	650	2
Chrysanthemum virus B (CVB)	685 — 690	6,7
= gynura latent virus		
Dandelion latent virus (DLV)	640	8
Elderberry carlavirus (ECV)	650	9
Helenium virus S (HeVS)	640	10
Hop latent virus (HoLV)	650	11
American hop latent virus (AHLV)	680	12
Hop mosaic virus (HoMV)	650	13,14
Kalanchoë virus 1 (KV1)	610	15
Lilac mottle virus (LiMV)	610	16
Lily symptomless virus (LSV)	640	17,18
Lonicera latent virus (LLV)	650	19,20
= honeysuckle latent virus		
Mulberry latent virus (MLV)	700	21
Muskmelon vein necrosis virus (MkVNV)	674	22
Narcissus latent virus (NLV)	650	23,24
Nerine latent virus (NeLV)	650	25,26
= hippeastrum latent virus		
Passiflora latent virus (PaLV)	648	27
Pea streak virus (PeSV)	619 — 630	3,4,28 — 31
= alfalfa latent virus		
Pepino latent virus (PeLV)	660 — 680	32
Poplar mosaic virus (PopMV)	670 — 685	19,33 — 38
Potato virus M (PVM)	651	39
Potato virus S (PVS)	657	40
Red clover vein mosaic virus (RCVMV)	654 — 670	30,31,41
Shallot latent virus (SLV)	650	42

Table 2
POSSIBLE MEMBERS OF THE CARLAVIRUS GROUP[a]

Virus	Modal length (nm)
Caper vein banding virus	678
Cassia mild mosaic virus	650
Chicory blotch virus	657
Cole latent virus	654
Cowpea mild mottle virus	650
Cynodon mosaic virus	650
Eggplant mild mottle virus	677 — 693
Fuchsia latent virus	650
Garlic mosaic virus	610 — 640
Groundnut crinkle virus	650
Nasturtium mosaic virus	650 — 710
White bryony mosaic virus	650

[a] Data modified from Matthews[1] and Wetter and Milne.[2]

not compared directly in the same laboratory. Wetter and colleagues,[45] for example, found that the modal lengths of a PeSV and an RCVMV isolate were 619 and 654 nm, respectively.

Bos and his colleagues[30] also compared PeSV and RCVMV isolates and found their modal lengths to be 630 and 670 nm, respectively. Thus, while absolute values varied somewhat, it is likely that PeSV and RCVMV do actually differ in modal length by 30 to 40 nm. Where direct comparisons are lacking, conclusions as to real difference cannot be drawn, but it is interesting to note that whereas most Carlaviruses have recorded modal lengths near 650 nm, a few appear to be appreciably shorter and some longer (Table 1). Accurate direct comparisons should show whether these differences are significant. If real differences are established, their value to taxonomy must be assessed critically. (See Chapter 14 for a discussion on the taxonomic value of Potyvirus particle lengths.)

Cowpea mild mottle virus (CPMMV) is listed by Matthews[1] as a Carlavirus but its status is now in doubt.[46] Like a Carlavirus, it has filamentous particles about 13 nm wide and 650 nm long, has one coat protein of M_r about 34.5 × 10^3, and possesses single-stranded RNA. However, the virus is vectored in a semi-persistent manner by the whitefly *Bemisia tabaci*, has a fairly wide host range, is seed transmitted to a high level in some hosts, and has a rather distinct cytopathology. In addition, the virus seems to be serologically unrelated to any true Carlavirus tested.[46,47] Groundnut crinkle virus (Table 2) is the name given to a virus isolate with properties similar to those of CPMMV[48] and serologically related to it.[49]

II. VIRUS STRUCTURE AND COMPOSITION

A. Particle Structure

The particles of Carlaviruses are slightly flexuous rods often appearing curved to one side, both in negatively stained and shadowed preparations (Figure 1). Generally, Potyviruses are more flexuous and Closteroviruses much more flexuous (see Chapter 14 and 15). The modal lengths within the group range between 610 and 700 nm.[1,2] The width of the particles is 12 to 13 nm. Usually, Carlaviruses, in purified preparations, exhibit clear longitudinal ribbing (Figure 1) due to the visibility of about 4 files of morphological subunits (higher order helices).[1,36,50] Usually the axial canal and cross-banding due to the primary helix are not visible. However, discs of the particles show a central hole indicating the axial canal.[1] Electron micrographs of CLV and RCVMV analyzed by optical diffraction have indicated that the basic pitch of the helix is 3.3 and 3.4 nm, respectively.[50] X-ray diffraction analysis of oriented gels of CLV have given a value for the pitch of 3.45 nm.[51]

Estimates have been made of the number of subunits per turn of the helix for the particles of several Carlaviruses, but the results vary widely. CLV and RCVMV were reported to have 12 and 10 subunits per turn, respectively,[50] while Richter[51] concluded that there were 10 subunits per turn for CLV. Veerisetty[52] suggested that CLV, CPMMV, PeSV, and RCVMV have only 6.8 subunits per turn, although for ALV (now thought to be synonymous with PeSV[4]), he proposed a different structure with 8.2 subunits per turn. It seems likely that all Carlaviruses have the same number of subunits per turn and this number is probably near 10. The untrastructure of Carlavirus particles, while revealing some detail in the electron microscope, have not been distinguished one from another, and the M_r (and hence physical size) of the coat protein subunits is similar for all the viruses for which estimates are available (Table 3).

Although the coat protein sizes of most of the Carlaviruses investigated are of M_r between 31 and 34 × 10^3 (Table 3), the estimates for ALV and PopMV are lower and higher than this, respectively, and should be checked. The only amino acid analysis reported, that of PVS coat protein,[53] indicated the presence of a minimum of 144 amino acids corresponding to M_r about 15.9 × 10^3. Doubling this value would bring it into close agreement with most of the values reported in Table 3.

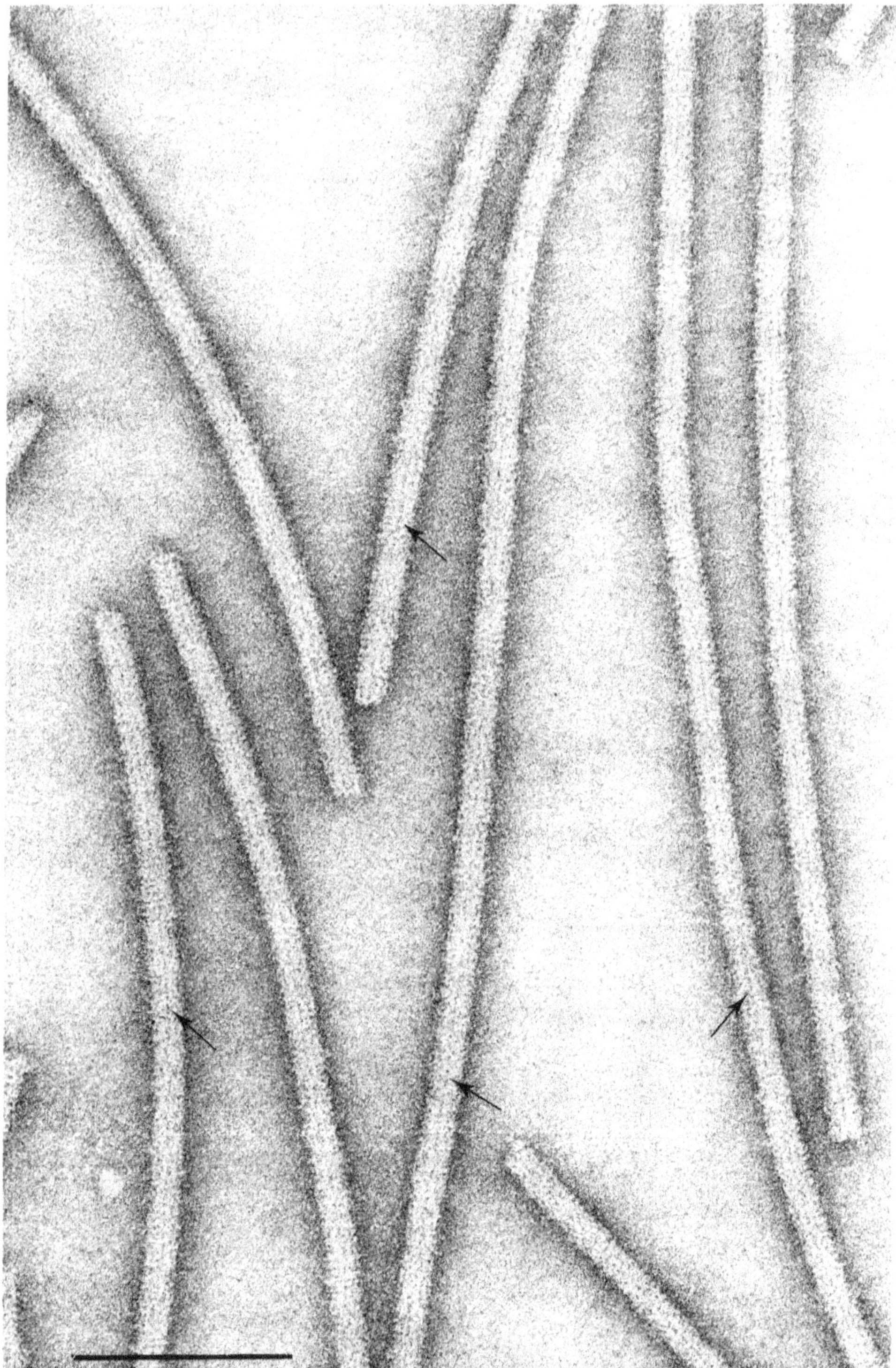

FIGURE 1. Purified carnation latent virus (CLV), stained in uranyl acetate (arrows point to the central canal resolved in some parts of the particles). Bar = 100 nm.

Table 3
CAPSID PROTEIN AND RNA SIZE OF SOME CARLAVIRUSES

Virus	M_r of RNA ($\times 10^6$)	M_r of coat protein ($\times 10^3$)	Ref.
Carnation latent virus (CLV)	ND[a]	32 — 34.8	51,54
Helenium virus S (HeVS)	ND	31	55
Nerine latent virus (NeLV)	ND	32	26
Lonicera latent virus (LLV)	2.8	31	20
Narcissus latent virus (NLV)	ND	32.6	23
Pea streak virus (PeSV)	2.55	33.5	31
= alfalfa latent-virus	2.45	27	31
Poplar mosaic virus (PopMV)	2.3	36.7	36
Red clover vein mosaic virus (RCVMV)	2.6[b]	33.5	31
Hop mosaic virus (HoMV)	3.0	34	13
Hop latent virus (HoLV)	2.9	33	11
American hop latent virus (AHLV)	3.0	33	12
Elderberry carlavirus (ECV)	ND	31	56

[a] ND, not determined.
[b] An additional small RNA component of 12 to 15S was detected in nucleic acid preparations of this virus.[31]

B. The Viral Genome

There is very little published information about the RNAs of Carlaviruses. However, it appears that they contain between 4 and 6% RNA which is single-stranded.[1,2] The few RNA size estimates that have been reported indicate values of M_r between 2.3 and 3.0 (Table 3). RNA preparations of RCVMV appear to contain a small RNA sedimenting at 12 to 15S in addition to the major RNA sedimenting at 39S, but only the 39S RNA was shown to be infectious, indicating that it is positive-sense. The significance of the smaller RNA is not clear. [31] There appear to be no reports on the molecular properties of Carlavirus RNAs or on their messenger activities in vitro.

III. CYTOPATHOLOGY

The cytopathology of most of the Carlaviruses has been examined[34-36,57-66] and reviewed.[67-69] The general conclusion is that Carlavirus-infected cells do not contain any virus-specific inclusions nor do they induce specific cytopathological effects by which infections can be recognized. The particles are found singly or in masses in the cytoplasm of infected cells (Figures 2 and 3), but have not been found in chloroplasts, nuclei or other organelles.

CPMMV, which as noted above, does not appear to be a true Carlavirus, also seems to be confined to the cytoplasm of infected cells. However, the particles aggregate in sheets or bundles, and most frequently, form brush-like inclusions.[46,70]

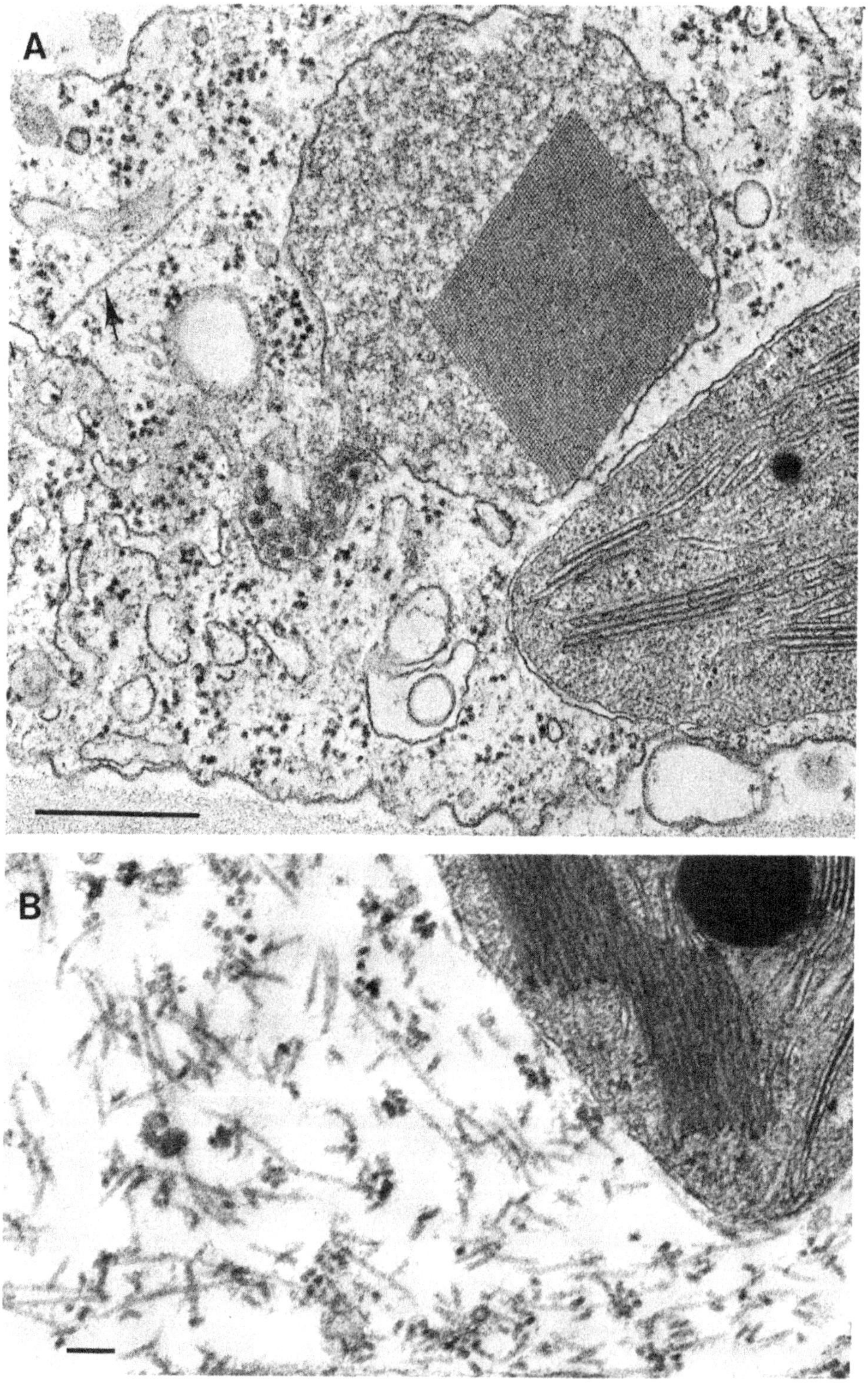

FIGURE 2. Thin section of *Nicotiana megalosiphon* cells infected with poplar mosaic virus (PopMV). A, Cell at an early stage of infection with the cytoplasm rich in ribosomes (arrow points to a virus particle); bar = 500 nm. B, Cell at a later stage of infection with numerous virus particles in the cytoplasm but few ribosomes and no endoplasmic reticulum; bar = 100 nm.

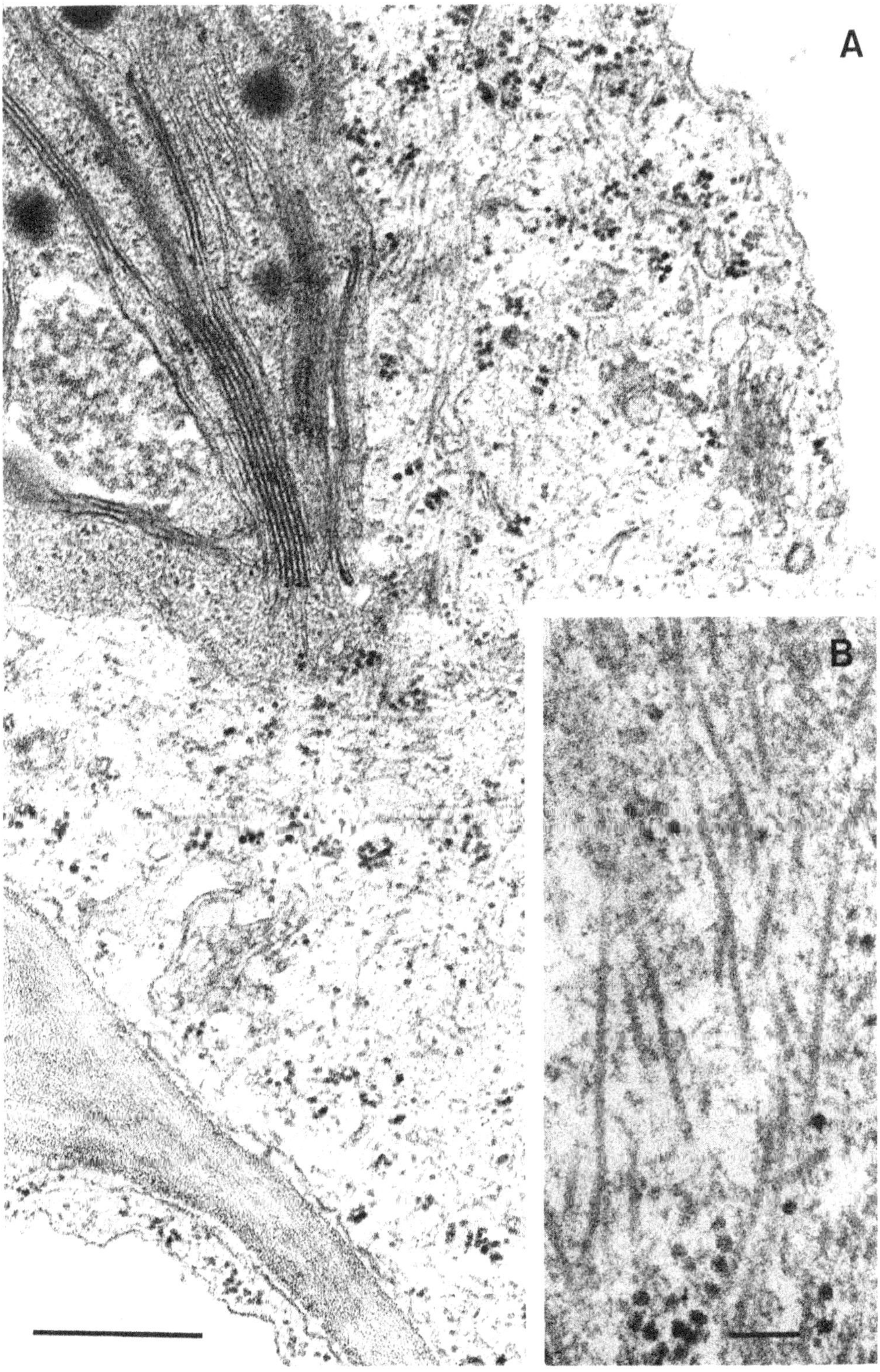

FIGURE 3. Thin sections of *Nicotiana megalosiphon* cells infected with PopMV. A, Cytoplasm of a cell heavily infected with virus; bar = 500 nm. B, Virus particles in the cytoplasm; bar = 100 nm.

REFERENCES

1. **Matthews, R. E. F.,** Classification and nomenclature of viruses. Fourth report of the International Committee on Taxonomy of Viruses, *Intervirology*, 17, 1, 1982.
2. **Wetter, C. and Milne, R. G.,** Carlaviruses, in *Handbook of Plant Virus Infections and Comparative Diagnosis*, Kurstak, E., Ed., Elsevier, Amsterdam, 1981, 695.
3. **Veerisetty, V. and Brakke, M. K.,** Alfalfa latent virus, a naturally occurring Carlavirus in alfalfa, *Phytopathology*, 67, 1202, 1977.
4. **Hampton, R. O.,** Evidence suggesting identity between alfalfa latent and pea streak viruses, *Phytopathology*, 71, 223, 1981 (abstract).
5. **Wetter, C.,** Carnation latent virus, *CMI/AAB Descriptions of Plant Viruses*, No. 61, 1971.
6. **Hollings, M. and Stone, O. M.,** Chrysanthemum virus B, *CMI/AAB Descriptions of Plant Viruses*, No. 110, 1972.
7. **Gumpf, D. J., Osman, F. M., and Weathers, L. G.,** Purification and some properties of a latent virus in *Gynura*, *Plant Dis. Rep.*, 61, 325, 1977.
8. **Johns, L. J.,** Purification and partial characterization of a carlavirus from *Taraxacum officinale*, *Phytopathology*, 72, 1239, 1982.
9. **Dijkstra, J. and van Lent, J. W. M.,** Elderberry carlavirus, *CMI/AAB Descriptions of Plant Viruses*, No. 263, 1983.
10. **Koenig, R. and Lesemann, D. -E.,** Helenium virus S, *CMI/AAB Descriptions of Plant Viruses*, No. 265, 1983.
11. **Barbara, D. J. and Adams, A. J.,** Hop latent virus, *CMI/AAB Descriptions of Plant Viruses*, No. 261, 1983.
12. **Barbara, D. J. and Adams, A. N.,** American hop latent virus, *CMI/AAB Descriptions of Plant Viruses*, No. 262, 1983.
13. **Adams, A. N. and Barbara, D. J.,** Host range, purification and some properties of hop mosaic virus, *Ann. Appl. Biol.*, 96, 201, 1980.
14. **Barbara, D. J., and Adams, A. N.,** Hop mosaic virus, *CMI/AAB Descriptions of Plant Viruses*, No. 241, 1981.
15. **Hearon, S. S.,** A carlavirus from *Kalanchoë blossfeldiana*, *Phytopathology*, 72, 838, 1982.
16. **Waterworth, H. E.,** Purification, serology, and properties of a virus from lilac, *Syringa oblata affinis*, *Plant Dis. Rep.*, 56, 923, 1972.
17. **Allen, T. C.,** Lily symptomless virus, *CMI/AAB Descriptions of Plant Viruses*, No. 96, 1972.
18. **Asjes, C. J., De Vos, N. P., and Van Slogteren, D. H. M.,** Brown ring formation and streak mottle, two distinct syndromes in Lilies associated with complex infections of lily symptomless virus and tulip breaking virus, *Neth. J. Plant Pathol.*, 79, 23, 1973.
19. **Van der Meer, F. A., Maat, D. Z., and Vink, J.,** Lonicera latent virus, a new carlavirus serologically related to poplar mosaic virus: some properties and inactivation in vivo by heat treatment, *Neth. J. Plant Pathol.*, 86, 69, 1980.
20. **Brunt, A. A., Phillips, S., and Thomas, B. J.,** Honeysuckle latent virus, a Carlavirus infecting *Lonicera periclymenum* and *L. japonica (Caprifoliaceae)*, *Acta Hortic.*, 110, 205, 1980.
21. **Tsuchizaki, T.,** Mulberry latent virus isolated from mulberry, *Ann. Phytopathol. Soc. Jpn.*, 42, 304, 1976.
22. **Freitag, J. H. and Milne, K. S.,** Host range, aphid transmission and properties of muskmelon vein necrosis virus, *Phytopathology*, 60, 166, 1970.
23. **Brunt, A. A.,** Some hosts and properties of narcissus latent virus, a Carlavirus commonly infecting narcissus and bulbous iris, *Ann. Appl. Biol.*, 87, 355, 1977.
24. **Brunt, A. A.,** Narcissus latent virus, *CMI/AAB Descriptions of Plant Viruses*, No. 170, 1976.
25. **Maat, D. Z., Huttinga, H., and Hakkaart, F. A.,** Nerine latent virus: some properties and serological detectability in *Nerine bowdenii*, *Neth. J. Plant Pathol.*, 84, 47, 1978.
26. **Brunt, A. A. and Phillips, S.,** Hippeastrum. *Hippeastrum hybridum* Hort., *Glasshouse Crops Res. Inst., Annu. Rep.*, p.144, 1978.
27. **Brandes, J. and Wetter, C.,** Untersuchungen über Eigenschaften und Verwandtschaftsbeziehungen des Latenten *Passiflora*-Virus (*Passiflora latent virus*), *Phytopathol. Z.*, 49, 61, 1963/64.
28. **Bos, L.,** Pea streak virus, *CMI/AAB Descriptions of Plant Viruses*, No. 112, 1973.
29. **Veerisetty, V.,** Alfalfa latent virus, *CMI/AAB Descriptions of Plant Viruses*, No. 211, 1979.
30. **Bos, L., Maat, D. Z., and Markov, M.,** A biologically highly deviating strain of red clover vein mosaic virus, usually latent in pea (*Pisum sativum*), and its differentiation from pea streak virus, *Neth. J. Plant Pathol.*, 78, 125, 1972.
31. **Veerisetty, V. and Brakke, M. K.,** Differentiation of legume carlaviruses based on their biochemical properties, *Virology*, 83, 226, 1977.
32. **Thomas, W., Mohamed, N. A., and Fry, M. E.,** Properties of a Carlavirus causing a latent infection of pepino *(Solanum muricatum)*, *Ann. Appl. Biol.*, 95, 191, 1980.

33. **Biddle, P. G. and Tinsley, T. W.,** Poplar mosaic virus, *CMI/AAB Descriptions of Plant Viruses*, No. 75, 1971.
34. **Atkinson, M. A. and Cooper, J. I.,** Ultrastructural changes in leaf cells of poplar naturally infected with poplar mosaic virus, *Ann. Appl. Biol.*, 83, 395, 1976.
35. **Brunt, A. A., Stace-Smith, R., and Leung, E.,** Cytological evidence supporting the inclusion of poplar mosaic virus in the Carlavirus group of plant viruses, *Intervirology*, 7, 303, 1976.
36. **Boccardo, G. and Milne, R. G.,** Poplar mosaic virus: electron microscopy and polyacrylamide gel analysis, *Phytopathol. Z.*, 87, 120, 1976.
37. **Luisoni, E., Boccardo, G., and Milne, R. G.,** Purification and some properties of an Italian isolate of poplar mosaic virus, *Phytopathol. Z.*, 85, 65, 1976.
38. **Van der Meer, F. A., Maat, D. Z., and Vink, J.,** Poplar mosaic virus: purification, antiserum preparation, and detection in poplars with the enzyme-linked immunosorbent assay (ELISA) and with infectivity tests on *Nicotiana megalosiphon*, *Neth. J. Plant Pathol.*, 86, 99, 1980.
39. **Wetter, C.,** Potato virus M, *CMI/AAB Descriptions of Plant Viruses*, No. 87, 1972.
40. **Wetter, C.,** Potato virus S, *CMI/AAB Descriptions of Plant Viruses*, No. 60, 1971.
41. **Varma, A.,** Red clover vein mosaic cirus, *CMI/AAB Descriptions of Plant Viruses*, No. 22, 1970.
42. **Bos, L., Huttinga, H., and Maat, D. Z.,** Shallot latent virus, a new Carlavirus, *Neth. J. Plant Pathol.*, 84, 227, 1978.
43. **Van der Meer, F. A., Maat, D. Z., and Vink, J.,** Lonicera latent virus, a new carlavirus serologically related to poplar mosaic virus. some properties and inactivation *in vivo* by heat treatment, *Neth. J. Plant Pathol.*, 86, 69, 1980.
44. **Adams, A. N. and Barbara, D. J.,** Host range, purification and some properties of two carlaviruses from hop (*Humulus lupulus*): hop latent and American hop latent, *Ann. Appl. Biol.*, 101, 483, 1982.
45. **Wetter, C., Quantz, L., and Brandes, J.,** Vergleichende Untersuchungen über das Rotkleeadernmosaik-Virus und das Erbsenstrichel-Virus, *Phytopathol. Z.*, 44, 151, 1962.
46. **Brunt, A. A., Atkey, P. T., and Woods, R. D.,** The occurrence properties and intracellular location of cowpea mild mottle virus, *Proc. 4th Int. Congr. Plant Pathol., Melbourne*, Abstr., No. 443, 1983.
47. **Brunt, A. A. and Kenten, R. H.,** Cowpea mild mottle virus, *CMI/AAB Descriptions of Plant Viruses*, No. 140, 1974.
48. **Dubern, J. and Dollet, M.,** Groundnut crinkle, a new virus disease observed in Ivory Coast, *Phytopathol. Z.*, 95, 279, 1979.
49. **Fauquet, C.,** personal communication, 1983.
50. **Varma, A., Gibbs, A. J., Woods, R. D., and Finch, J. T.,** Some observations on the structure of the filamentous particles of several plant viruses, *J. Gen. Virol.*, 2, 107, 1968.
51. **Richter, W.,** Untersuchungen über das Carnation Latent Virus Und über den aufban seiner Proteinhülle, Thesis, University of Saarbrücken, West Germany, 1976.
52. **Veersitty, V.,** Relationships among structural parameters of virions of helical symmetry, *Virology*, 84, 523, 1978.
53. **Tremaine, J. H. and Goldsack, D. E.,** The structure of regular viruses in relation to their subunit amino acid composition, *Virology*, 35, 227, 1968.
54. **Paul, H. L.,** SDS polyacrylamide gel electrophoresis of virion proteins as a tool for detecting the presence of virus in plants, *Phytopathol. Z.*, 80, 330, 1974.
55. **Kuschki, G. H., Koenig, R., Düvel, D., and Kühne, H.,** Helenium virus S and Y — two new viruses from commercially grown *Helenium* hybrids, *Phytopathology*, 68, 1407, 1978.
56. **Van Lent, J. W. M., Wit, A. J., and Dijkstra, J.,** Characterization of a Carlavirus in elderberry (*Sambucus* spp.), *Neth. J. Plant Pathol.*, 86, 117, 1980.
57. **Lyons, A. R. and Allen, T. C.,** Electron microscopy of viruslike particles associated with necrotic fleck of *Lilium longiflorum*, *J. Ultrastruct. Res.*, 27, 198, 1969.
58. **Bos, L. and Rubio-Huertos, M.,** Intracellular accumulation of *Passiflora* latent virus in *Chenopodium quinoa*, *Neth. J. Plant Pathol.*, 77, 145, 1971.
59. **Bos, L. and Rubio Huertos, M.,** Light and electron microscopy of pea streak virus in crude sap and tissues of pea *(Pisum sativum)*, *Neth. J. Plant Pathol.*, 78, 247, 1972.
60. **Tu, J. C. and Hiruki, C.,** Ultrastructure of potato infected with potato virus M, *Virology*, 42, 238, 1970.
61. **Tu, J. C. and Hiruki, C.,** Electron microscopy of cell wall thickening in local lesions of potato virus-M infected red kidney bean, *Phytopathology*, 61, 862, 1971.
62. **Hiruki, C. and Tu, J. C.,** Light and electron microscopy of potato virus M lesions and marginal tissue in red kidney bean, *Phytopathology*, 62, 77, 1972.
63. **De Bokx, J. A. and Waterreus, H. A. J. I.,** Electron microscopy of plant tissues infected wtih potato viruses A and S, *Neth. J. Plant Pathol.*, 77, 106, 1971.
64. **Hiruki, C. and Shukla, P.,** Intracellular location of potato virus S in leaf tissue of *Chenopodium quinoa*, *Can. J. Bot.*, 51, 1699, 1973.

65. **Shukla, P. and Hiruki, C.,** Ultrastructural changes in leaf cells of *Chenopodium quinoa* infected with potato virus S. *Physiol. Plant Pathol.*, 7, 189, 1975.
66. **Biddle, P. G. and Tinsley, T. W.,** Poplar mosaic virus in Great Britain. *New Phytol.*, 70, 61, 1971.
67. **Martelli, G. P. and Russo, M.,** Plant virus inclusion bodies. *Adv. Virus Res.*, 21, 175, 1977.
68. **Christie, R. G. and Edwardson, J. R.,** Light and electron microscopy of plant virus inclusions. *Fla. Agric. Exp. Sta. Monogr. Ser.*, 9, 1977, 150 pp.
69. **Edwardson, J. R. and Christie, R. G.,** Use of virus-induced inclusions in classification and diagnosis. *Annu. Rev. Phytopathol.*, 16, 31, 1978.
70. **Brunt, A. A., Atkey, P. T., and Woods, R. D.,** Intracellular occurrence of cowpea mild mottle virus in two unrelated plant species. *Intervirology*, 20, 137, 1983.

Chapter 14

POTYVIRUS GROUP

This, the largest group of plant viruses, derives its name from potato virus Y (PVY), the type member.[1] It includes many viruses which are poorly characterized and hence are of uncertain relationship to each other. However, varying degrees of serological relationship have been reported among more than 30 apparently distinct group members.[2]

Potyviruses have flexuous rod-shaped particles between 680 and 900 nm long and about 12 nm in diameter, sedimenting at about 150S. Each particle is constructed with helical symmetry from a single species of protein of M_r between 32 and 36 $\times$ 10^3 encapsidating a single molecule of positive-sense, single-stranded RNA of M_r between 3.0 and 3.5 $\times$ 10^6. The viruses induce very characteristic inclusions in the cytoplasm of infected cells, usually referred to as cylindrical inclusions or pinwheels.[3-6]

Most Potyviruses have fairly restricted host ranges and are transmitted non-persistently by aphids and in some cases through seeds. They can also be transmitted by mechanical inoculation.[4-6]

I. MEMBERS OF THE GROUP AND THEIR RELATIONSHIPS

In 1959, Brandes and Wetter[7] suggested the establishment of a number of groups of plant viruses with elongated particles based on their modal lengths and certain other characters. One of the proposed groups included PVY and thirteen other viruses. By 1971, the group was defined more precisely and the name Potyvirus was suggested.[8] A description of the group appeared in the first report of the ICTV with 12 definite and 13 possible members.[9] The membership of the group has grown by leaps and bounds over the years (Table 1) but judging by the discrepancies in the lists prepared by various authors, there remain differences of opinion about the exact membership of the group.

With so many members and possible members of the Potyvirus group and the continual description of new candidates (Table 1), it seems imperative to define satisfactory taxonomic parameters for the group. Firstly, the minimum essential characters for assigning viruses to the group must be agreed and secondly, suitable criteria must be defined which will distinguish viruses from strains. Although the first requirement is probably easier, even here all difficulties have not been resolved.

It is generally agreed that to qualify for inclusion in the Potyvirus group, a virus isolate must have particles with the characteristic morphology and be able to induce typical cylindrical inclusions in the cytoplasm of infected cells.[1] Most viruses with these properties are transmitted non-persistently by aphids, and this property is also considered by some as essential for inclusion in the group.[6] However, there are some viruses with Potyvirus-like particles and which induce Potyvirus-like inclusions but which are transmitted by soil fungi, mites, or whiteflies (Table 2). At present they are considered by the ICTV as possible members of the group.[1] None of these viruses appear to be serologically related to any of the aphid-transmitted Potyviruses. It seems that their taxonomic position will not be resolved until more data about them become available and agreement is reached about the significance of vector specificity as a Potyvirus group characteristic. In relation to the latter point, it should be noted that there are other groups of viruses whose members are transmitted by different types of vectors; for example the Geminivirus group and the Rhabdoviridae (see Chapters 3 and 5, Volume I).

There is another instance where it has been difficult to determine if a virus should be included in the Potyvirus group. Maclura mosaic virus (MaMV) is transmitted by aphids in

Table 1
INCREASE IN THE NUMBER OF DESCRIBED POTYVIRUSES WITH TIME

	Number of viruses[a]			
Year	Definite members	Possible members	Total members[a]	Ref.
1959	14	0	14	7
1965	20	0	20	10
1971[b]	12	13	25	9
1974	—	—	92[c]	4
1976[b]	34	11	45	11
1979[b]	38	35	73	12
1981[d]	34	51	85	5
1982[b]	48	56	104	1

[a] Does not include viruses with demonstrated vectors other than aphids.
[b] Successive Reports of the ICTV.
[c] Viruses were not distinguished as definite and possible members.
[d] In another review,[6] the same authors listed 60 viruses as members of the Potyvirus group, 32 as probable strains of the members or viruses of uncertain relationship to the members, and 8 as possible members (a total of 100 virus isolates).

Table 2
VIRUSES WITH POTYVIRUS-LIKE PARTICLES AND CYTOPATHOLOGY BUT NON-APHID VECTORS

Virus[a]	Modal length (nm)	Vector	Ref.
Barley yellow mosaic virus (BaYMV)[b]	275 and 550[c]	Fungus	13
Oat mosaic virus (OMV)	600—750	Fungus	14
= soilborne oat mosaic virus			
Rice necrosis mosaic virus (RNMV)	275 and 550[c]	Fungus	15
Wheat spindle streak mosaic virus (WSSMV)	275—300 and 600—625[c]	Fungus	16,17
Wheat yellow mosaic virus (WYMV)[b]	275—300 and 575—600	Fungus	18
Agropyron mosaic virus (AgMV)[d]	~717	Mite	19
= agropyron green mosaic virus			
= agropyron streak mosaic virus			
= agropyron yellow mosaic virus			
= couch grass streak mosaic virus			
Hordeum mosaic virus (HorMV)[d]	650—750	Mite	20
Ryegrass mosaic virus (RGMV)	703	Mite	21
= ryegrass streak mosaic virus			
= ryegrass streak virus			
Wheat streak mosaic virus (WSMV)[d]	700	Mite	22
Sweet potato mild mottle virus (SPMMV)	~950	Whitefly	23

[a] Generally accepted names are given together with synonyms.
[b] Viruses vectored by the fungus *Polymyxa graminis* and serologically related but whose degree of relationship is uncertain.
[c] Particle length distribution appears to be bimodal.
[d] Viruses vectored by mites and serologically related but only remotely so.

a non-persistent manner and induces typical Potyvirus-like cell inclusions but has particles with modal length of only 672 nm. Furthermore, the particles are a little thicker than those of Potyvirus and the virus does not appear to be serologically related to any of a number of viruses tested within the Potyvirus, Carlavirus, and Closterovirus groups.[24,25] At present MaMV is looked upon as a possible Potyvirus[1] and its exact taxonomic position needs clarification.

Difficulties regarding the delineation of viruses within the Potyvirus group are immense. They stem from the apparently inexhaustible, continuous variation within the group and from technical difficulties associated with purifying the viruses in sufficient amounts for meaningful physical, chemical, and serological studies.[5] Consequently, it is very difficult to decide which of all the isolates being described warrant consideration as distinct viruses and which as virus strains. The use of differences in their physical, chemical, antigenic, and cytopathological properties for taxonomic purposes have been considered and it appears that some of these are more useful than others.[3-6,26-31]

Although considerable variation has been reported in the modal lengths of Potyviruses,[3] it seems unlikely that most of the differences are real and of taxonomic significance. For those viruses whose modal lengths have been determined repeatedly, the values obtained vary considerably (Table 3). At least some of this variation stems from measurement and calibration errors[27] but it has been shown that the length of Potyvirus particles is dependent on the composition of suspending media.[27,32] Nevertheless, some differences in the reported modal lengths appear to be significant, even between closely related viruses. For example, it has been demonstrated that very closely related virus isolates of bean yellow mosaic virus (BYMV) as judged by serological cross-reactivity and coat protein amino acid composition had significantly different lengths when examined using identical methods.[27]

The antigenic properties of Potyviruses have been used extensively for their differentiation but there are a number of contradictory reports regarding some relationships.[3] It seems that there are a number of factors concerned with the use of serology for comparing Potyviruses which could lead to spurious results. The most obvious are the use of antisera of widely different titers, and different assay techniques.

Although much work on the antigenic properties of Potyviruses has been done with intact virus particles as antigen, some has also been done with dissociated viral proteins. The latter method has been especially popular when using immunodiffusion tests in agar to avoid problems with the poor diffusion of the intact particles in gels.[73-76] Reactions with intact particles and their dissociated proteins appear to involve different populations of antibodies and it has been shown that antisera raised to intact virus particles can react poorly or not at all with dissociated virus, and vice versa.[26,77] It has been observed that antisera to dissociated Potyvirus proteins are better at recognizing more remote serological relationships among the viruses than antisera to intact particles.[77] Furthermore, it has been demonstrated that there are differences in the specificity of antisera raised against the protein of a Potyvirus prepared by different chemical procedures.[78]

Another complication associated with serological studies of Potyviruses may arise from the instability of their coat proteins. It has been reported that during purification and storage of virus preparations, the coat protein subunits may undergo partial proteolysis.[76,79-83] It seems likely that coat proteins in different state of digestion will differ in their antigenic specificity.

A potentially useful method for the differentiation of Potyviruses is the comparison of the RNA nucleotide sequence homologies as determined by the amount of annealing between the RNA of one virus isolate with the complementary DNA prepared to another.[29] The exact value of this method will only become apparent when more data become available. However, preliminary results with BYMV and clover yellow vein virus (ClYVV) RNAs indicate that molecular hybridization tests may constitute a better method for discriminating between Potyvirus isolates than serology.[29,84]

Table 3
MEMBERS OF THE POTYVIRUS GROUP

Virus[a]	Modal length (nm)	Ref.
Potato virus Y (PVY)	730 (705—759)[b]	33
= tobacco vein-banding virus		
= potato virus C		
Bean common mosaic virus (BCMV)	750 (738—751)	26,27,34
= mungbean mosaic virus		
Bean yellow mosaic virus (BYMV)	750 (743—842)	26,27,35
= pea mosaic virus		
= lupin mottle virus		
Bearded iris mosaic virus (BIMV)	750 (753—761)	36
Beet mosaic virus (BtMV)	730 (688—739)	37
Bidens mottle virus (BiMV)	720 (710—730)	38
Carnation vein mottle virus (CVMV)	790 (735—790)	39
Carrot thin leaf virus (CTLV)	736	40
Celery mosaic virus (CeMV)	780 (750—792)	41
Clover yellow vein virus (ClYVV)	760 (670—767)	42
= pea necrosis virus		
Cocksfoot streak virus (CSV)	752 (740—760)	43
cocksfoot mosaic virus		
Cowpea aphid-borne mosaic virus (CABMV)	750	44
= cowpea common mosaic virus		
= Azuki bean mosaic virus		
Dasheen mosaic virus (DaMV)	750	45
Henbane mosaic virus (HMV)	800 (724—850)	46
= atropa mild mosaic virus		
Hippeastrum mosaic virus (HiMV)	750	47
= amaryllis mosaic virus		
Iris mild mosaic virus (IMMV)	760	48
= iris mosaic virus		
= iris latent mosaic virus		
Leek yellow stripe virus (LYSV)	820	49
Lettuce mosaic virus (LMV)	750 (746—778)	26,27,50
Narcissus yellow stripe virus (NYSV)	755 (750—800)	51
Onion yellow dwarf virus (OYDV)	775 (772—800)	52
Papaya ringspot virus (PRSV)	800 (780—800)	53
Parsnip mosaic virus (ParMV)	736	54
Passionfruit woodiness virus (PWV)	750 (670—737)	26,27,55
Pea seed-borne mosaic virus (PSMV)	770 (750—770)	56
= pea leafroll mosaic virus		
= pea fizzletop virus		
Peanut mottle virus (PeMoV)	745 (704—812)	57
= peanut mild mosaic virus		
= peanut severe mosaic virus		
= groundnut mottle virus		
Pepper mottle virus (PepMoV)	737	58
= arizona pepper virus		
Pepper veinal mottle virus (PVMV)	770 (750—770)	59
Plum pox virus (PPV)	764 (700—764)	60
= sharka virus		
Pokeweed mosaic virus (PokMV)	776	61
= phytolacca decandra mosaic virus		
= potato virus P		
Potato virus A (PVA)	730 (730—750)	62
= potato mild mosaic virus		
Soybean mosaic virus (SoyMV)	750 (685—750)	63
Sugarcane mosaic virus (SCMV)	750 (700—784)	64,65
= abaca mosaic virus		
= grass mosaic virus		

Table 3 (continued)
MEMBERS OF THE POTYVIRUS GROUP

Virus[a]	Modal length (nm)	Ref.
= maize dwarf mosaic virus		
= sorghum red stripe virus		
Tobacco etch virus (TEV)[b]	730 (715—780)	66
= datura virus Z		
= tomato etch virus		
Tobacco vein mottling virus (TVMV)	765	67
Tulip breaking virus (TBV)	740 (700—750)	68
= tulip mosaic virus		
= lily mottle virus		
Turnip mosaic virus (TuMV)	720 (680—800)	69
= cabbage black ringspot virus		
= horseradish mosaic virus		
Watermelon mosaic virus (WMV)[c]	750 (720—765)	70-72

[a] Generally accepted names are given together with synonyms.
[b] Values are those considered the most trustworthy, and those in brackets are the ranges within which values reported by different authors fall.[3]
[c] This name is confusing because there are two distinct viruses for which the same name has been used referred to as WMV 1 and WMV 2.[71]

Different Potyviruses induce a variety of cytopathic effects whose differences have been used for taxonomic purposes. Edwardson and Christie[3,30,31] have suggested that Potyviruses could be divided into three subgroups based on the morphology of their cytoplasmic cylindrical (pinwheel) inclusions whereas Moghal and Francki[27] concluded that only two would be more appropriate. Other cytopathological features resulting from Potyvirus infection, such as the presence and morphology of nuclear inclusions, can also be useful characters for distinguishing certain viruses within the group.

The above discussion indicates the difficulties of deciding whether a given Potyvirus isolate should be considered as a distinct virus or a strain. On the information available in the literature, we have concluded that the 37 aphid-transmitted Potyviruses listed in Table 3 should be considered as definitive members of the group. Many of them are serologically related[3] but the relationships are rather remote. Many Potyviruses described and named in the literature appear to be so similar to those listed in Table 3 that they can be considered as their strains or isolates. Their names are also included in Table 3 under the viruses to which they appear to be synonymous. Also, many of the isolates listed in Table 4 as possible members of the Potyvirus group are probably synonymous to some of the viruses listed in Table 3. For example, there has been considerable confusion about the identity of virus isolates initially described as cowpea aphid-borne mosaic virus (CABMV) and blackeye cowpea mosaic virus (BlCMV). BlCMV was described in 1955[85] as a distinct virus but was thought at one time to be a strain of BYMV.[86] In 1966, CABMV was described[87] as a new virus but was later considered as possibly synonymous with BlCMV.[88] It now appears that although the two viruses can be confused, their antigenic and biological properties are sufficiently dissimilar for them to be considered as distinct viruses.[89,90] However, it has been shown that BlCMV is serologically closely related to bean common mosaic virus (BCMV) and that the two could be looked on as strains of the same virus.[89] Thus we have included BlCMV as a possible Potyvirus (Table 4) because of insufficient data to conclude whether it is significantly different from BCMV.

Although the main problem in Potyvirus taxonomy is that many new viruses are described which later turn out to be synonymous with already existing ones, there are also some viruses whose recognized strains are so different that they should probably be considered as separate

Table 4
POSSIBLE MEMBERS OF THE POTYVIRUS GROUP[a]

Virus	Ref.
Alliaria mosaic	93
Alstroemeria mosaic	94
Amaranthus leaf mottle	
Anthoxanthum mosaic	
Anthurium	
Aquilegia	
Araujia mosaic	
Artichoke latent	
Asystasia gangetica	95
Bean western mosaic	
Bidens mosaic	
Blackeye cowpea mosaic	
Brinjal mild mosaic	96
Bromus mollis	97
Bryonia mottle	
Canavalia maritima mosaic	
Canna mosaic	
Carrot mosaic	
Carthamus latent	
Cassia yellow blotch	98
Celery latent	
Celery yellow mosaic	
Clover (Croatian)	
Colombian datura	
Commelina mosaic	
Crinum	
Daffodil mosaic	
Daphne Y	
Datura 437	
Datura mosaic	
Datura shoestring	
Dendrobium mosaic	
Desmodium mosaic	
Dioscorea green banding (= yam mosaic)	
Dioscorea trifida	
Dipsacus fullonum	
Dock mottling mosaic	
Eucharis mosaic	
Euphorbia ringspot	
Fern	
Freesia mosaic	
Freesia streak	
Garlic yellow streak	
Gloriosa stripe mosaic	
Groundnut eyespot	
Guar symptomless	
Guinea grass mosaic	
Helenium Y	
Holcus streak	
Hungarian datura innoxia mosaic	99
Hyacinth mosaic	
Iris fulva mosaic	
Iris severe mosaic	
Iris yellow mosaic	
Isachne mosaic	
Kennedya Y	
Lilium speciosum streak mottle	
Maclura mosaic	
Malva vein clearing	
Marigold mottle	
Melilotus mosaic	100
Mungbean mottle	
Narcissus degeneration	
Narcissus late season yellows (= jonquil mild mosaic)	
Narcissus white streak	
Nerine	
New burley tobacco	
Nanhoscordum mosaic	
Oat necrotic mottle	
Ornithogalum mosaic	
Palm mosaic	
Passionfruit ringspot	
Pepper mild mottle	101
Pepper severe mosaic	
Perilla mottle	102
Peru tomato	103
Pigweed mosaic	
Plantain 7	104
Poa palustris mosaic	
Populus	105
Primula mosaic	
Rape mosaic	
Red clover	
Reed canary mosaic	
Spartina mottle	106
Statice Y	
Sunflower	
Sweet potato A	
Sweet potato feathery mottle	
Sweet potato russet crack	
Sweet vetch crinkly mosaic	
Teasel mosaic	
Tigrida latent	
Tigrida mosaic	
Tamarillo mosaic	
Tradescantia	
Ullucus mosaic	107
Vallota mosaic	108
Vicia faba virus	
Wheat streak	
White bryony mosaic	
Wild potato mosaic	
Wisteria vein mosaic	
Zucchini yellow fleck[b]	109
Zucchini yellow mosaic[b]	110

[a] Data obtained from reviews[1,3,5,6] and research papers where individual references are quoted.

[b] Both viruses reported from Italy independently at about the same time. They have been sufficiently well characterized to indicate that they are not synonymous with any of the viruses listed in Table 3.

viruses. For example, it is now generally recognized that two distinct viruses have been described as watermelon mosaic virus (WMV); they are sometimes referred to as WMV 1 and WMV 2. These viruses have been shown to be serologically distinct, to induce different cytopathic effects, and to differ significantly in their host ranges.[3,71,72,76] Furthermore, their RNAs appear to have no detectable base sequence homology.[91] Similarly, a number of sugarcane mosaic virus (SCMV) strains have been described, some of which differ so much from each other that they probably merit the status of distinct viruses.[3,65,92]

Table 4 lists all the remaining viruses which have been described as members or possible members of the Potyvirus group. Some have been characterized relatively thoroughly but their assignment as definite members of the group should wait until it is confirmed that they are not synonymous with any of the viruses listed in Tables 2 and 3. Others will need to be characterized much more thoroughly before their taxonomic status can be assessed.

II. VIRUS STRUCTURE AND COMPOSITION

A. Particle Structure

The morphology of Potyvirus particles is probably much more uniform than one would deduce from the published reports.[3] (Table 3). They are flexuous rods about 11 to 12 nm wide and although the modal lengths can vary between 680 and 900 nm, most members of the group have particles about 750 nm long. Most negatively stained preparations of the particles show little fine structure (Figure 1). However, in some preparations particles with narrow central canals have been seen [5] and optical diffraction studies indicate that the protein subunits are arranged in a helix with a pitch of 3.3 to 3.5 nm.[111]

Much of the reported variation in Potyvirus modal lengths must stem from the instability of the particles and from preparative artifacts. The particle lengths of several viruses have been shown to depend largely on the presence or absence of divalent metal ions in the suspending medium.[27,32,112] For example, particles of pepper veinal mottle virus (PVMV) were flexuous and 750 nm long when extracted in EDTA solution but straight and 850 nm long in $MgCl_2$ solution.[32] Also particles of cocksfoot streak virus (CSV) were longer and straighter in extracts from plants grown under conditions of high levels of Mg^{2+}.[112] Most of the reported differences in the width of Potyvirus particles are probably due to problems in measurment because of differences in the thickness and distribution of the negative stain around the particles. It has been claimed that particles of some Potyviruses swell or partially disrupt in certain negative stains.[113,114] However, the possibilities that the increased width of the particles is due to stretching of the supporting film, flattening of the particles, or movement of the particles during drying, cannot be ruled out.

Potyvirus particles can be dissociated into coat protein subunits and RNA by relatively mild treatments such as exposure to 2 *M* $LiCl_2$ or $CaCl_2$.[26,115] Such preparations have been repolymerized to form filamentous structures resembling virus particles but were found to be noninfectious.[116]

Polyacrylamide-gel electrophoretic studies on dissociated Potyvirus coat proteins indicate that each virus is constructed from a single protein species. Minor, faster migrating protein components are sometimes observed in the gels but they appear to be degradation products[26,78,80,81,83] although in some cases, heterogeneiety of protein preparations may also be due to variation in the disulfide-bridge patterns of the protein.[78] The estimated size of apparently undegraded Potyvirus coat proteins range in M_r from about 32×10^3 to more than 40×10^3.[65,79] Although much of this variation may be due to the methods used in different laboratories, some differences between viruses and virus strains may also exist.[65] Earlier estimates of the coat protein sizes of some Potyviruses, derived from their amino acid compositions and peptide maps, were somewhat lower, and were almost certainly underestimates.[13,117-124]

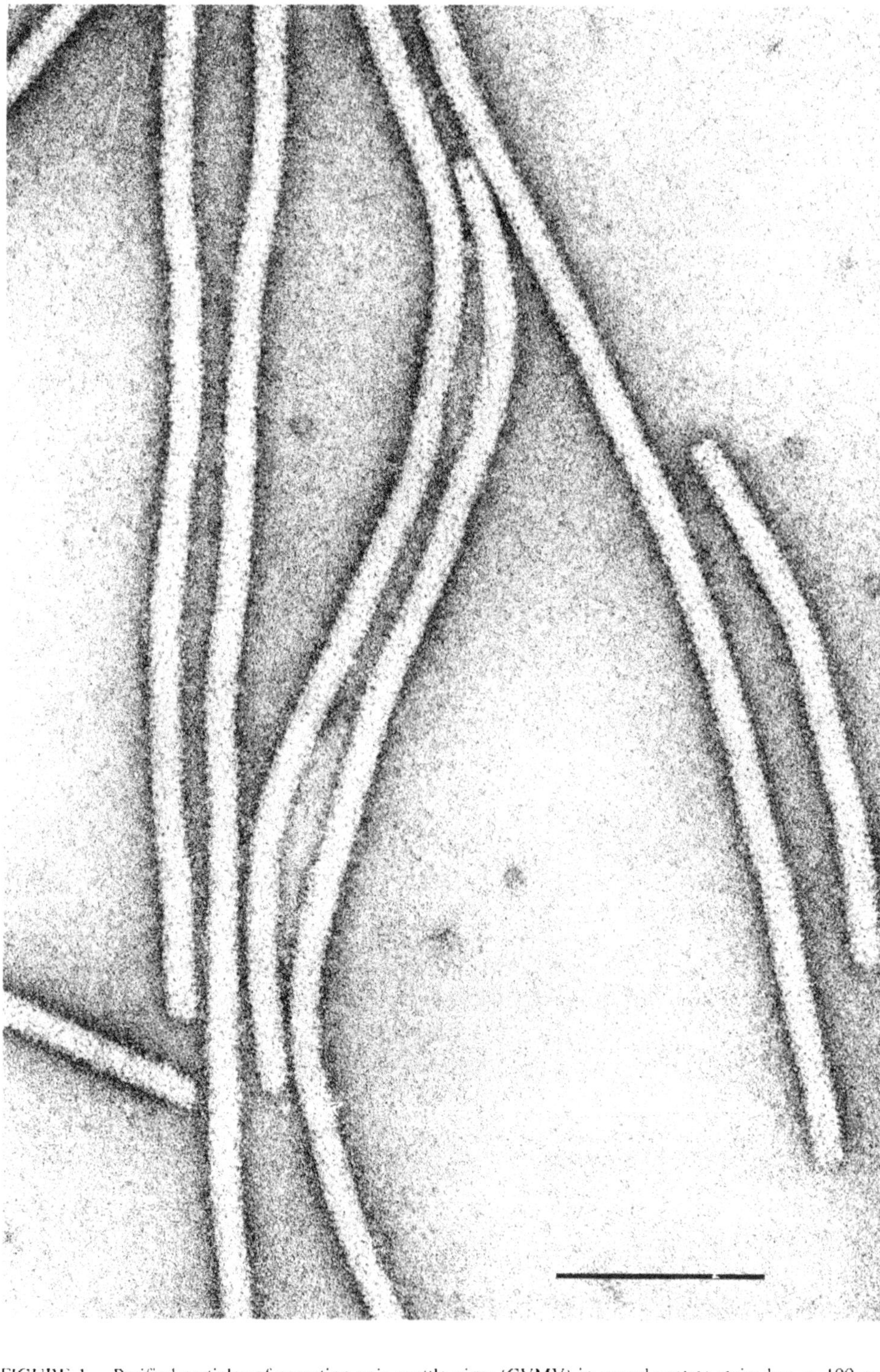

FIGURE 1. Purified particles of carnation vein mottle virus (CVMV) in uranyl acetate stain; bar = 100 nm.

The sizes of some Potyvirus RNAs have been estimated from electrophoretic mobilities in polyacrylamide gels and sedimentation rates; they range from M_r about 2.9 to 3.5 × 10^6.[122-125] From the RNA and coat protein sizes and their arrangement[111] it would seem that each Potyvirus particle, which sediments at about 150S,[5] consists of a single RNA molecule coated by 1700 to 2000 protein subunits with a total M_r of 60 to 70 × 10^6.[6]

B. The Viral Genome

Of the Potyviruses whose RNAs have been isolated, that of tobacco etch virus (TEV) appears to have been the most thoroughly studied.[125,126] Preparations of TEV RNA have been separated into two fractions, those whose 3′ ends contain poly A tracts of various lengths and those that do not. However, both fractions were shown to be equally infective.[125,126] It was also demonstrated that TEV RNA has a protein (VPg) of M_r about 6 × 10^3 covalently linked to its 5′ end.[126] It will be interesting to see if the RNAs of other Potyviruses have similar properties.

The RNAs of three Potyviruses, TEV, pepper mottle virus (PepMoV), and tobacco vein mottling virus (TVMV), have been expressed in in vitro translation systems.[127-133] Dougherty and Hiebert[127-129] detected and identified four specific proteins in rabbit reticulocyte lysates programmed with TEV or PepMoV RNA as messengers. Viral coat protein, two virus-specific nuclear inclusion, and one cytoplasmic cylindrical (pinwheel) inclusion proteins were identified by immunoprecipitation. However, some larger proteins were also precipitated by the antisera indicating that they were read-through products. On the basis of these experiments a genetic map was proposed for the two viruses.[129] It was suggested that the genome consists of six adjacent genes coding from the 5′ terminus for proteins of M_r about 78 to 87 × 10^3 (unidentified), 49 × 10^3 (nuclear inclusion protein 1),41 to 50 × 10^3 (unidentified), 68 to 70 × 10^3 (cytoplasmic cylindrical inclusion protein), 54 to 56 × 10^3 (nuclear inclusion protein 2), and 30 to 33 × 10^3 (viral coat protein).

When Hari and his colleagues[132,133] translated highly purified, full-length TEV RNA in the rabbit reticulocyte and wheat germ in vitro systems, they detected only one polypeptide. The product was of M_r about 39 × 10^3 in the reticulocyte lysate and 41 × 10^3 in the wheat germ extract. They suggested that there may be a terminator which is recognized in the reticulocyte but not in the wheat germ system. However, they detected a number of subgenomic RNAs in purified preparations of TEV and some additional ones in infected tissues with messenger activities for a variety of protein products which, however, were not identified.[133] They concluded that the translational strategy of TEV involves the selective transcription of genomic RNA into subgenomic RNAs some of which can be encapsidated.[133] The data on TEV RNA translation from the two laboratories appears inconsistent and needs clarification.

In vitro translation studies on TVMV RNA in reticulocyte lysates revealed that the RNA coded for a number of proteins.[130,131] The RNA used as messenger was purified by sucrose density-gradient centrifugation but was not checked for the presence of subgenomic RNAs. The translation products were identified with appropriate antisera as proteins of the viral coat, cytoplasmic cylindrical (pinwheel) inclusions, two nuclear inclusions, and the helper component required for aphid transmission of the virus.[131] Specificity of immunoprecipitation was shown by competing with the various proteins to which antisera were raised and by sequentially precipitating with two different antisera. It was concluded that the TVMV genome consists of at least five different nonoverlapping coding regions.[131]

III. CYTOPATHOLOGY

Infection of plant cells by Potyviruses induces characteristic changes in the cytoplasm and sometimes also in the nuclei. These ctyopathic effects have been studied in plants infected by many Potyviruses including those which have vectors other than aphids (Table 2) and a vast literature on the subject has accumulated over the past two decades. However, all the investigations have been primarily descriptive and little effort appears to have been made to reveal the composition and functions of the wide variety of intricate structures associated with infection by this group of viruses. The field has been reviewed periodically.[3,5,30,134-136]

A. Cylindrical Inclusions in the Cytoplasm

All Potyviruses induce complex structures in the cytoplasm of infected cells, unlike those associated with infection by viruses of any other group. These structures have been referred to be various terms such as ''dense bands'', ''curled inclusions'', ''pinwheels'', ''bundles'', ''laminated inclusions'', and ''circular inclusions'' to describe them when viewed in two dimensions. When referring to them as three-dimensional structures, terms such as ''plate-like inclusions'', ''tubes'', ''scrolls'', ''cylindrical inclusions'', and ''lamellar inclusions'' have been used.[137-141]However, the most commonly used terms are ''cylindrical inclusions'' when referring to them as three-dimensional structures; ''pinwheels'', ''scrolls'', or ''laminated inclusions'', depending on their shape when viewed in transverse section; and ''bundles'' in longitudinal section.

The detailed structure of cytoplasmic inclusions has been deduced from conventional ultrastructural studies including serial sectioning, freeze-etch electron microscopy and computer-assisted analytic geometry supported by data from electron microscopy of thick sections.[142-144] The inclusions appear to consist of a central core tubule from which radiate a series of plates with a finely striated substructure with a periodicity of 5 nm (Figures 2 and 3). In some cases the plates are larger and more curved than in others (compare pinwheel inclusions in Figures 3, 4, and 5). In embedded material, cylindrical inclusions appear as pinwheels in transverse section and as ''bundles'' in longitudinal sections (Figures 2 and 5). Sometimes the plates are not attached to the pinwheel cores,and then may appear in transverse section as ''laminated inclusions''; several such inclusions, when closely associated, have been referred to as ''laminated aggregates'' (Figures 6, 7A and 8A). When the plates are curled or rolled up they give the appearance of ''scrolls'' or ''tubes'' (Figure 9). In three-dimensions, pinwheel inclusions are usually cylindrical[138] (Figures 2B and 5A) but may be conicol.[142] It has also been suggested that they may be elliptic hyperboloids.[144] In most instances, however, they appear to be more or less cylindrical.

Cylindrical inclusions induced by different Potyviruses vary considerably; this led Edwardson[3] to use the differences for subdividing the Potyviruses into three subgroups. He assigned viruses which induce scroll-like inclusions to subdivision I, those inducing laminated aggregates to subdivision II, and viruses inducing both scrolls and laminated aggregates to subdivision III. In our experience the cytopathology of subdivision II viruses is reasonably easy to recognize (Figures 4A, 7, and 8). We have had much more difficulty in distinguishing viruses which should be assigned to subgroups I and III. For example, we were unable to distinguish convincingly between the morphology of cytoplasmic inclusions induced by BCMV and passionfruit woodiness virus (PWV) which Edwardson[3] assigned to subgroups I and III, respectively. These viruses also appear to be distantly related serologically.[26] There are a number of other Potyviruses which have been shown to be serologically related and yet are assigned to different subgroups on the basis of their cylindrical inclusion morphology.[3] For example BCMV (Subgroup I), BYMV (Subgroup II), and soybean mosaic virus (SoyMV, Subgroup III) have all been reported to be serologically related.[3,26] Furthermore, it would appear that on the basis of cylindrical inclusion morphology, different strains of some viruses fall into different subgroups; as would strains of ryegrass mosaic virus (RGMV)[145] and those of SCMV.[3] It seems that although cylindrical inclusion morphology may be of use in helping

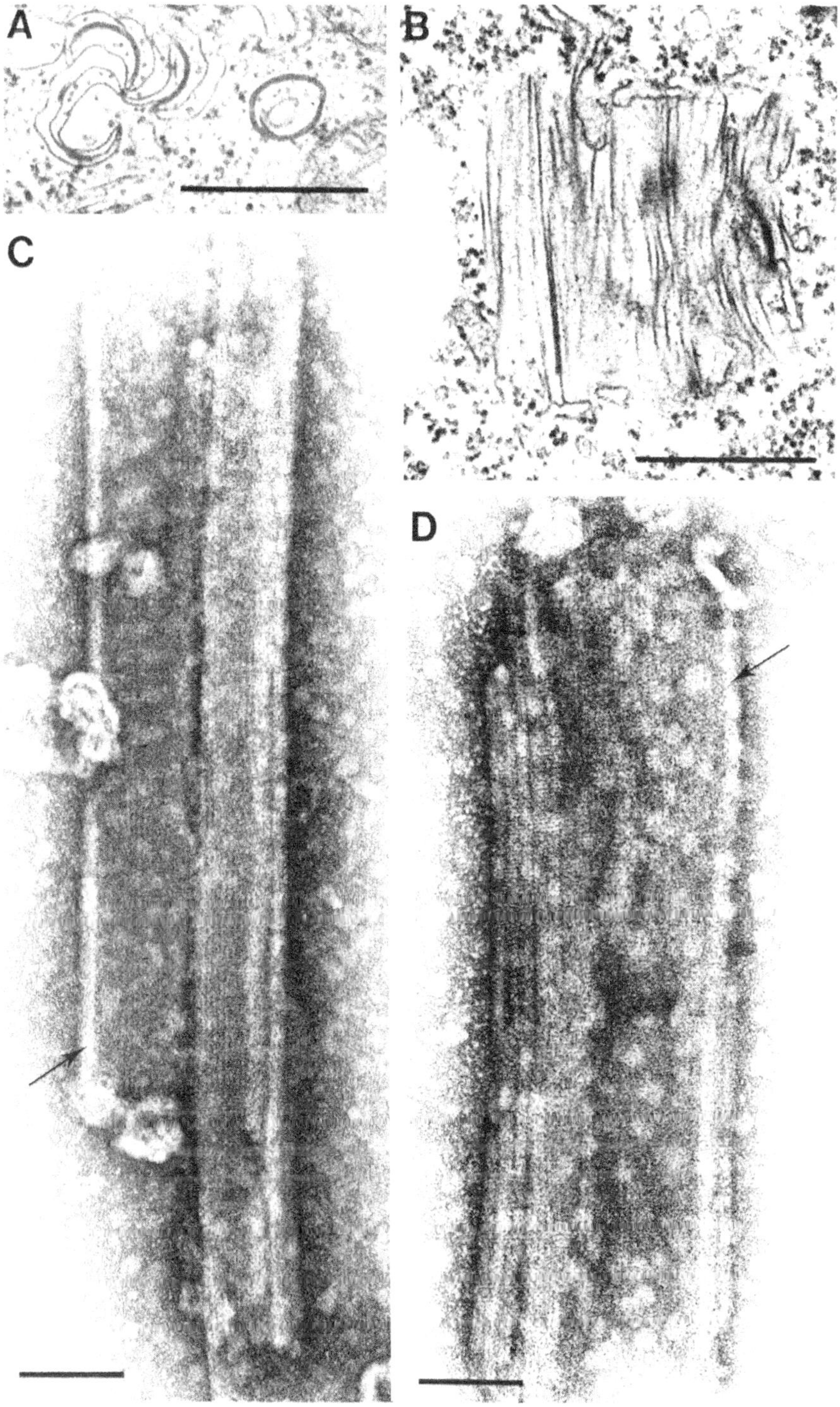

FIGURE 2. The structure of Potyvirus-induced cylindrical inclusions. A, Transverse section of potato virus Y (PVY) induced inclusions in the cytoplasm of a *Nicotiana glutinosa* leaf cell seen in transverse sections as a pinwheel (left) and a scroll (right). B, Similar structures to those in (A) but in longitudinal section seen as a bundle. Bars = 500 nm. C and D, Cylindrical inclusions isolated from *Bryonia cretica* infected with an unidentified Potyvirus showing the plate-like structures whose surfaces and edges (arrows) can be seen; uranyl acetate staining. Bars = 100 nm.

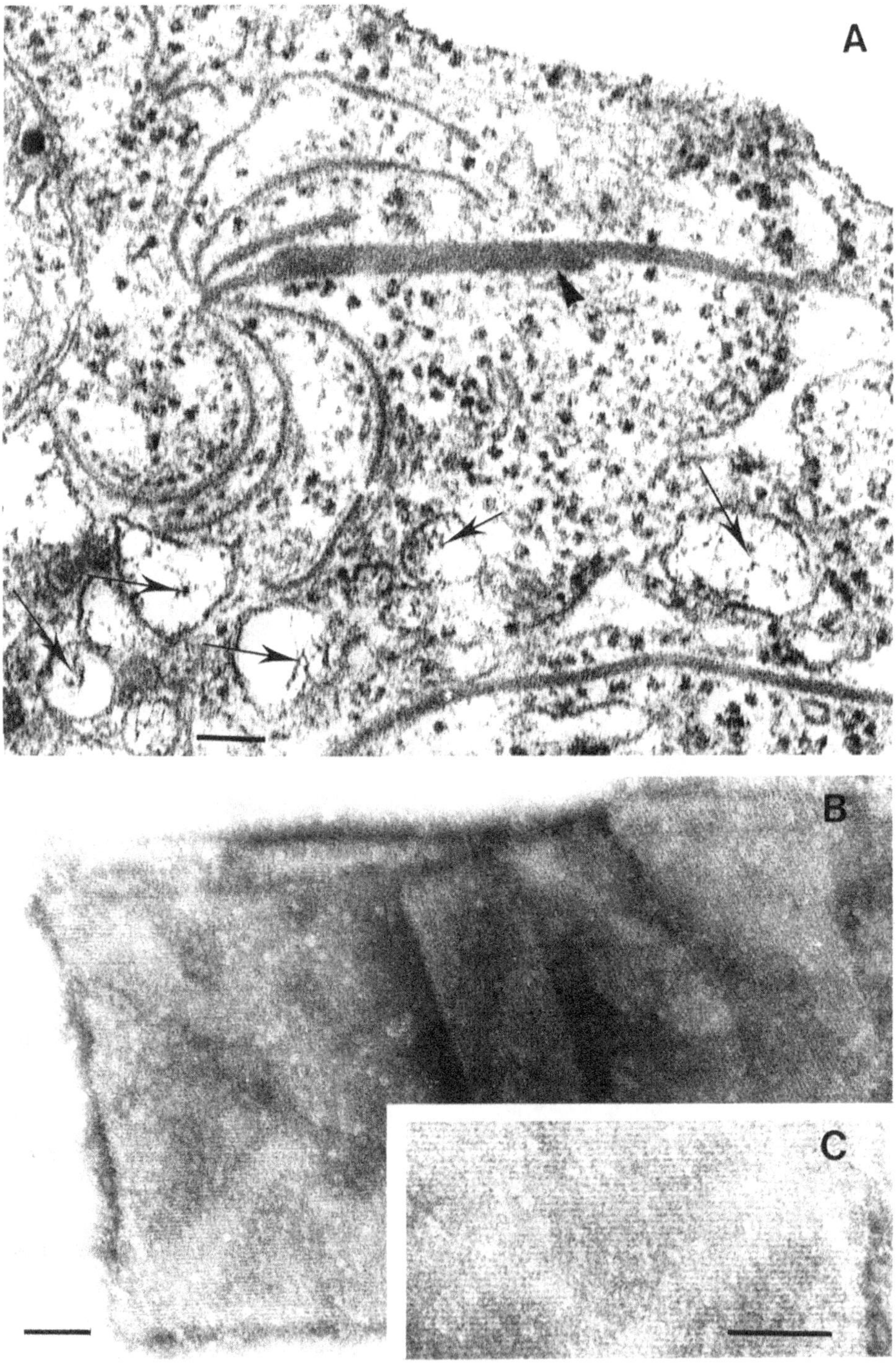

FIGURE 3. A, Thin section of a *Beta vulgaris* (silverbeet) leaf cell infected with beet mosaic virus (BtMV) showing a cylindrical inclusion cut transversely with radiating plates, one of which has other plates attached to it (arrowhead). Vesicles containing fibrils reminiscent of nucleic acid (arrows) can also be seen. B and C, Cylindrical inclusion plates isolated from BtMV-infected leaves and stained with uranyl acetate, showing their finely striated surface structure. Bars = 100 nm. (Courtesy of Mr. C. Davies, Department of Plant Pathology, Biochemistry, University of Adelaide.)

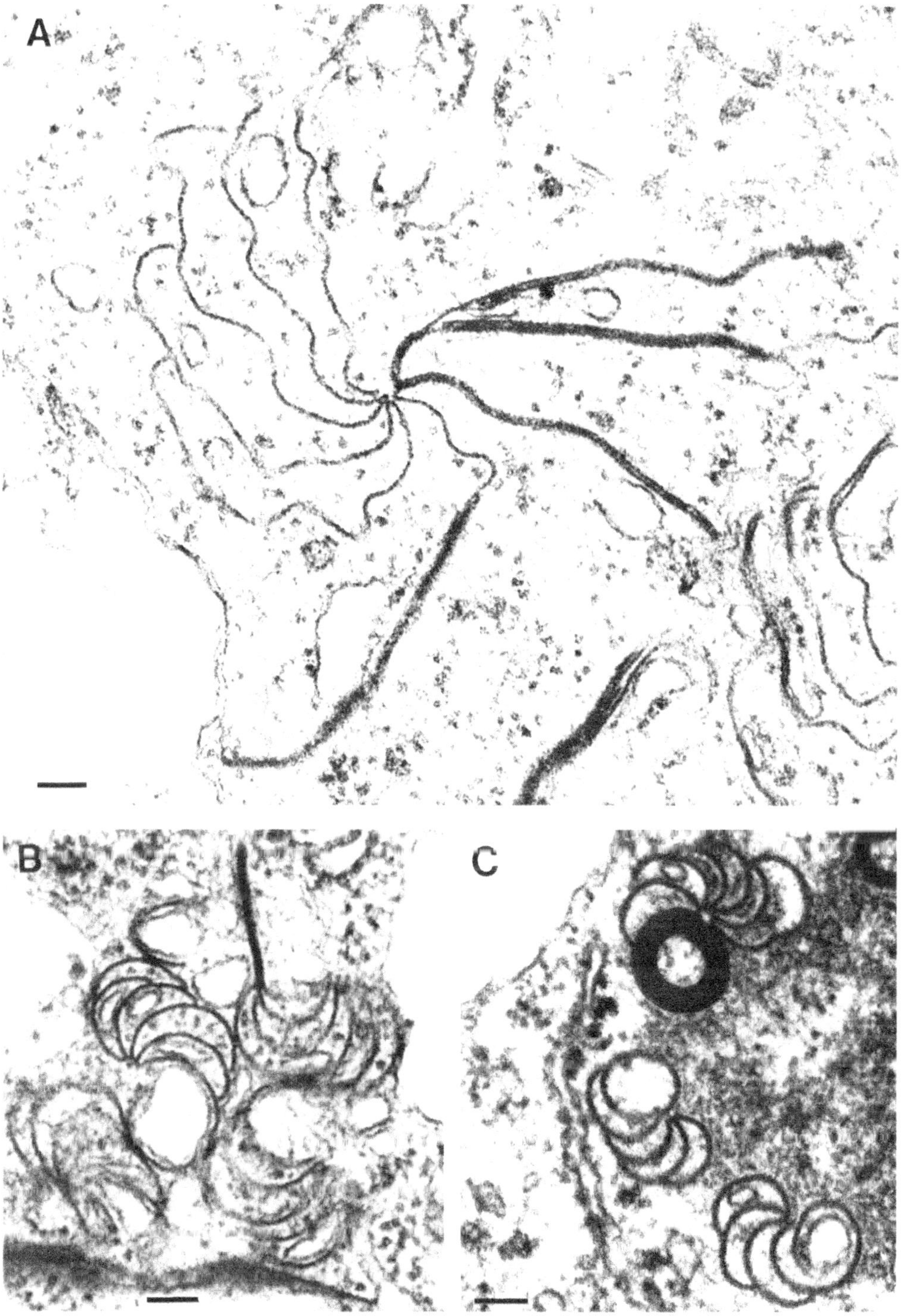

FIGURE 4. Thin sections of cylindrical inclusions cut transversely (pinwheels) in leaf cells of A, *Vicia faba* (broad bean) infected with bean yellow mosaic virus (BYMV); B, *Zea mays* (maize) infected with sugarcane mosaic virus (SCMV); C, *Phaseolus vulgaris* (French bean) infected with bean common mosaic virus (BCMV). Bars = 100 nm.

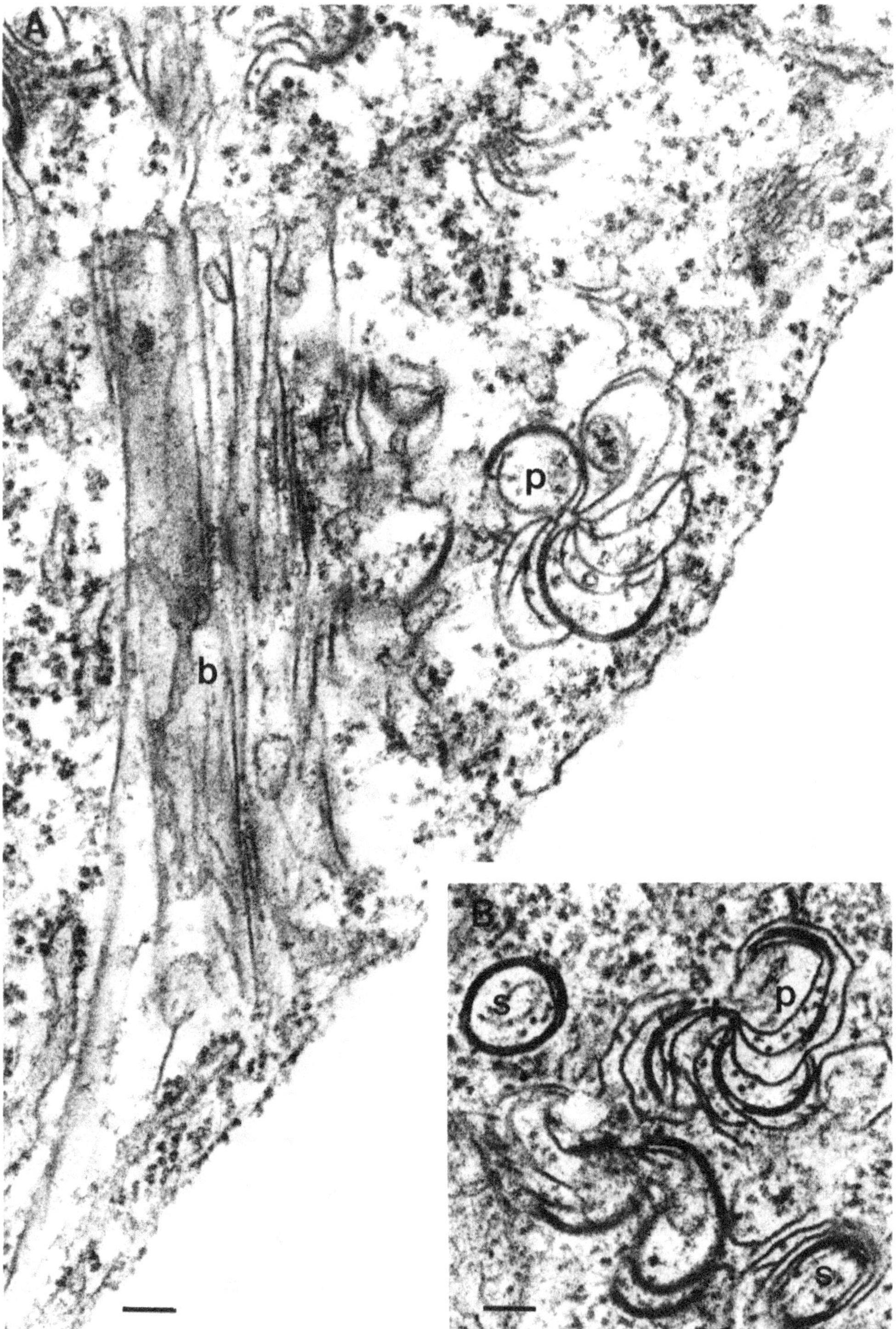

FIGURE 5. Thin sections of cylindrical inclusions in the cytoplasm of *Nicotiana glutinosa* leaf cells infected with PVY. A, Inclusions in longitudinal section appearing as a bundle (b) and in transverse section as a pinwheel (p); B, a pinwheel (p) and scrolls (s). Bars = 100 nm.

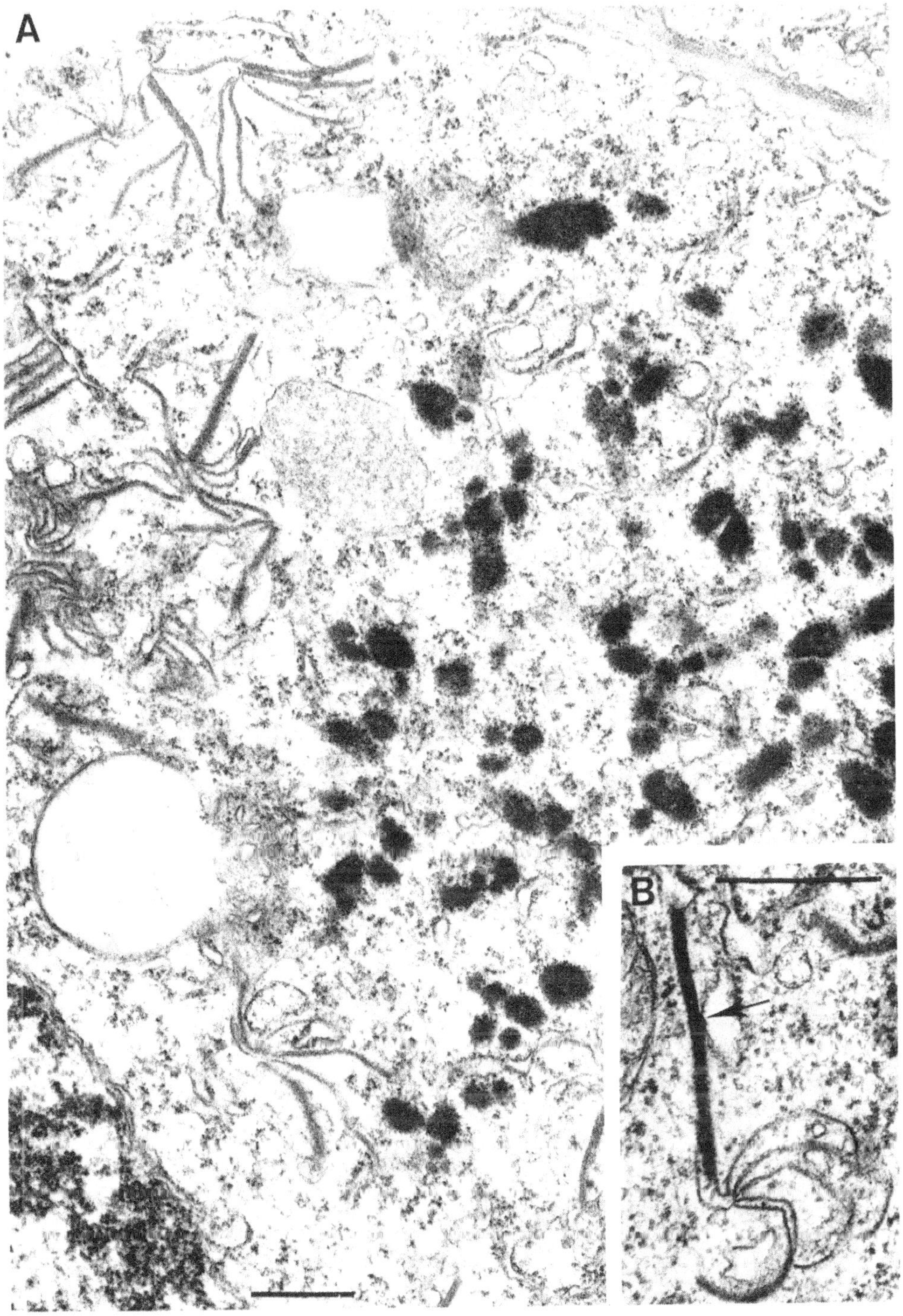

FIGURE 6. Thin sections of *Vicia faba* (broad bean) leaf cells infected with BYMV. A, Numerous cylindrical inclusions mostly in transerve section and densely stained virus-specific crystalline inclusions scattered throughout the cytoplasm; B, a pinwheel to which is attached a laminated aggregate (arrow). Bars = 500 nm.

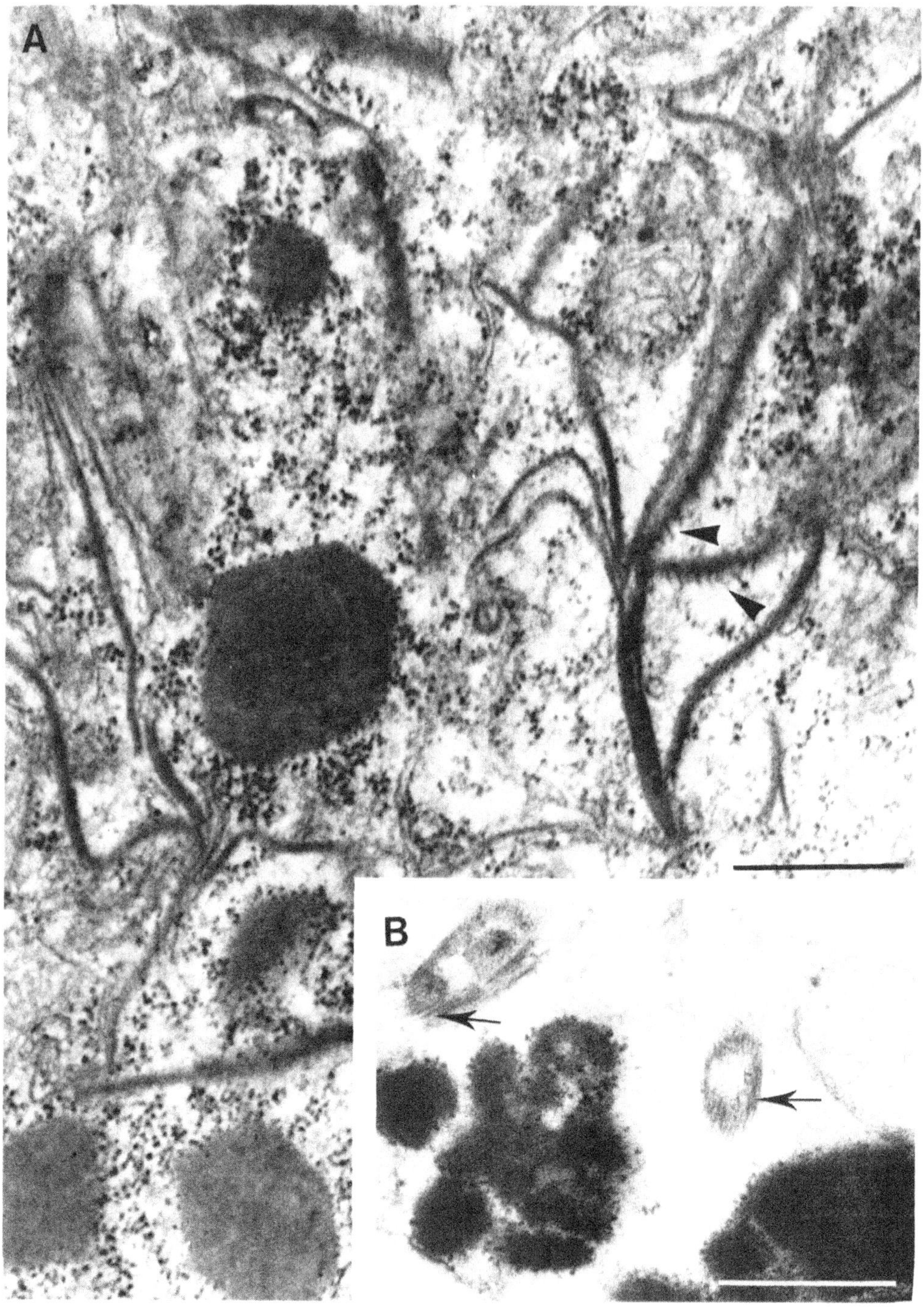

FIGURE 7. Thin sections of *Pisum arvense* (sweet pea) leaf cells infected with BYMV. A. A relatively young cell with numerous cylindrical inclusions and densely stained virus-specific crystalline inclusions in cytoplasm rich in ribosomes (arrowheads point to plates of cylindrical inclusion cut slightly obliquely and showing attached virus particles). (Courtesy of Dr. S. M. Moghal, Department of Plant Pathology, Agricultural Research Institute, Tandojam, Sind, Pakistan.) B. An older cell with ribosomes surrounding only the virus-specific crystalline inclusions, and uniseriate bundles of virus particles cut obliquely (arrows). Bars = 500 nm.

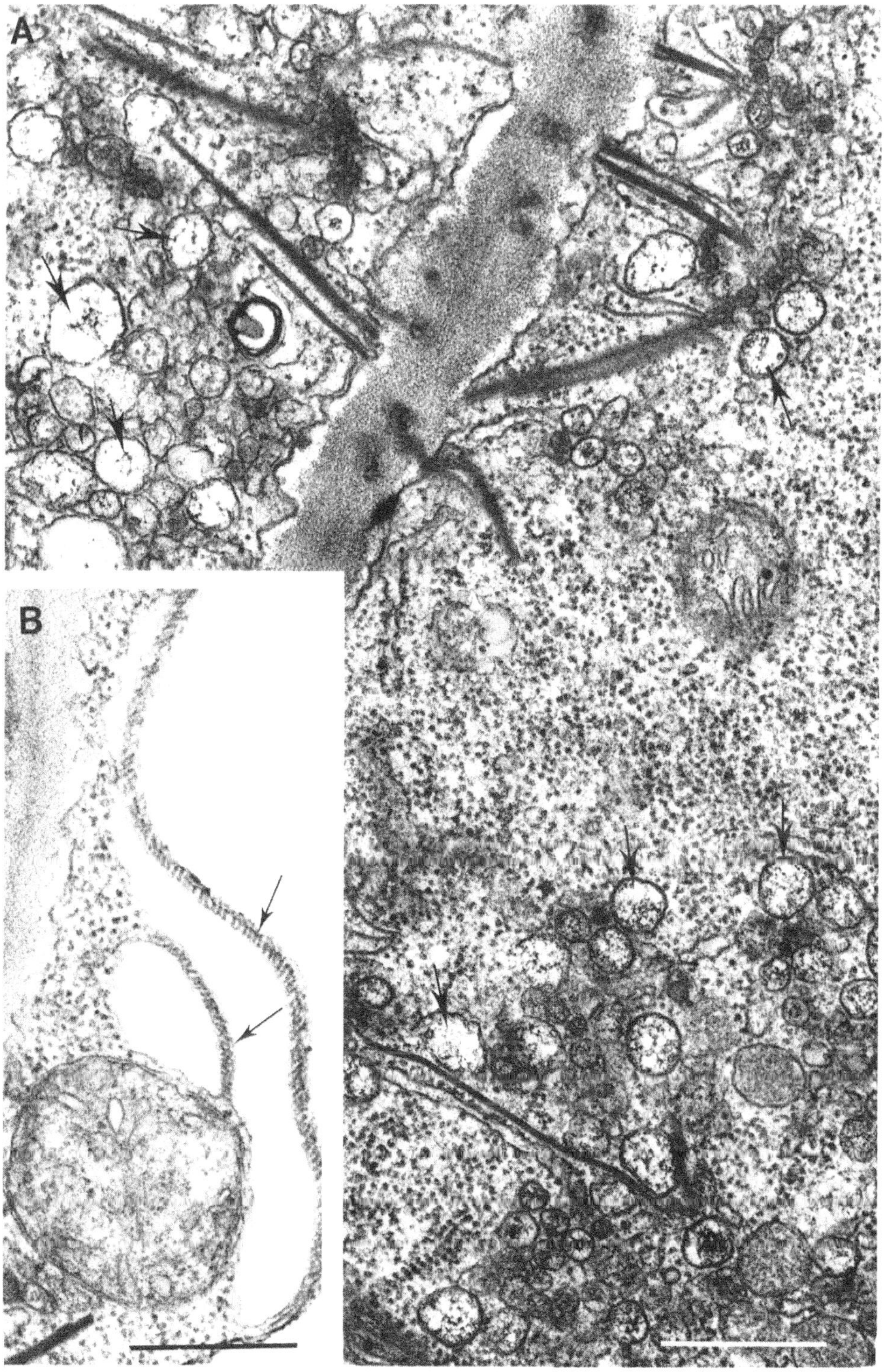

FIGURE 8. Thin sections of *Beta vulgaris* (silver beet) leaf cells infected with BtMV. A, Showing cylindrical inclusions in the cytoplasm near the cell wall with plasmodesmata and numerous vesicles with fibrils reminiscent of nucleic acid (arrows); B, vacuolated cytoplasm with uniseriate rows of virus particles cut slightly obliquely (arrows) adjacent to the tonoplast. Bars = 500 nm.

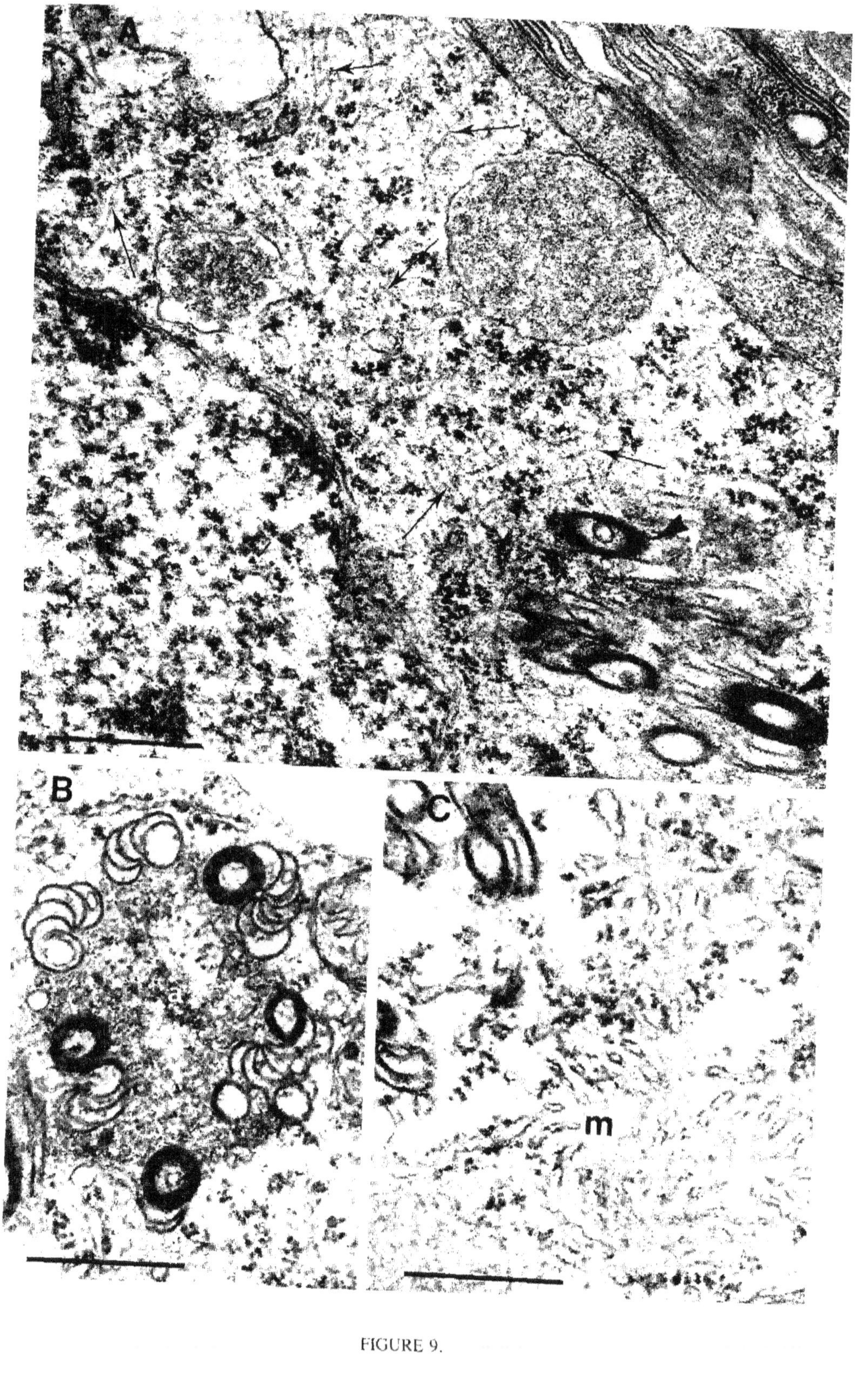

FIGURE 9.

to identify some individual Potyviruses, their use for hierarchical subdivisions within the group could be hazardous.

In Potyvirus-infected cells, cylindrical inclusions develop during early stages of infection at the cell periphery where they increase in size and number. They usually originate in association with the plasma membrane near plasmodesmata (Figure 8A).[142,146] The central tubules of the cylindrical inclusions are located directly over the plasmodesmata and it seems that their membranes may be continuous from one cell to another. The cores and plates of the inclusions extend into the cytoplasm as the inclusions grow and later are released from the plasmodesmata into the cytoplasm. At later stages of infection, large masses of cylindrical inclusions appear scattered throughout the cytoplasm (Figures 6, 10, and 11). Virus particles may be intimately associated with the pinwheel arms at all times (Figure 7A), but especially at early stages of infection.[142] Later, tubular and laminate rather than the characteristic pinwheel inclusions predominate. It seems that they arise by detachment of the arms from pinwheels.[147]

Cylindrical inclusions (mostly fragmented) have been purified from plants infected by several Potyviruses and their chemical and antigenic properties have been examined.[79,82,148] Such preparations contained only one protein of M_r about 67 to 70 $\times$ 10^3 when analyzed by polyacrylamide gel electrophoresis.[79] This is about twice the size of the viral coat protein (see Section II.A) indicating that the two proteins are distinct. This conclusion is supported by serological data which established that the inclusion protein is antigenically unrelated to its respective viral coat protein, or to any host plant proteins.[148,149] It has been shown that proteins of inclusions induced by different viruses are antigenically distinct, although in some cases they are related.[150] These observations indicate that cylindrical inclusion proteins are virus coded, a conclusion also supported by experiments in which it was established that one of the products of Potyvirus RNA translation in vitro is a protein with chemical and antigenic properties of the inclusion protein (see Section II.B).

It is interesting that in spite of the overwhelming evidence that cylindrical inclusions are the products of the Potyvirus genome, similar structures have been observed in apparently virus-free plant cells. Wilson and his colleagues[151,152] observed pinwheel inclusions like those associated with Potyvirus infection in carrot tissue cultured in vitro. All efforts to demonstrate the presence of virus in the tissue were unsuccessful and some evidence was obtained indicating that the inclusions developed in response to specific cultural conditions.[152] These observations raise interesting questions regarding the origin and function of pinwheel inclusions in plant cells which, however, remain unanswered.

B. Other Cytoplasmic Effects

Whereas the presence of cylindrical inclusions is the most conspicuous and universal effect of Potyvirus infection, a number of other structures have been reported to be associated with infection by some viruses of the group.[3,5,30,136] Unfortunately, the literature concerning these inclusions can be confusing because of the large number of descriptive terms used. Sometimes different structures are referred to by similar names and in other cases, different terms are used to describe similar structures.

In spite of the spectacular ultrastructural changes in the cytoplasm of Potyvirus-infected cells, it is unusual for any of the organelles other than nuclei to be visibly affected. Perhaps the most interesting exception is the large number of crystals observed in the peroxisomes of PVY-infected cells (Figure 10).[3,153]

Infection by Potyviruses usually results in greatly increased vacuolation and vesiculation

FIGURE 9. Thin sections of *Phaseolus vulgaris* (French bean) leaf cells infected with bean common mosaic virus (BCMV). A, Showing virus particles scattered in the cytoplasm (arrows) and cylindrical inclusions cut in oblique section (arrowheads); B, an amorphous body (a) surrounded by cylindrical inclusions appearing as pinwheels and scrolls; C, a membranous body (m). Bars = 500 nm.

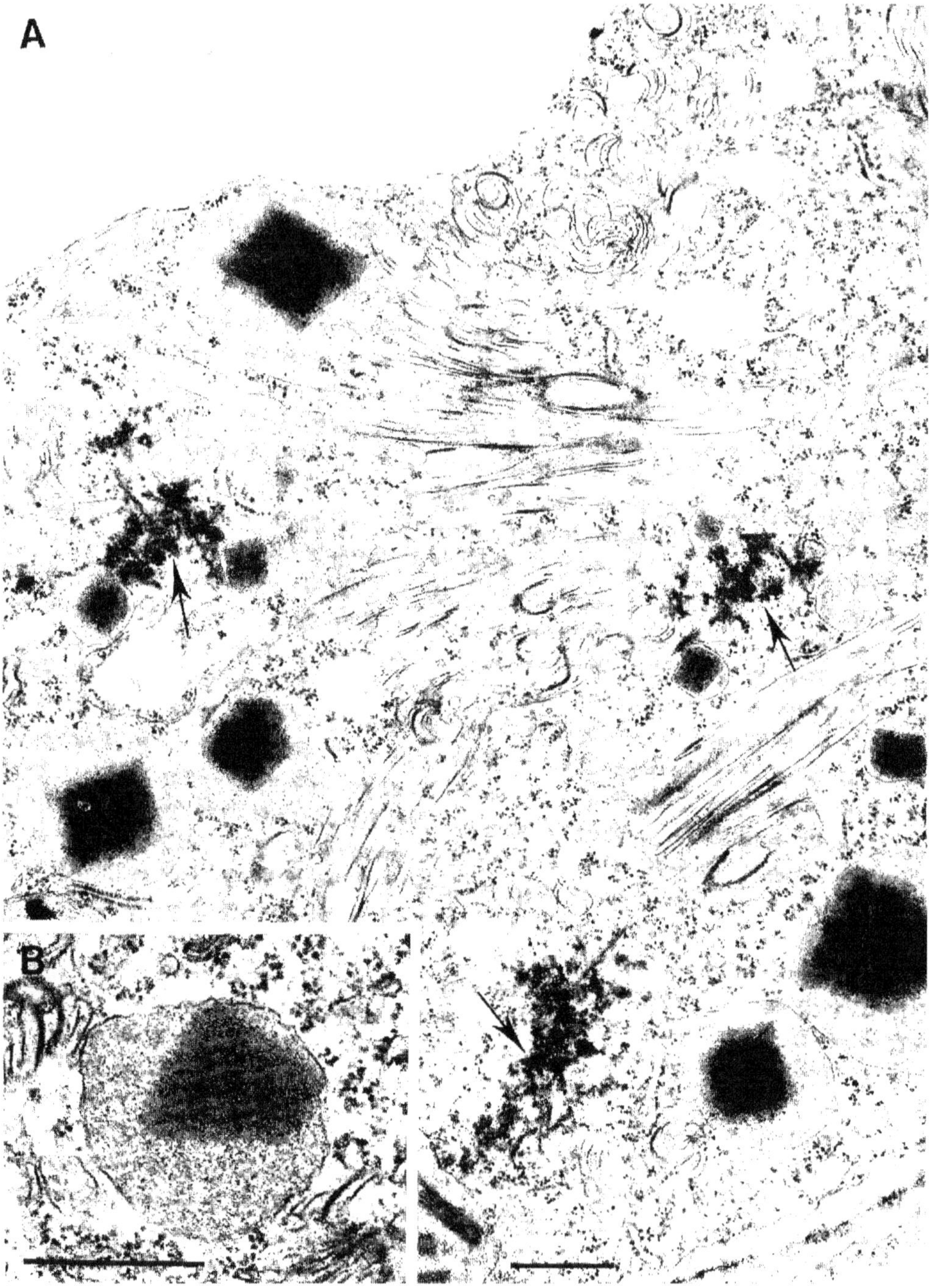

FIGURE 10. Thin sections of *Nicotiana glutinosa* leaf cells infected with PVY. A, Showing cylindrical inclusions cut at various angles, peroxisomes with virus-induced crystals and amorphous bodies (arrows) all scattered throughout the cytoplasm; B, a peroxisome containing a virus-induced crystal. Bars = 500 nm.

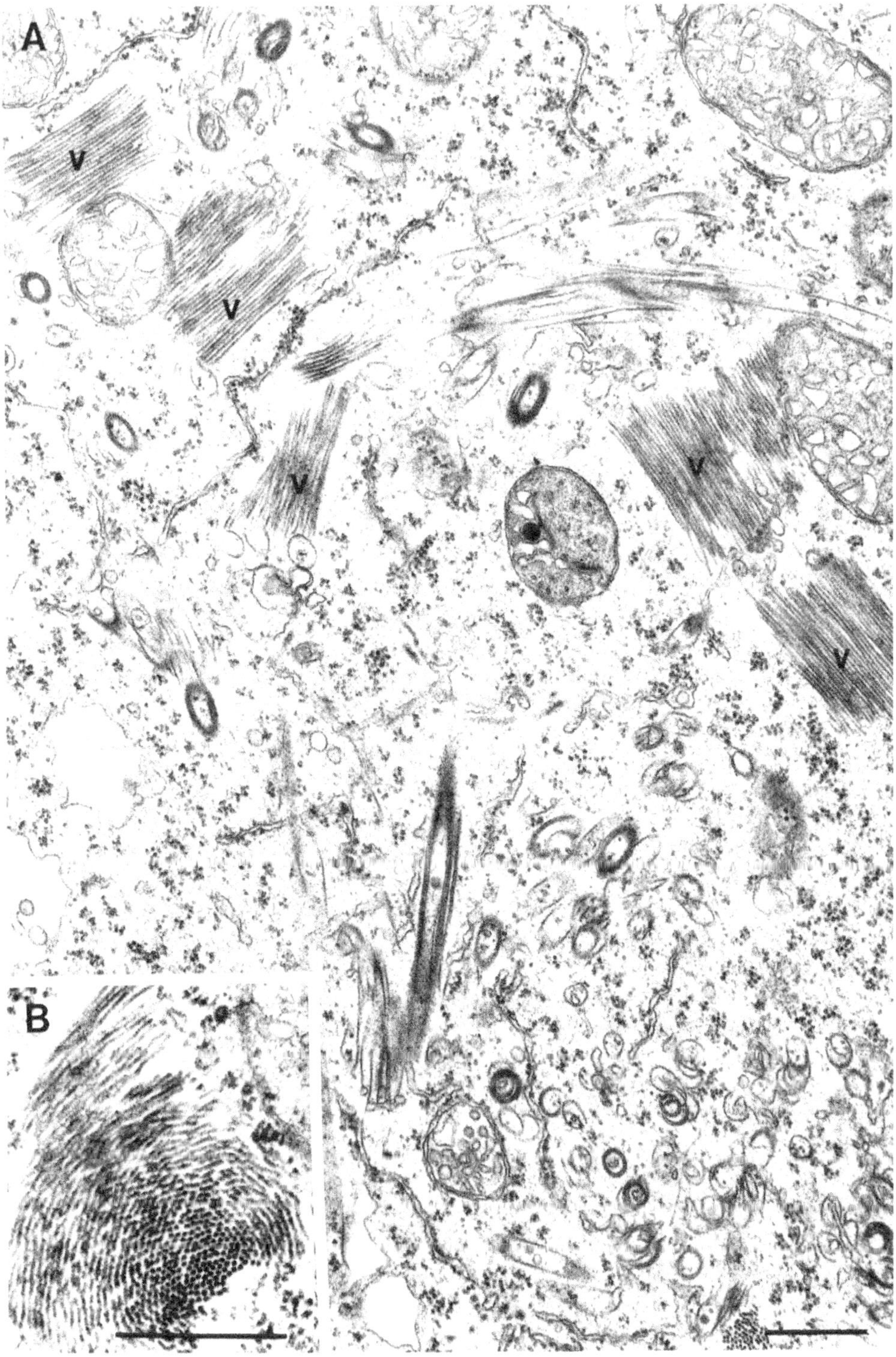

FIGURE 11. Thin sections of *Cucumis sativus* (cucumber) leaf cells infected with watermelon mosaic virus 1 (WMV 1). A, Showing numerous cylindrical inclusions cut at various angles and bundles of virus particles scattered through the cytoplasm (v); B, a bundle of virus particles cut so that the particles are seen in various aspects. Bars = 500 nm.

in the cytoplasm (Figures 8 and 11). The virus-induced vesicles seem to originate from the endoplasmic reticulum or the golgi bodies. [154-160] Some of the vesicles appear empty but others contain electron-dense material, often in the form of fine fibrils reminiscent of nucleic acid (Figures 3A and 8A). It is tempting to speculate that these vesicles are the sites of viral RNA synthesis like those in cells infected by Comoviruses (see Chapter 1). With WMV 2, it was observed that the virus-specific vesicles appeared early in infection,[160] which would be expected if they were the sites of viral RNA synthesis.

The secondary vacuolation observed in cells infected by many Potyviruses is often associated with the formation of characteristic structures consisting of monolayers of virus particles sandwiched between two membranes. These sheets of particles often line the vacuoles and may bridge secondary vacuoles or extend in a convoluted form into the central vacuole (Figure 8B).[154,155,159-163] Uniseriate arrays of Potyvirus particles attached to a single membrane can also sometimes be seen in the body of the cytoplasm (Figure 12A). The membranes often appear to be in the form of cylinders with the particles attached on the outside (Figure 12A). It seems that these structures may be the same as the uniseriate arrays of virus particles adjoining the tonoplasts already mentioned. If a sheet of particles attached to a single membrane in the cytoplasm migrated and made contact with the tonoplast, the particles would become sandwiched between two membranes.

In cells infected by many Potyviruses the particles are rather inconspicuous. When scattered in the cytoplasm without aggregating, they are not easy to detect[158] because of their thin morphology and rather poor stain uptake (Figures 9A and 12B). They are often randomly distributed in the cytoplasm but sometimes can also be seen in the plasmodesmata.[153,164] However, in many cases the particles occur as aggregates forming conspicuous bundles in the cytoplasm (Figures 11, 13B, and 14A).[145,155,160,165-168] In some instances the particles have been seen as uniseriate aggregates between the outer membranes of clumped masses either of chloroplasts, as in the case of turnip mosaic virus (TuMV) infection,[169] or mitochondria, as with PVY[3] and henbane mosaic virus (HMV).[170] In none of these instances did the organelles themselves appear to be adversely affected.

A few Potyviruses, notably BYMV, induce electron-dense, crystalline inclusions in the cytoplasm (Figures 6 and 7) and sometimes also in the nuclei and nucleoli.[3,134,171] These inclusions in the cytoplasm are usually surrounded by ribosomes (Figure 7). The crystalline structures consist of parallel striations with electron-opaque strands 1 to 3 nm wide separated by electron-transparent striations about 7 nm wide.[171] They were shown to be susceptible to proteolytic enzymes but not to either RNase or DNase indicating that they are composed entirely or primarily of protein.[172]

Some Potyviruses such as wheat spindle streak mosaic virus (WSSMV) and zucchini yellow fleck virus (ZYFV) have been shown to induce large and conspicuous membranous bodies.[173,174] Similar structures have occasionally been seen in cells infected with BCMV (Figure 9C). They appear to consist of variously interconnected tubes. It was suggested by Langenberg and Schroeder[173] that membranous bodies may be involved in the development of cylindrical inclusions, but there seems little evidence to support this. It is interesting to note that structures similar to the Potyvirus-induced membranous bodies have been observed in the protophloem of some apparently virus-free plants.[175]

Some Potyviruses, notably WMV 2, have been reported to induce the formation of amorphous bodies in the cytoplasm.[160] These structures consist of masses of finely granular electron-dense material (Figures 9B and 10) and should not be confused with some Potyvirus-induced structures referred to by the same name which appear to be composed of fibrous or tubular material observed in cells infected by PVY and PVMV.[3] It has been shown that the amorphous bodies in WMV 2-infected cells consist of a uniformly dense matrix without any obvious substructure but later on in infection, they became less compact and many thread-like structures, probably virus particles, were seen to permeate through the inclusions.[160]

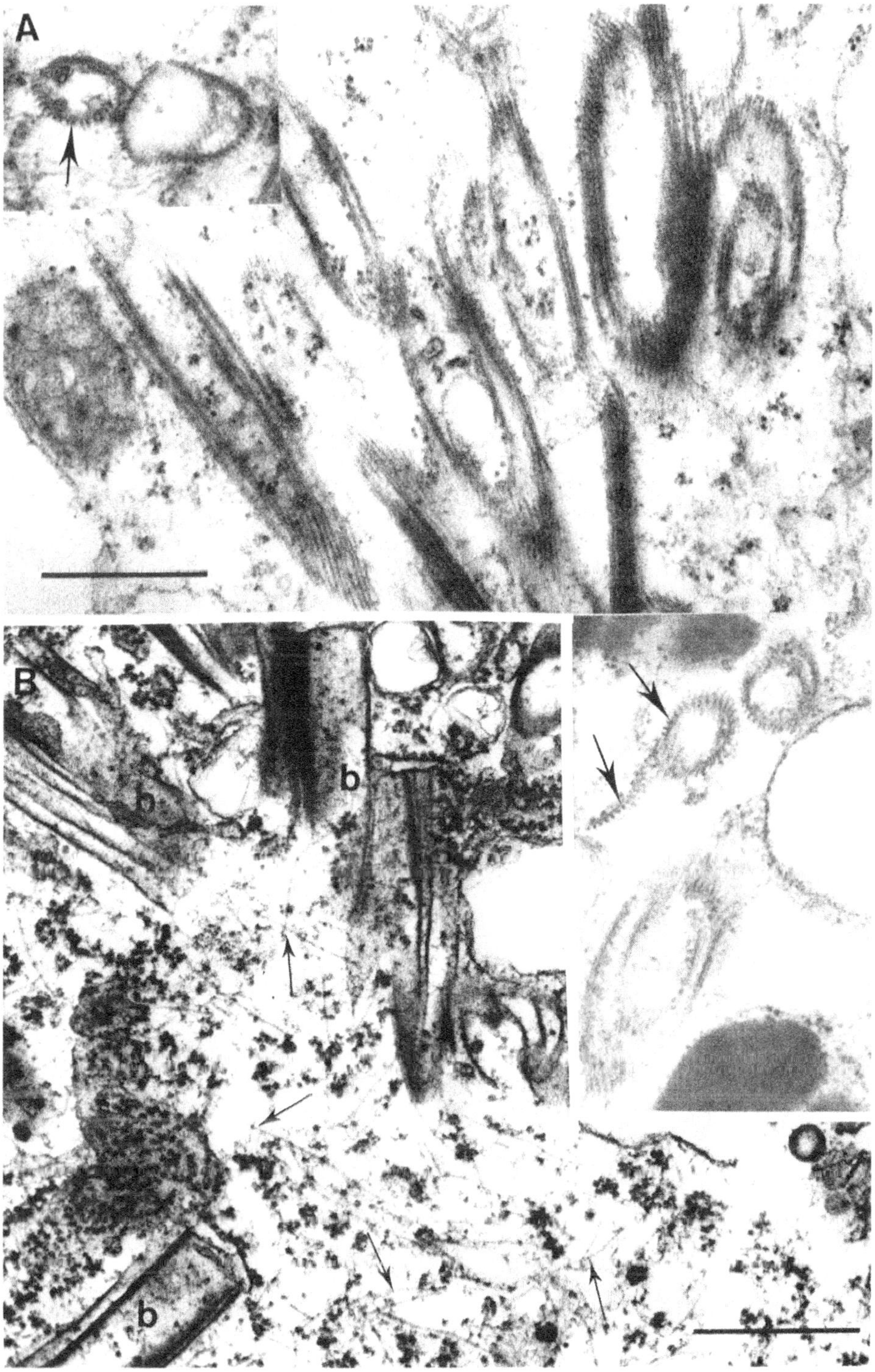

FIGURE 12. A, Thin sections of *Pisum arvense* (sweet pea) leaf cells infected with BYMV showing cytoplasm containing uniseriate sheets of virus particles cut at various angles; sheets cut approximately in transverse section are marked with arrows. B, Thin section of a *Zea mays* (maize) leaf cell infected with sugarcane mosaic virus (SCMV), showing cytoplasm with scattered virus particles (arrows) and cylindrical inclusions appearing as bundles (b). Bars = 500 nm.

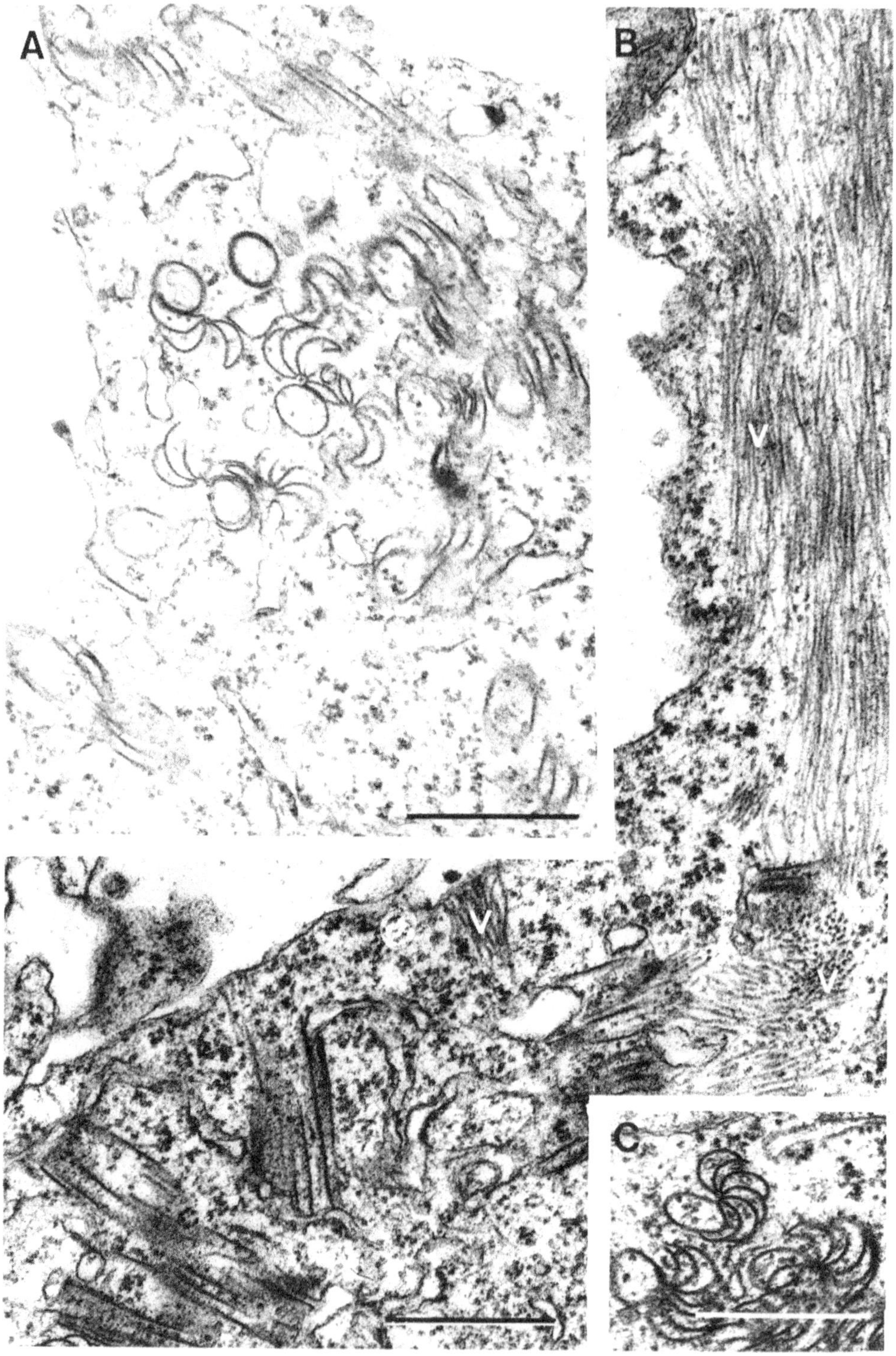

FIGURE 13. Thin sections of *Cucumis sativus* (cucumber) leaf cells infected with watermelon mosaic virus 2 (WMV 2). A, Cytoplasm with scattered cylindrical inclusions sectioned transversely (pinwheels) and obliquely; B, same tissue as in A but also showing bundles of virus particles (V) sectioned in various planes; C, a group of cylindrical inclusions sectioned transversely (pinwheels). Bars = 500 nm.

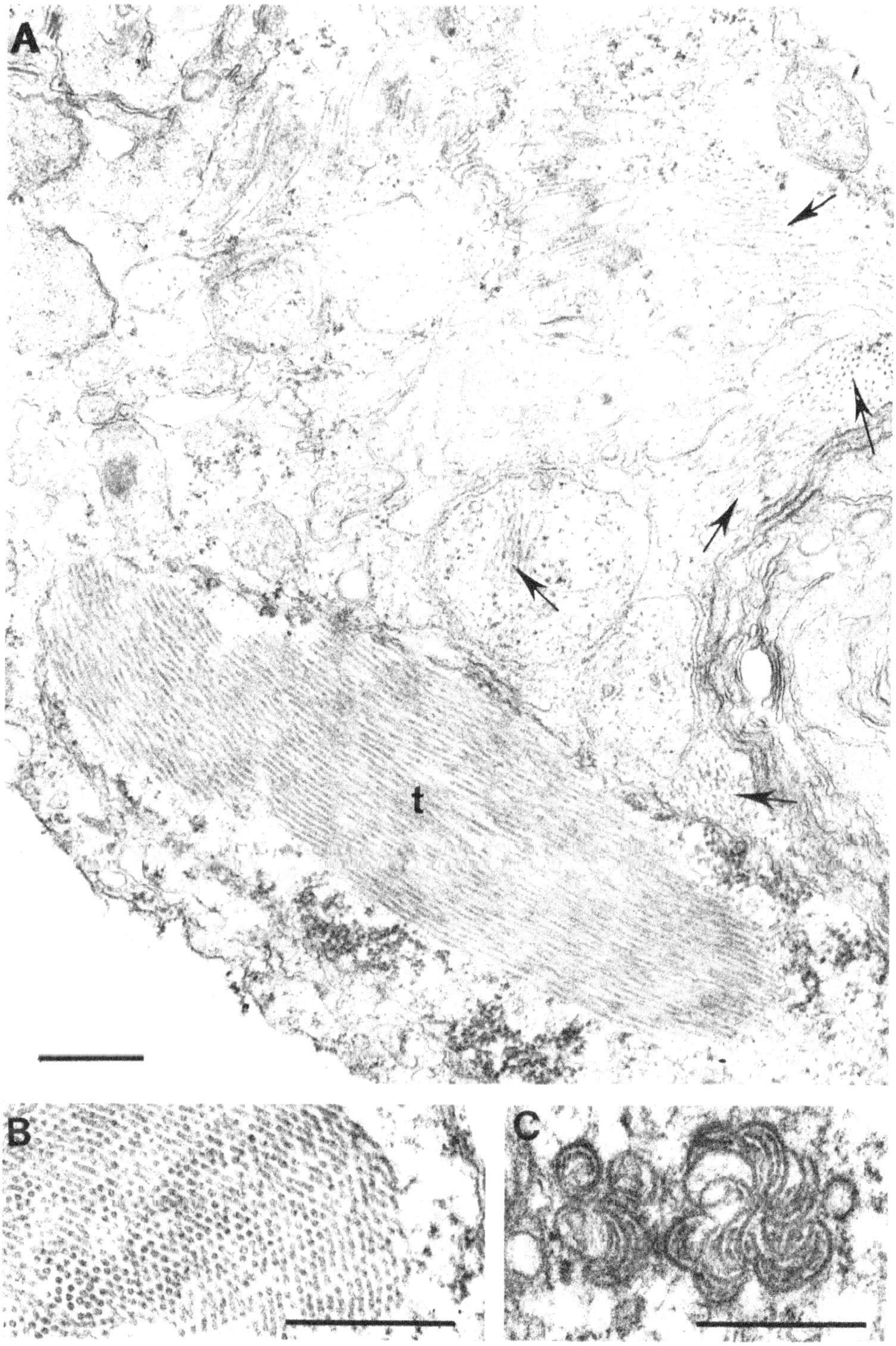

FIGURE 14. Thin sections of *Coriandrum sativum* (coriander) leaf cells infected with an unidentified Potyvirus (similar to celery yellow mosaic virus[183]), showing in A, extensive membrane development and bundles of virus particles (arrows) in the cytoplasm and bundles of virus-specific tubules in the nucleus (t); B, a bundle of virus-specific tubules in the nucleoplasm with some of the tubules cut in transverse section and some obliquely; C, cylindrical inclusions in the cytoplasm cut transversely (pinwheels). Bars = 500 nm.

Recently, Martelli and his colleagues[174] have reported the presence of structures to which they refer as fimbriate bodies in cells infected with ZYFV. These structures appear to consist of scattered flakes of densely staining material with a finely granular or fibrous texture. In older infections, the flakes increase in size, becoming elongated and curved to form aggregates occupying large areas of the ground cytoplasm. The fimbriate bodies in ZYFV-infected cells have at least a superficial resemblance to some of the less-common structures associated with infection by some Potyviruses, for example, the "amorphous inclusions" observed in PVY- and PVMV-infected cells by Edwardson.[3]

C. Nuclear Inclusions

Although many Potyviruses induce inclusions in the nuclei of the cells they infect, this is not a general characteristic of the group. Furthermore, a number of different types of inclusions have been observed, described as crystalline, paracrystalline, satellite, and fibrous.[3] Of these, the crystalline inclusions induced by TEV are probably the most thoroughly studied (Figure 15), but similar structures are induced by a few other Potyviruses such as some strains of BYMV (Figure 16).[3,134]

The nuclear inclusions induced by TEV can be detected by light microscopy.[134] In the electron microscope their appearance differs somewhat in plants infected with different virus strains.[66] With most strains, they are truncated pyramids which can appear as squares or trapezoids depending on the plane of section (Figure 15). Electron microscopy of negatively stained and freeze-etched inclusions, both *in situ* and after isolation, revealed that the inclusions consist of subunits packed in a rectangular array and organised into laminae which assemble into crystals by stacking vertically.[176,177] Cytochemical, enzyme-digestion and autoradiographical studies on the inclusions *in situ* indicate that they consist of protein.[178-180] Chemical analysis of purified preparations of the inclusions showed that they consist of two polypeptides of M_r 49.8 and 54.5 $\times$ 10^3.[177] Immunochemical studies have established that the inclusion proteins are antigenically unrelated to viral coat and cytoplasmic cylindrical inclusion proteins.[181] In vitro translation studies with TEV RNA, PepMoV and TVMV have confirmed that the nuclear inclusion proteins are coded for by the virus, and shown that they are the products of two distinct viral genes (see Section II.B). These two genes account for nearly half the coding potential of the viral RNAs and hence it would be surprising if they were not essential for virus multiplication. It is interesting to note that somewhat similar nuclear crystals have been observed in two apparently virus-free *Dianthus* species.[182]

The nuclei of beet infected with beet mosaic virus (BtMV) undergo unique ultrastructural changes including the development of satellite inclusions (Figure 17).[156] The origin of these inclusions appears to be associated with the nucleoli near which develop granular, ribosome-like particles. The satellite bodies are large rounded structures apparently consisting of protein but it is not known if this is of host of viral origin.[156] As the satellite bodies enlarge, they become vacuolated (Figure 17B) and paracrystalline aggregates of tubules develop in the vacuoles (Figure 17C). The tubules are about 18 to 20 nm in diameter and also appear to consist of protein.[156]

Another type of nuclear inclusion has been observed in cells of plants infected with celery yellow mosaic virus.[183] These inclusions consist of bundles of fibrous, thread-like structures about 25 nm in diameter (Figure 14). They also seem to be made of protein, but their origin is not known.[183,184]

The variety of cell inclusions induced by Potyviruses is bewildering. However, only the presence of the characteristic cylindrical inclusions in the cytoplasm appears to be a cytopathic effect common to all Potyviruses and is an excellent diagnostic character of the group. All the other cytoplasmic and nuclear inclusions that have been observed are only induced by some Potyviruses and can be useful in recognizing individual members of the group.

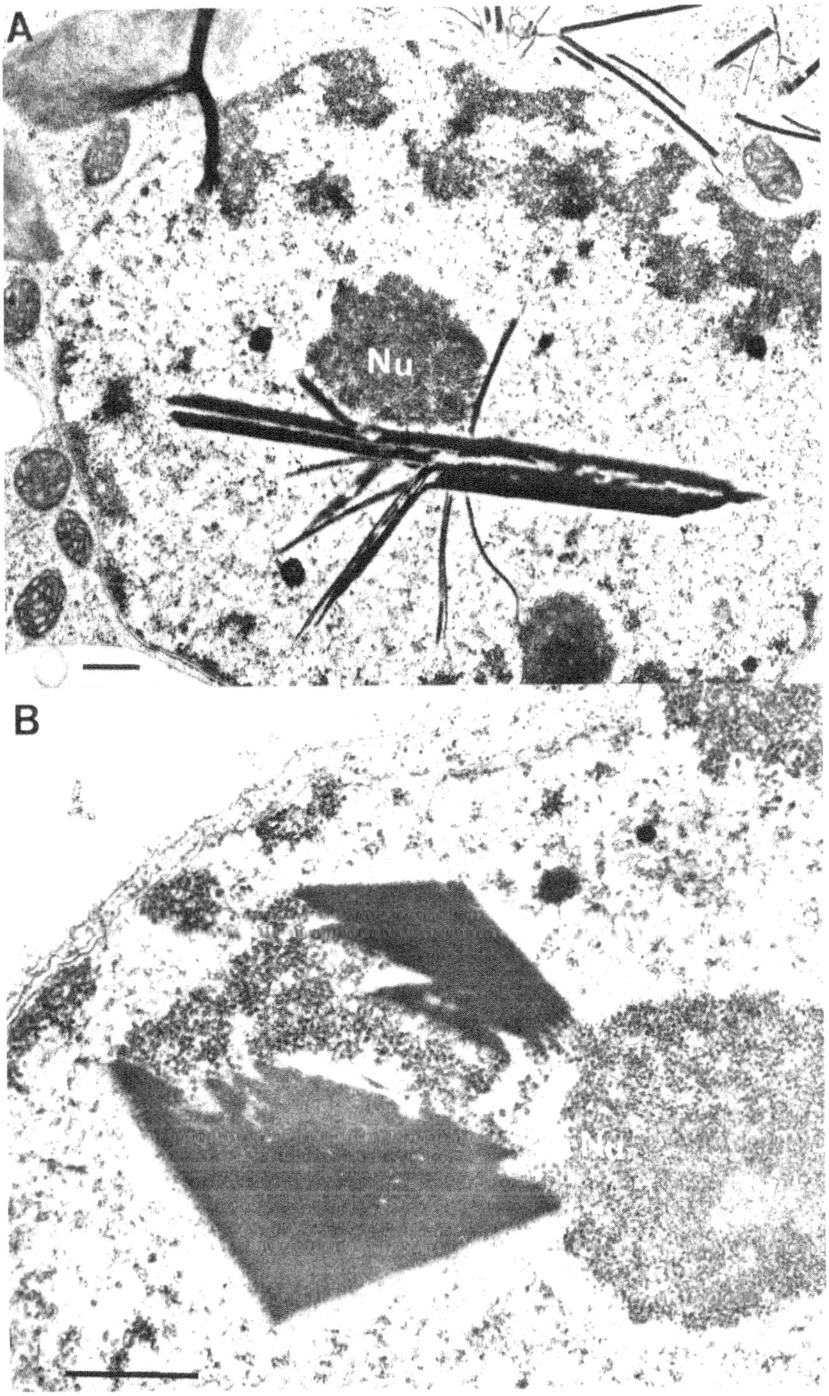

FIGURE 15. Thin sections of *Nicotiana tabacum* (tobacco) leaf cells infected with tobacco etch virus (TEV), showing characteristic virus-induced crystalline inclusions in the nuclei. A, Section is cut at right angles to the inclusion plates; B, section is cut more or less in the plane of two bipyramidal inclusion plates. The inclusions are in close association with the nucleolus (nu). Bars = 500 nm. (Courtesy of Dr. J. R. Edwardson, Department of Agronomy, University of Florida, Gainesville.)

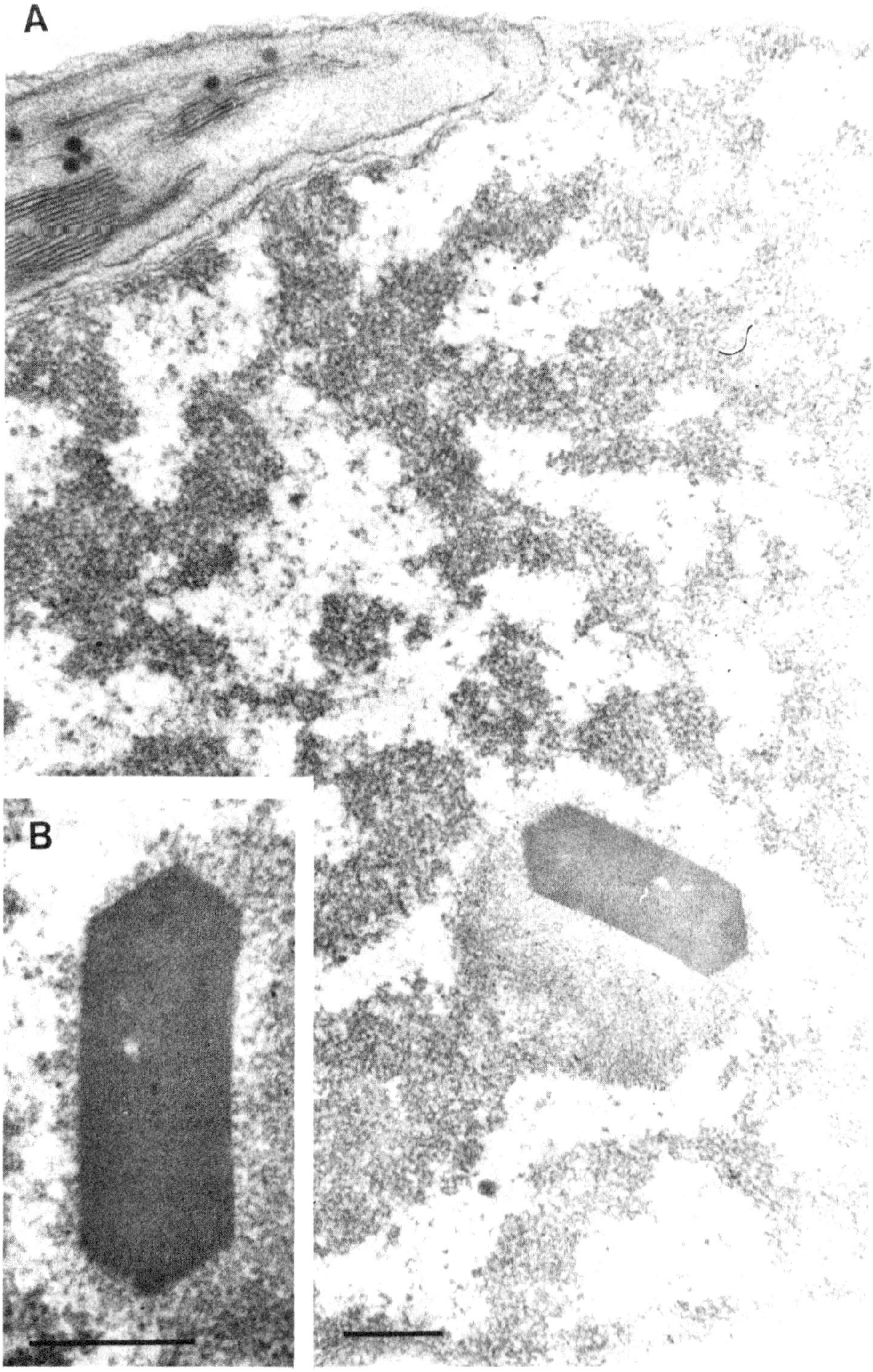

FIGURE 16. Thin section of a *Vicia faba* (broad bean) leaf cell infected with BYMV. A, Showing nucleus containing a virus-induced crystal; B, crystal seen at a higher magnification. Bars = 500 nm.

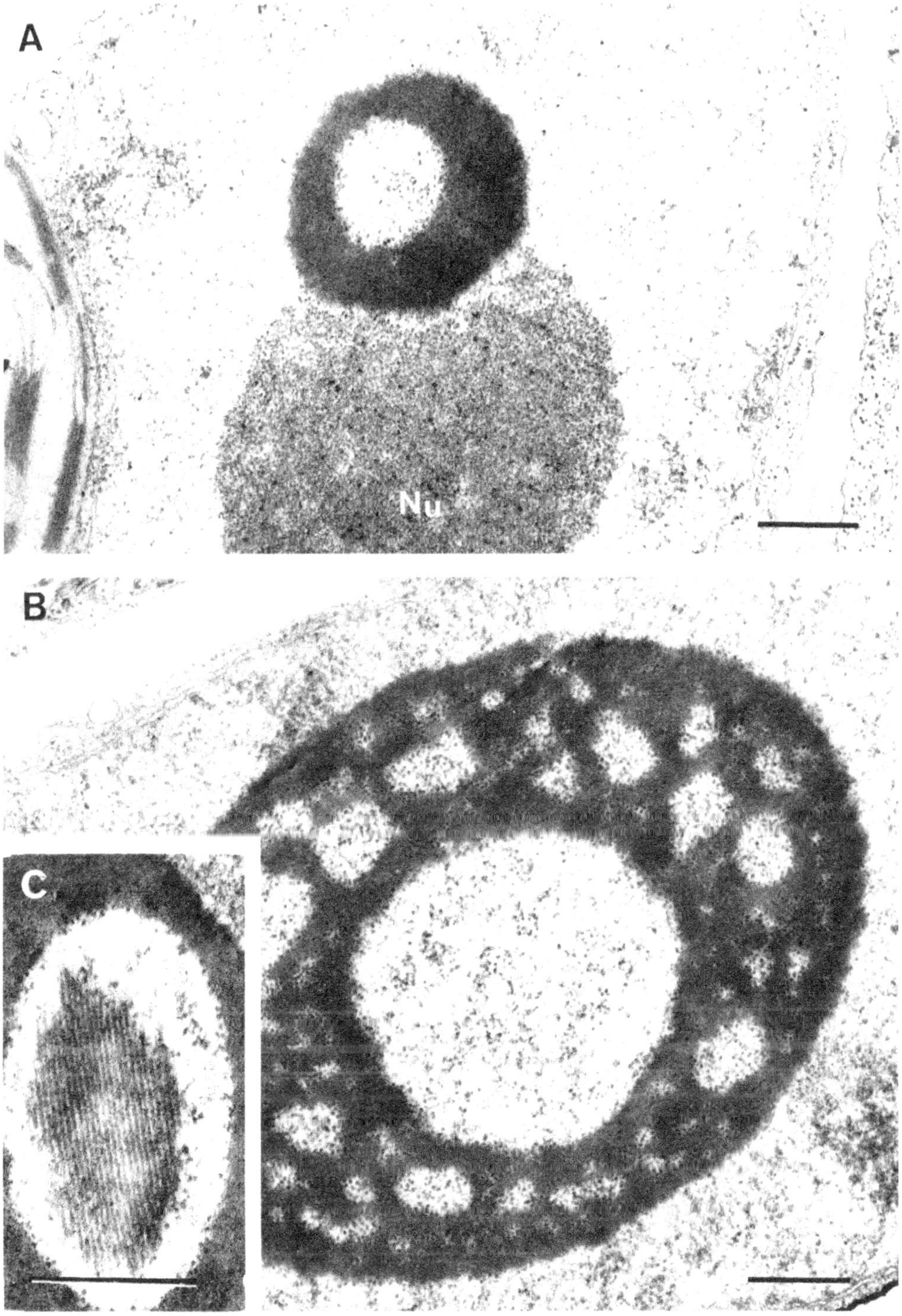

FIGURE 17. Thin sections of *Beta vulgaris* (silver beet) leaf cells infected with BtMV. A, Showing a developing virus-induced satellite body near the nucleolus (Nu); B, older satellite body with characteristic vacuolation; C, paracrystalline aggregates of tubules associated with a satellite body. Bars = 500 nm. (Courtesy of Mr. C. Davies, Department of Biochemistry, University of Adelaide, Australia.)

REFERENCES

1. **Matthews, R. E. F.,** Classification and nomenclature of viruses. Fourth report of the International Committee on Taxonomy of Viruses, *Intervirology*, 17, 1, 1982.
2. **van Regenmortel, M. H. V.,** *Serology and Immunochemistry of Plant Viruses*, Academic Press, New York, 1982, 302 pp.
3. **Edwardson, J. R.,** Some properties of the potato virus Y-group, *Fla. Agric. Expt. Stn. Monogr. Ser. No.* 4, 1974, 398 pp.
4. **Edwardson, J. R.,** Host ranges of viruses in the PVY-group, *Fla. Agric. Expt. Stn. Monogr. Ser. No.* 5, 1974, 225 pp.
5. **Hollings, M. and Brunt, A. A.,** Potyviruses, in *Handbook of Plant Virus Infections and Comparative Diagnosis*, Kurstak, E., Ed., Elsevier, Amsterdam, 1981, 731.
6. **Hollings, M. and Brunt, A. A.,** Potyvirus group, *CMI/AAB Descriptions of Plant Viruses*, No. 245, 1981.
7. **Brandes, J. and Wetter, C.,** Classification of elongated plant viruses on the basis of particle morphology, *Virology*, 8, 99, 1959.
8. **Harrison, B. D., Finch, J. T., Gibbs, A. J., Hollings, M., Shepherd, R. J., Valenta, V., and Wetter, C.,** Sixteen groups of plant viruses, *Virology*, 45, 356, 1971.
9. **Wildy, P.,** Classification and nomenclature of viruses. First report of the International Committee of Nomenclature of Viruses, *Monogr. Virol.*, 5, 1, 1971.
10. **Brandes, J. and Bercks, R.,** Gross morphology and serology as a basis for classification of elongated plant viruses, *Adv. Virus Res.*, 11, 1, 1965.
11. **Fenner, F.,** Classification and nomenclature of viruses. Second report of the International Committee on Taxonomy of Viruses, *Intervirology*, 7, 1, 1976.
12. **Matthews, R. E. F.,** Classification and nomenclature of viruses. Third report of the International Committee on Taxonomy of Viruses, *Intervirology*, 12, 129, 1979.
13. **Inouye, T. and Saito, Y.,** Barley yellow mosaic virus, *CMI/AAB Descriptions of Plant Viruses*, No. 143, 1975.
14. **Hebert, T. T. and Panizo, C. H.,** Oat mosaic virus, *CMI/AAB Descriptions of Plant Viruses*, No. 145, 1975.
15. **Inouye, T.,** Rod shaped particles associated with necrosis mosaic of rice, *Ann. Phytopathol. Soc. Jpn.*, 34, 301, 1968.
16. **Slykhuis, J. T. and Polák, Z.,** Factors affecting manual transmission, purification, and particle lengths of wheat spindle streak mosaic virus, *Phytopathology*, 61, 569, 1971.
17. **Hooper, G. R. and Wiese, M. V.,** Cytoplasmic inclusions in wheat affected by wheat spindle streak mosaic, *Virology*, 47, 664, 1972.
18. **Inouye, T.,** Filamentous particles as the causal agent of yellow mosaic disease of wheat, *Nogaku Kenkyu Ohara Inst. Agric. Biol. Okayama Univ.*, 53, 61, 1969.
19. **Slykhuis, J. T.** Agropyron mosaic virus, *CMI/AAB Descriptions of Plant Viruses*, No. 118, 1973.
20. **Slykhuis, J. T. and Bell, W.,** Differentiation of agropyron mosaic, wheat streak mosaic and a hitherto unrecognized hordeum mosaic virus in Canada, *Can. J. Bot.*, 44, 1191, 1966.
21. **Slykhuis, J. T. and Paliwal, Y. C.,** Ryegrass mosaic virus, *CMI/AAB Descriptions of Plant Viruses*, No. 86, 1972.
22. **Brakke, M. K.,** Wheat streak mosaic virus, *CMI/AAB Descriptions of Plant Viruses*, No. 48, 1971.
23. **Hollings, M., Stone, O. M., and Bock, K. R.,** Sweet potato mild mottle virus, *CMI/AAB Descriptions of Plant Viruses*, No. 162, 1976.
24. **Pleše, N., Koenig, R., Lesemann, D. E., and Bozarth, R. F.,** Maclura mosaic virus — an elongated plant virus of uncertain classification, *Phytopathology*, 69, 471, 1979.
25. **Koenig, R., Lesemann, D.-E., and Pleše, N.,** Maclura mosaic virus, *CMI/AAB Descriptions of Plant Viruses*, No. 239, 1981.
26. **Moghal, S. M. and Francki, R. I. B.,** Towards a system for the identification and classification of potyviruses. I. Serology and amino acid composition of six distinct viruses, *Virology*, 73, 350, 1976.
27. **Moghal, S. M. and Francki, R. I. B.,** Towards a system for the identification and classification of potyviruses. II. Virus particle length, symptomatology, and cytopathology of six distinct viruses, *Virology*, 112, 210, 1981.
28. **Randles, J. W., Davies, C., Gibbs, A. J., and Hatta, T.,** Amino acid composition of capsid protein as a taxonomic criterion for classifying the atypical S strain of bean yellow mosaic virus, *Aust. J. Biol. Sci.*, 33, 245, 1980.
29. **Abu-Samah, N. and Randles, J. W.,** A comparison of the nucleotide sequence homologies of three isolates of bean yellow mosaic virus and their relationship to other potyviruses, *Virology*, 110, 436, 1981.
30. **Christie, R. G. and Edwardson, J. R.,** Light and electron microscopy of plant virus inclusions, *Fla. Agric. Expt. Stn. Monogr. Ser. No.* 9, 1977, 150 pp.

31. **Edwardson, J. R. and Christie, R. G.,** Use of virus-induced inclusions in classification and diagnosis, *Annu. Rev. Phytopathol.*, 16, 31, 1978.
32. **Govier, D. A. and Woods, R. D.,** Changes induced by magnesium ions in the morphology of some plant viruses with filamentous particles, *J. Gen. Virol.*, 13, 127, 1971.
33. **de Bokx, J. A. and Huttinga, H.,** Potato virus Y, *CMI/AAB Descriptions of Plant Viruses*, No. 242, 1981.
34. **Bos, L.,** Bean common mosaic virus, *CMI/AAB Descriptions of Plant Viruses*, No. 73, 1971.
35. **Bos, L.,** Bean yellow mosaic virus, *CMI/AAB Descriptions of Plant Viruses*, No. 40, 1970.
36. **Barnett, O. W. and Brunt, A. A.,** Bearded iris mosaic virus, *CMI/AAB Descriptions of Plant Viruses*, No. 147, 1975.
37. **Russell, G. E.,** Beet mosaic virus, *CMI/AAB Descriptions of Plant Viruses*, No. 53, 1971.
38. **Purcifull, D. E., Christie, S. R., and Zitter, T. A.,** Bidens mottle virus, *CMI/AAB Descriptions of Plant Viruses*, No. 161, 1976.
39. **Hollings, M. and Stone, O. M.,** Carnation vein mottle virus, *CMI/AAB Descriptions of Plant Viruses*, No. 78, 1971.
40. **Howell, W. E. and Mink, G. I.,** Carrot thin leaf virus, *CMI/AAB Descriptions of Plant Viruses*, No. 218, 1980.
41. **Shepard, J. F. and Grogan, R. G.,** Celery mosaic virus, *CMI/AAB Descriptions of Plant Viruses*, No. 50, 1971.
42. **Hollings, M. and Stone, O. M.,** Clover yellow vein virus, *CMI/AAB Descriptions of Plant Viruses*, No. 131, 1974.
43. **Catherall, P. L.,** Cocksfoot streak virus, *CMI/AAB Descriptions of Plant Viruses*, No. 59, 1971.
44. **Bock, K. R. and Conti, M.,** Cowpea aphid-borne mosaic virus, *CMI/AAB Descriptions of Plant Viruses*, No. 134, 1974.
45. **Zettler, F. W., Abo El-Nil, M. M., and Hartman, R. D.,** Dasheen mosaic virus, *CMI/AAB Descriptions of Plant Viruses*, No. 191, 1978.
46. **Govier, D. A. and Plumb, R. T.,** Henbane mosaic virus, *CMI/AAB Descriptions of Plant Viruses*, No. 95, 1972.
47. **Brunt, A. A.,** Hippeastrum mosaic virus, *CMI/AAB Descriptions of Plant Viruses*, No. 117, 1973.
48. **Brunt, A. A.,** Iris mild mosaic virus, *CMI/AAB Descriptions of Plant Viruses*, No. 116, 1973.
49. **Bos, L.,** Leek yellow stripe virus, *CMI/AAB Descriptions of Plant Viruses*, No. 240, 1981.
50. **Tomlinson, J. A.,** Lettuce mosaic virus, *CMI/AAB Descriptions of Plant Viruses*, No. 9, 1970.
51. **Brunt, A. A.,** Narcissus yellow stripe virus, *CMI/AAB Descriptions of Plant Viruses*, No. 76, 1971.
52. **Bos, L.,** Onion yellow dwarf virus, *CMI/AAB Descriptions of Plant Viruses*, No. 158, 1976.
53. **Purcifull, D. E.,** Papaya ringspot virus, *CMI/AAB Descriptions of Plant Viruses*, 84, 1972.
54. **Murant, A. F.,** Parsnip mosaic virus, *CMI/AAB Descriptions of Plant Viruses*, No. 91, 1972.
55. **Taylor, R. H. and Greber, R. S.,** Passionfruit woodiness virus, *CMI/AAB Descriptions of Plant Viruses*, No. 122, 1973.
56. **Hampton, R. O. and Mink, G. I.,** Pea seed-borne mosaic virus, *CMI/AAB Descriptions of Plant Viruses*, No. 146, 1975.
57. **Bock, K. R. and Kuhn, C. W.,** Peanut mottle virus, *CMI/AAB Descriptions of Plant Viruses*, No. 141, 1975.
58. **Nelson, M. R., Wheeler, R. E., and Zitter, T. A.,** Pepper mottle virus, *CMI/AAB Descriptions of Plant Viruses*, No. 253, 1982.
59. **Brunt, A. A. and Kenten, R. H.,** Pepper veinal mottle virus, *CMI/AAB Descriptions of Plant Viruses*, No. 104, 1972.
60. **Kegler, H. and Schade, C.,** Plum pox virus, *CMI/AAB Descriptions of Plant Viruses*, No. 70, 1971.
61. **Shepherd, R. J.,** Pokeweed mosaic virus, *CMI/AAB Descriptions of Plant Viruses*, No. 97, 1972.
62. **Bartels, R.,** Potato Virus A, *CMI/AAB Descriptions of Plant Viruses*, No. 54, 1971.
63. **Bos, L.,** Soybean mosaic virus, *CMI/AAB Descriptions of Plant Viruses*, No. 93, 1972.
64. **Pirone, T. P.,** Sugarcane mosaic virus, *CMI/AAB Descriptions of Plant Viruses*, No. 88, 1972.
65. **Gough, K. H. and Shukla, D. D.,** Coat protein of potyviruses. I. Comparison of the four Australian strains of sugarcane mosaic virus, *Virology*, 111, 455, 1981.
66. **Purcifull, D. E. and Hiebert, E.,** Tobacco etch virus, *CMI/AAB Descriptions of Plant Viruses*, No. 258, 1982.
67. **Sun, M. K. C., Gooding, G. V., Pirone, T. P., and Tolin, S. A.,** Properties of tobacco vein-mottling virus, a new pathogen of tobacco, *Phytopathology*, 64, 1133, 1974.
68. **van Slogteren, D. H. M.,** Tulip breaking virus, *CMI/AAB Descriptions of Plant Viruses*, No. 71, 1971.
69. **Tomlinson, J. A.,** Turnip mosaic virus, *CMI/AAB Descriptions of Plant Viruses*, No. 8, 1970.
70. **Van Regenmortel, M. H. V.,** Watermelon mosaic virus, *CMI/AAB Descriptions of Plant Viruses*, No. 63, 1971.

71. **Russo, M., Martelli, G. P., Vovlas, C., and Ragozzino, A.,** Comparative studies on mediterranean isolates of watermelon mosaic virus, *Phytopathol. Mediterr.*, 18, 94, 1979.
72. **Webb, R. E. and Scott, H. A.,** Isolation and identification of watermelon mosaic viruses 1 and 2, *Phytopathology*, 55, 895, 1965.
73. **Purcifull, D. E. and Gooding, G. V.,** Immunodiffusion tests for potato Y and tobacco etch viruses, *Phytopathology*, 60, 1036, 1970.
74. **Shepard, J. F., Jutila, J. W., Catlin, J. E., Newman, F. S., and Hawkins, W. H.,** Immunodiffusion assay for potato virus M infection, *Phytopathology*, 61, 873, 1971.
75. **Purcifull, D. E. and Batchelor, D. L.,** Immunodiffusion tests with sodium dodecyl sulfate (SDS)-treated plant viruses and plant viral inclusions, *Fla. Agric. Exp. Stn. Tech. Bull.*, No.788, 1977, 39 pp.
76. **Purcifull, D. E. and Hiebert, E.,** Serological distinction of watermelon mosaic virus isolates, *Phytopathology*, 69, 112, 1979.
77. **Shepard, J. F., Secor, G. A., and Purcifull, D. E.,** Immunochemical cross reactivity between the dissociated capsid proteins of PVY group plant viruses, *Virology*, 58, 464, 1974.
78. **Hill, J. H. and Benner, H. I.,** Properties of soybean mosaic virus and its isolated protein, *Phytopathol. Z.*, 97, 272, 1980.
79. **Hiebert, E. and McDonald, J. G.,** Characterization of some proteins associated with viruses in the potato Y group, *Virology*, 56, 349, 1973.
80. **Huttinga, H. and Mosch, W. H. M.,** Properties of viruses of the potyvirus group. II. Buoyant density, S value, particle morphology, and molecular weight of the coat protein subunit of bean yellow mosaic virus, pea mosaic virus, lettuce mosaic virus and potato virus Y^N, *Neth. J. Plant Pathol.*, 80, 19, 1974.
81. **Huttinga, H.,** Properties of viruses of the potyvirus group. III. A comparison of buoyant density, S value, particle morphology and molecular weight of the coat protein subunit of 10 viruses and virus isolates, *Neth. J. Plant Pathol.*, 81, 58, 1975.
82. **McDonald, J. G. and Hiebert, E.,** Characterization of the capsid and cylindrical inclusion proteins of three strains of turnip mosaic virus, *Virology*, 63, 295, 1975.
83. **Michelin-Lausarot, P. and Papa, G.,** The coat protein of the *Alliaria* strain of turnip mosaic virus: molecular weight and degradation products formed during purification and upon storage, *J. Gen. Virol.*, 29, 121, 1975.
84. **Barnett, O. W. and Randles, J. W.,** personal communication, 1983.
85. **Anderson, C. W.,** Vigna and Crotalaria viruses in Florida. II. Notations concerning cowpea mosaic virus *(Marmor vignae)*, *Plant Dis. Rep.*, 39, 349, 1955.
86. **Kuhn, C. W.,** Separation of cowpea virus mixtures, *Phytopathology*, 54, 739, 1964.
87. **Lovisolo, O. and Conti, M.,** Identification of an aphid-transmitted cowpea mosaic virus, *Neth. J. Plant Pathol.*, 72, 265, 1966.
88. **Martyn, E. B.,** Plant virus names, *Phytopathol. Pap.*, No. 9, Commonwealth Mycological Institute, Kew, Surrey, p. 41, 1968.
89. **Taiwo, M. A. and Gonsalves, D.,** Serological grouping of isolates of blackeye cowpea mosaic and cowpea aphidborne mosaic viruses, *Phytopathology*, 72, 583, 1982.
90. **Taiwo, M. A., Gonsalves, D., Provvidenti, R., and Thurston, H. D.,** Partial characterization and grouping of isolates of blackeye cowpea mosaic and cowpea aphidborne mosaic viruses, *Phytopathology*, 72, 590, 1982.
91. **Abu-Samah, N. and Randles, J. W.,** personal communication, 1983.
92. **Teakle, D. S. and Grylls, N. E.,** Four strains of sugar-cane mosaic virus infecting cereals and other grasses in Australia, *Aust. J. Agric. Res.*, 24, 465, 1973.
93. **Papa, G., Michelin-Lausarot, P., and Casetta, A.,** Purification and properties of alliaria mosaic virus, *Phytopathol. Z.*, 78, 344, 1973.
94. **Phillips, S. and Brunt, A. A.,** *Glasshouse Crops Research Inst. Ann. Rpt.* Littlehampton, Sussex, 192, 1981.
95. **Thouvenel, J. -C., Fauquet, C., and Monsarrat, A.,** Un nouveau Potyvirus: le virus de la marbrure d'Asystasia gangetica, *C. R. Acad. Sci. Paris*, 295, 213, 1982.
96. **Naqvi, Q. A. and Mahmood, K.,** Purification and properties of Brinjal mild mosaic virus, *Phytopathol. Z.*, 93, 86, 1978.
97. **Miličić, D., Kujundžić, M., Wrischer, M., and Plavšić, B.,** A potyvirus isolated from *Bromus mollis*, *Acta Bot. Croatica*, 39, 27, 1980.
98. **Paguio, O. R. and Kitajima, E. W.,** Isolamento de um potyvirus de *Cassia hoffmannseggi*. Mart. *Fitopatol. Brasileira*, 6, 187, 1981.
99. **Peralta, E. -L., Beczner, L., and Dezséry, M.,** Characterization of the Hungarian *Datura innoxia* mosaic virus, *Acta Phytopathol. Acad. Sci. Hung.*, 16, 85, 1981.
100. **Naqvi, Q. A. and Mahmood, K.,** Studies on *Melilotus* mosaic virus, *Phytopathol. Z.*, 93, 249, 1978.
101. **Ladera, P., Lastra, R., and Debrot, E. A.,** Purification and partial characterization of a potyvirus infecting peppers in Venezuela, *Phytopathol. Z.*, 104, 97, 1982.

102. **Lee, J. T., Yamashita, S., Doi, Y., and Yora, K.,** A potyvirus found in mosaic plants of perilla, *Ann. Phytopathol. Soc. Jpn.*, 46, 672, 1980.
103. **Fribourg, C. E.,** Host plant reactions, some properties and serology of Peru tomato virus, *Phytopathology*, 69, 441, 1979.
104. **Hammond, J.,** Viruses occurring in *Plantago* species in England, *Plant Pathol.*, 30, 237, 1981.
105. **Martin, R. R., Berbee, J. G., and Omuemu, J. O.,** Isolation of a Potyvirus from declining clones of *Populus*, *Phytopathology*, 72, 1158, 1982.
106. **Jones, P.,** Leaf mottling of *Spartina* species caused by a newly recognized virus, spartina mottle virus, *Ann. Appl. Biol.*, 94, 77, 1980.
107. **Brunt, A. A., Phillips, S., Jones, R. A. C., and Kenten, R. H.,** Viruses detected in *Ullucus tuberosus* (Basellaceae) from Peru and Bolivia, *Ann. Appl. Biol.*, 101, 65, 1982.
108. **Inouye, N. and Hakkaart, F. A.,** Preliminary description of a potyvirus from *Vallota speciosa*, *Neth. J. Plant Pathol.*, 86, 265, 1980.
109. **Vovlas, C., Hiebert, E., and Russo, M.,** Zucchini yellow fleck virus, a new potyvirus of Zucchini squash, *Phytopathol. Mediterr.*, 20, 123, 1981.
110. **Lisa, V., Boccardo, G., D'Agostino, G., Dellavalle, G., and d'Aquilio, M.,** Characterization of a potyvirus that causes zucchini yellow mosaic, *Phytopathology*, 71, 667, 1981.
111. **Varma, A., Gibbs, A. J., Woods, R. D., and Finch, J. T.,** Some observations on the structure of the filamentous particles of several plant viruses, *J. Gen. Virol.*, 2, 107, 1968.
112. **Chamberlain, J. A. and Catherall, P. L.,** Electron microscopy of cocksfoot streak virus and its differentiation from ryegrass mosaic virus in naturally infected *Dactylis glomerata* plants, *Ann. Appl. Biol.*, 85, 105, 1977.
113. **Hill, J. H., Ford, R. E., and Benner, H. I.,** Purification and partial characterization of maize dwarf virus strain B (sugarcane mosaic virus), *J. Gen. Virol.*, 20, 327, 1973.
114. **Barnett, O. W., and Alper, M.,** Characterization of iris fulva mosaic virus, *Phytopathology*, 67, 448, 1977.
115. **McDonald, J. G., Beveridge, T. J., and Bancroft, J. B.,** Self-assembly of protein from a flexuous virus, *Virology*, 69, 327, 1976.
116. **McDonald, J. G. and Bancroft, J. B.,** Assembly studies on potato virus Y and its coat protein, *J. Gen. Virol.*, 35, 251, 1977.
117. **Miki, T. and Oshima, N.,** On the size of the protein subunits in potato virus Y, *J. Gen. Virol.*, 15, 179, 1972.
118. **Makkouk, K. M. and Gumpf, D. J.,** Characterization of the protein and nucleic acid of potato virus Y strains isolated from pepper, *Virology*, 63, 336, 1975.
119. **Stace-Smith, R. and Tremaine, J. H.,** Purification and composition of potato virus Y, *Phytopathology*, 60, 1785, 1970.
120. **Damirdagh, I. S. and Shepherd, R. J.,** Some of the chemical properties of the tobacco etch virus and its protein and nucleic acid components, *Virology*, 40, 84, 1970.
121. **Hill, J. H. and Shepherd, R. J.,** Biochemical properties of turnip mosaic virus, *Virology*, 47, 807, 1972.
122. **Pring, D. R. and Langenberg, W. G.,** Preparation and properties of maize dwarf mosaic virus ribonucleic acid, *Phytopathology*, 62, 253, 1972.
123. **Brakke, M. K. and van Pelt, N.,** Properties of infectious ribonucleic acid from wheat streak mosaic virus, *Virology*, 42, 699, 1970.
124. **Hill, J. H. and Benner, H. I.,** Properties of potyvirus RNAs: turnip mosaic, tobacco etch, and maize dwarf mosaic viruses, *Virology*, 75, 419, 1976.
125. **Hari, V., Siegel, A., Rozek, C., and Timberlake, W. E.,** The RNA of tobacco etch virus contains poly(A), *Virology*, 92, 568, 1979.
126. **Hari, V.,** The RNA of tobacco etch virus: further characterization and detection of protein linked to RNA, *Virology*, 112, 391, 1981.
127. **Dougherty, W. G. and Hiebert, E.,** Translation of potyvirus RNA in a rabbit reticulocyte lysate: reaction conditions and identification of capsid protein as one of the products of *in vitro* translation of tobacco etch and pepper mottle viral RNAs, *Virology*, 101, 466, 1980.
128. **Dougherty, W. G. and Hiebert, E.,** Translation of potyvirus RNA in a rabbit reticulocyte lysate: identification of nuclear inclusion proteins as products of tobacco etch virus RNA translation and cylindrical inclusion protein as a product of the potyvirus genome, *Virology*, 104, 174, 1980.
129. **Dougherty, W. G. and Hiebert, E.,** Translation of potyvirus RNA in a rabbit reticulocyte lysate: cell-free translation strategy and a genetic map of the potyviral genome, *Virology*, 104, 183, 1980.
130. **Hellmann, G. M., Shaw, J. G., Lesnaw, J. A., Chu, L-Y., Pirone, T. P., and Rhoads, R. E.,** Cell-free translation of tobacco vein mottling virus RNA, *Virology*, 106, 207, 1980.
131. **Hellmann, G. M., Thornbury, D. W., Hiebert, E., Shaw, J. G., Pirone, T. P., and Rhoads, R. E.,** Cell-free translation of tobacco vein mottling virus RNA. II. Immunoprecipitation of products by antisera to cylindrical inclusion, nuclear inclusion, and helper component proteins, *Virology*, 124, 434, 1983.

132. **Koziel, M. G., Hari, V., and Siegel, A.,** *In vitro* translation of tobacco etch virus RNA, *Virology*, 106, 177, 1980.
133. **Otal, T. and Hari, V.,** Detection and cell-free translation of subgenomic RNAs of tobacco etch virus, *Virology*, 125, 118, 1983.
134. **Teakle, D. S. and Pares, R. D.,** Potyvirus (potato virus Y) group, in *The Atlas of Insect and Plant Viruses*, Maramorosch, K., Ed., Academic Press, New York, 1977, 311.
135. **Israel, H. W. and Wilson, H. J.,** Pinwheel inclusions and plant viruses, in *The Atlas of Insect and Plant Viruses*, Maramorosch, K., Ed., Academic Press, New York, 1977, 405.
136. **Martelli, G. P. and Russo, M.,** Plant virus inclusion bodies, *Adv. Virus Res.*, 21, 175, 1977.
137. **Matsui, C. and Yamaguchi, A.,** Electron microscopy of host cells infected with tobacco etch virus. I. Fine structures of leaf cells at later stages of infection, *Virology*, 22, 40, 1964.
138. **Edwardson, J. R.,** Electron microscopy of cytoplasmic inclusions in cells infected with rod-shaped viruses, *Am. J. Bot.*, 53, 359, 1966.
139. **Edwardson, J. R., Purcifull, D. E., and Christie, R. G.,** Structure of cytoplasmic inclusions in plants infected with rod-shaped viruses, *Virology*, 34, 250, 1968.
140. **Lee, P. E.,** Electron microscopy of inclusions associated with wheat streak mosaic virus, *J. Ultrastruct. Res.*, 13, 359, 1965.
141. **Kitajima, E. W., Camargo, I. J. B., and Costa, A. S.,** Intranuclear crystals and cytoplasmic membranous inclusions associated with infection by two Brazilian strains of potato virus Y, *J. Electron Micros.*, 17, 144, 1968.
142. **Andrews, J. H. and Shalla, T. A.,** The origin, development, and conformation of amorphous inclusion body components in tobacco etch virus-infected cells, *Phytopathology*, 64, 1234, 1974.
143. **McDonald, J. G. and Hiebert, E.,** Ultrastructure of cylindrical inclusions induced by viruses of the potato Y group as visualized by freeze-etching, *Virology*, 58, 200, 1974.
144. **Mernaugh, R. L., Gardner, W. S., and Yocom, K. L.,** Three-dimensional structure of pinwheel inclusions as determined by analytic geometry, *Virology*, 106, 273, 1980.
145. **Chamberlain, J. A., Catherall, P. L., and Jellings, A. J.,** Symptoms and electron microscopy of ryegrass mosaic virus in different grass species, *J. Gen. Virol.*, 36, 297, 1977.
146. **Lawson, R. H. and Hearon, S. S.,** The association of pinwheel inclusions with plasmodesmata, *Virology*, 44, 454, 1971.
147. **Nome, S. F., Shalla, T. A., and Petersen, L. J.,** Comparison of virus particles and intracellular inclusions associated with vein mosaic, feathery mottle, and russet crack diseases of sweet potato, *Phytopathol. Z.*, 79, 169, 1974.
148. **Hiebert, E., Purcifull, D. E., Christie, R. G., and Christie, S. R.,** Partial purification of inclusions induced by tobacco etch virus and potato virus Y, *Virology*, 43, 638, 1971.
149. **Shepard, J. F. and Shalla, T. A.,** Tobacco etch virus cylindrical inclusions: antigenically unrelated to the causal virus, *Virology*, 35, 185, 1969.
150. **Purcifull, D. E., Hiebert, E., and McDonald, J. G.,** Immunochemical specificity of cytoplasmic inclusions induced by viruses in the potato Y group, *Virology*, 55, 275, 1973.
151. **Wilson, H. J., Israel, H. W., and Steward, F. C.,** Morphogenesis and the fine structure of cultured carrot cells, *J. Cell. Sci.*, 15, 57, 1974.
152. **Wilson, H. J., Goodman, R. M., and Israel, H. W.,** Pinwheel inclusions in morphogenesis: a possible alternative to induction by viruses, *Arch. Virol.*, 51, 347, 1976.
153. **Weintraub, M., Ragetli, H. W. J., and Lo, E.,** Potato virus Y particles in plasmodesmata of tobacco leaf cells, *J. Ultrastruct. Res.*, 46, 131, 1974.
154. **Hoefert, L. L.,** Proteinaceous and virus-like inclusions in cells infected with beet mosaic virus, *Virology*, 37, 498, 1969.
155. **Kim, K. S. and Fulton, J. P.,** Electron microscopy of pokeweed leaf cells infected with pokeweed mosaic virus, *Virology*, 37, 297, 1969.
156. **Martelli, G. P. and Russo, M.,** Nuclear changes in mesophyll cells of *Gomphrena globosa* L. associated with infection by beet mosaic virus, *Virology*, 38, 297, 1969.
157. **Camargo, I. J. B., Kitajima, E. W., and Costa, A. S.,** Microscopia electrônica de inclusões citoplasmáticas e alteracões celulases associadas ao virus do mosaico de Cenoura, *Bragantia*, 30, 31, 1971.
158. **Krass, C. J. and Ford, R. E.,** Ultrastructure of corn systemically infected with maize dwarf mosaic virus, *Phytopathology*, 59, 431, 1969.
159. **Russo, M. and Martelli, G. P.,** Cytology of *Gomphrena globosa* L. plants infected with beet mosaic virus, *Phytopathol. Mediterr.*, 8, 65, 1969.
160. **Martelli, G. P. and Russo, M.,** Unusual cytoplasmic inclusions induced by watermelon mosaic virus, *Virology*, 72, 352, 1976.
161. **Lawson, R. H., Hearon, S. S., and Smith, F. F.,** Development of pinwheel inclusions associated with sweet potato russet crack virus, *Virology*, 46, 453, 1971.

162. **Hampton, R. O., Phillips, S., Knesek, J. E., and Mink, G. I.,** Ultrastructural cytology of pea leaves and roots infected by pea seedborne mosaic virus, *Arch. Gesamte Virusforsch.*, 42, 242, 1973.
163. **Weintraub, M. and Ragetli, H. W. J.,** Distribution of viruslike particles in leaf cells of *Dianthus barbatus* infected with carnation vein mottle virus, *Virology*, 40, 868, 1970.
164. **McMullen, C. R. and Gardner, W. S.,** Cytoplasmic inclusions induced by wheat streak mosaic virus, *J. Ultrastruct. Res.*, 72, 65, 1980.
165. **Shepard, J. F. and Carroll, T. W.,** Electron microscopy of wheat streak mosaic virus particles in infected plant cells, *J. Ultrastruct. Res.*, 21, 145, 1967.
166. **Weintraub, M., Agrawal, H. O., and Ragetli, H. W. J.,** Cytoplasmic and nuclear inclusions in leaf cells infected with *Datura* shoestring virus (DSV), *Can. J. Bot.*, 51, 855, 1973.
167. **Langenberg, W. G. and Schroeder, H. F.,** The appearance of hordeum mosaic virus in wheat leaf cells after fixation in glutaraldehyde or chromate, *J. Gen. Virol.*, 23, 51, 1974.
168. **Koenig, R. and Lesemann, D.,** A potyvirus from *Gloriosa rothschildiana*, *Phytopathol. Z.*, 80, 136, 1974.
169. **Kitajima, E. W. and Costa, A. S.,** Aggregates of chloroplasts in local lesions induced in *Chenopodium quinoa* Wild. by turnip mosaic virus, *J. Gen. Virol.*, 20, 413, 1973.
170. **Kitajima, E. W. and Lovisolo, O.,** Mitochondrial aggregates in *Datura* leaf cells infected with henbane mosaic virus, *J. Gen. Virol.*, 16, 265, 1972.
171. **Weintraub, M. and Ragetli, H. W. J.,** Fine structure of inclusions and organelles in *Vicia faba* infected with bean yellow mosaic virus, *Virology*, 28, 290, 1966.
172. **Weintraub, M. and Ragetli, H. W. J.,** Intracellular characterization of bean yellow mosaic virus-induced inclusions by differential enzyme digestion, *J. Cell Biol.*, 38, 316, 1968.
173. **Langenberg, W. G. and Schroeder, H. F.,** Endoplasmic reticulum derived pinwheels in wheat infected with wheat spindle streak mosaic virus, *Virology*, 55, 218, 1973.
174. **Martelli, G. P., Russo, M., and Vovlas, C.,** Ultrastructure of zucchini yellow fleck virus infections, *Phytopathol. Mediterr.*, 20, 193, 1981.
175. **Oparka, K. J., Johnson, R. P. C., and Bowen, I. D.,** Sites of acid phosphatase in differentiating root protophloem of *Nymphoides peltata* (S.G. Gmel.) O. Küntze, *Plant Cell. Environ.*, 4, 27, 1981.
176. **McDonald, J. G. and Hiebert, E.,** Ultrastructure of the crystalline inclusion induced by tobacco etch virus visualized by freeze-etching, *J. Ultrastruct. Res.*, 48, 138, 1974.
177. **Knuhtsen, H., Hiebert, E., and Purcifull, D. E.,** Partial purification and some properties of tobacco etch virus induced intranuclear inclusions, *Virology*, 61, 200, 1974.
178. **Hooker, W. J. and Summanwar, A. S.,** Intracellular acridine orange fluorescence in plant virus infections, *Exp. Cell Res.*, 33, 609, 1964.
179. **Shepard, J. F.,** Electron microscopy of subtilisin treated tobacco etch virus nuclear and cytoplasmic inclusions, *Virology*, 36, 20, 1968.
180. **Hayashi, T. and Matsui, C.,** Electron microscopy of host cells infected with tobacco etch virus. IV. Tritiated uridine and leucine uptake of intracellular virus and intranuclear crystalline inclusions, *Virology*, 33, 47, 1967.
181. **Shepard, J. F., Gaard, G., and Purcifull, D. E.,** A study of tobacco etch virus-induced inclusions using indirect immunoferritin procedures, *Phytopathology*, 64, 418, 1974.
182. **Weintraub, M., Ragetli, H. W. J., and Veto, M.,** Fine structure of nuclear crystals in leaves of two *Dianthus* species, *Am. J. Bot.*, 55, 214, 1968.
183. **Kitajima, E. W. and Costa, A. S.,** The fine structure of the intranuclear, fibrous inclusions associated with the infection of celery yellow mosaic virus, *Fitopathol. Brasileira*, 3, 287, 1978.
184. **Da Cruz, N. D., Medina, D. M., Kitajima, E. W., Costa, A. S., and Landin, C. C.,** Citoquímica de inclusões intranucleares associadas ao virus do mosaico amarelo do salsão, *Bragantia*, 31, 217, 1972.

Chapter 15

CLOSTEROVIRUS GROUP

Members of the group have long, thin, very flexuous thread-like particles, and the group name is derived from the Greek word "kloster", meaning a spindle thread.[1] The type member of the group, beet yellows virus (BYV), has particles 1250 to 1450 nm long and about 12 nm wide, with a very characteristic open flexible structure. Other members have particles as long as 2000 nm or as short as 600 nm. Each BYV particle consists of a single molecule of positive-sense, single stranded RNA of between M_r 4.0 and 4.7 × 10^6, encapsidated in a coat of identical protein subunits of M_r about 22 × 10^3.[1-4]

BYV is confined to the phloem and phloem parenchyma of its host plants. The host ranges of individual Closteroviruses are typically rather narrow and the diseases produced are generally of the yellows type due to phloem necrosis and severe derangement of the chloroplasts.[5] The virus particles appear in the cytoplasm and are accompanied by characteristic accumulations of membranous vesicles.[2] Most group members are experimentally transmissible by mechanical inoculation, but often only with difficulty. BYV, like many Closteroviruses, is semipersistently transmitted by aphids. Recently, a mealybug has been confirmed as the vector of one Closterovirus, and whiteflies have been reported as the vectors of several Clostero-like viruses.[6-8]

I. MEMBERS OF THE GROUP AND THEIR RELATIONSHIPS

As knowledge of the Clostero-like viruses has developed, the outlines of the group have become less rather than more distinct. Members and possible members include several viruses resembling BYV, but there are others that deviate in one or more characters. For example, some viruses have typical particle structure but are only 600 to 800 nm in length; others have particles, either short or long, that lack the characteristic open structure and have coat protein subunits of M_r about 27 × 10^3 rather than 22 to 25 × 10^3. Some Closteroviruses are not confined to the phloem but invade parenchyma cells, and several do not appear to induce in infected cells the membranous vesicles characteristic of typical members. Some Closteroviruses are apparently not aphid transmitted, and at least one, grapevine virus A (GVA), has a mealybug vector.[6] Other Clostero-like viruses are transmitted by whiteflies.[7,8]

Because of this diversity, we have divided the members and possible members into three subgroups (I, II, and III) shown in Tables 1, 2, and 3. The viruses placed by Bar-Joseph and colleagues[2] in subgroup B become our subgroup III, while our subgroups I and II are the remaining viruses divided rather arbitrarily into those with long and short particles, respectively.

In subgroup I there are only three well-studied viruses: BYV, carnation necrotic fleck (CNFV), and citrus tristeza (CTV). CNFV is serologically related to BYV[20,62] and more distantly to wheat yellow leaf virus (WYLV)[33] None of the other subgroup I viruses appear to be interrelated serologically.

CNFV from Japan, carnation yellow fleck virus from Israel,[63] and a carnation virus from Australia were all independently described but are now known to be closely related and are included under the name CNFV.[20] A virus called carnation streak[17,18] also appears to be the same as CNFV.[3]

Beet yellow stunt virus (BYSV) is transmitted, like BYV, by *Myzus presicae* and both cause similar diseases in sugar beet and *Chenopodium capitatum*. However, their experimental host ranges differ, and no serological relationship has been detected between the two viruses[64] despite a suggestion to the contrary.[2]

Table 1
SOME PROPERTIES OF SUBGROUP I CLOSTEROVIRUSES

Virus	Modal length (nm)	Pitch of helix (nm)	Vectors	Location in tissues	Type of vesicles induced	Ref.
Members						
Beet yellows virus (BYV)	1250—1450	3.0—3.4	*Myzus persicae*, *Aphis fabae*, and others (aphids)	Phloem (epidermis and parenchyma in local lesions)	BYV-type (see text)	9—15
Carnation necrotic fleck virus (CNFV) (including carnation yellow fleck and carnation streak)	1100—1250, 1400—1500	3.4	*Myzus persicae* (aphid)	Phloem, epidermis, parenchyma, and xylem	BYV-type	16—20
Citrus tristeza virus (CTV)	~2000	3.7	*Toxoptera citricidus*, *Aphis gossypii*, and others (aphids)	Phloem	BYV-type	21—24
Beet yellow stunt virus (BYSV)	~1400	—[a]	*Nasonova lactucae* and others (aphids)	Phloem, phloem parenchyma	BYV-type	25—28
Burdock yellows virus (BuYV)	1600	3.6	*Macrosiphum gobonis* (aphid)	Phloem	BYV-type	29,30
Carrot yellow leaf virus (CYLV)	1600	3.7	*Semiaphis heraclei* (aphid)	Phloem	BYV-type	31
Clover yellows virus (CYV)	1700—1800	3.7	*Aphis craccivora* (aphid)	Phloem	BYV-type	2,3,32
Wheat yellow leaf virus (WYLV)	1600—1850	3.4	*Rhopalosiphum maidis*, *R. padi* (aphids)	Phloem	BYV-type	33,34
Possible members						
Alligatorweed stunting virus (AWSV)	1717	—	—	Phloem	BYV-type	2,35
Dendrobium vein necrosis virus (DVNV)	1865	—	—	Mesophyll	Mitochondrial vesicles	36
Festuca necrosis virus (FNV)	1725	—	*Rhopalosiphum padi* (aphid)	—	—	37
Heracleum virus 6 (HV6)	1600	—	*Cavariella* spp. (aphids)	—	—	38,39

[a] Not determined.

Table 2
SOME PROPERTIES OF SUBGROUP II CLOSTEROVIRUSES

Virus	Modal length (nm)	Pitch of helix (nm)	Vectors	Location in Tissues	Type of vesicles induced	Ref.
Members						
Apple chlorotic leafspot virus (ACLV)	600	3.8	ND[a]	Parenchyma	None found	2,3,40,41
Grapevine virus A (GVA) (formerly grapevine stem-pitting associated)	800	3.6—4.0	*Pseudococcus longispinus* (mealybug)	Parenchyma	None found	6,42,43
Heracleum latent virus (HLV)	730	3.8	*Cavariella spp.* (aphids) but only if helped by subgroup I Closterovirus HV6	Phloem, phloem parenchyma	ND	38,44—47
Possible Members						
Citrus tatter leaf virus (CiTLV)	650	ND	ND	ND	ND	48,49
Cucumber yellows virus (CuYV)	1000	ND	*Trialeurodes vaporariorum* (whitefly)	Phloem	BYV-type	7
Grapevine leafroll-associated virus (GLR-AV)	~1000	ND	ND	Phloem	BYV-type	50—52
Grapevine virus B (GVB)	ND	~3.8	ND	ND	ND	43
Nandina stem-pitting virus (NSPV)	755	3.9	ND	Phloem	None found	53

[a] ND, not determined.

Table 3
SOME PROPERTIES OF SUBGROUP III CLOSTEROVIRUSES

Virus	Modal length (nm)	Pitch of helix (nm)	Vectors	Location in tissues	Type of vesicles induced	Ref.
Apple stem grooving virus (ASGV)	600—700	3.4—3.8	ND[a]	ND	ND	54—56
Lilac chlorotic leafspot virus (LCLV)	~1540	3.7	ND	Phloem parenchyma, mesophyll	Dilated endoplasmic reticulum	57—59
Potato virus T (PVT)	610—640	3.4	ND	ND	ND	56,60,61

[a] ND, not determined.

The other viruses listed in Table 1 as possible members of subgroup I need further characterization before their relationships become clear. They infect quite different hosts but produce similar symptoms and cytopathology in them, accumulating in the phloem and inducing the proliferation of characteristic vesicles in the cytoplasm. These vesicles are further discussed in Section III and are referred to as "BYV-type" in Table 1. Festuca necrosis virus (FNV)[37] may be the same virus as WYLV though their serological relationship has not been studied. The viruses share several hosts and produce similar symptoms, though barley is infected by WYLV only.[33] The width of the FNV particle was reported as 18 nm[37] but the measurement was made on metal-shadowed preparations. The width of negatively stained particles might well be in line with that of bona fide Closteroviruses.

Table 2 lists the three known short Closteroviruses of subgroup II and several possible members. The subgroup II members share with those of subgroup I the same or closely similar particle fine structure (Figures 1 and 2) and have coat proteins of similar size (Table 4). Both of these characters differ in subgroup III (Figure 2). None of the subgroup II viruses are known to be serologically interrelated. Only heracleum latent virus (HLV) is known to be semipersistently aphid transmitted, but it depends for transmission on the helper virus heracleum virus 6 (HV6).[39] For the other viruses no vectors have been found.

Cucumber yellows virus (CuYV), despite its anomalous (whitefly) vector, will probably come to be included as a true subgroup II Closterovirus, but its nucleic acid, coat protein, and serological relationships need further study. Nandina stem-pitting virus (NSPV) may also be a new subgroup II Closterovirus, but its structure, chemical properties, and serological relationships are not yet clear enough for definitive assignment.

The possible and definite Closteroviruses associated with grapevines currently present a complicated picture. Clostero-like viruses occurring in the phloem and inducing "BYV-type" vesicles are associated with grapevine leafroll symptoms in Japan and California[50,52] and Italy;[51] these are listed in Table 2 as grapevine leafroll-associated virus (GLR-AV). The Japanese virus has a modal length of about 1000 nm.[50,52] There is a report of a Clostero-like virus in the phloem of grapevines with yellows symptoms in Germany,[73] and a Clostero-like virus has also been seen in negative-stain preparations from diseased grapevines in New Zealand[74] and Australia.[75]

A Closterovirus isolated from a stem-pitting diseased grapevine and multiplied in *Nicotiana clevelandii* was initially referred to as grapevine stem-pitting-associated virus,[42,72] but is now called grapevine virus A (GVA, Table 2). This virus has been serologically identified in a number of grapevines from Italy, Germany, and Israel, and appears to be correlated with leafroll symptoms more strongly than with those of stem-pitting.[43] The virus has been transmitted by mealybugs.[6]

Another Closterovirus-like particle serologically unrelated to GVA[43] has been detected in grapevines and it is suggested that it be provisionally named grapevine virus B (GVB, Table 2). Whether GLR-AV corresponds to GVA or GVB or a mixture of these remains to be determined.

The three viruses in subgroup III are listed in Table 3. They may eventually form the basis of a separate new virus group.[2] Apple stem grooving virus (ASGV) and potato virus T (PVT) are serologically closely related, differing by an SDI of between 1 and 2.[60] We include lilac chlorotic leafspot virus (LCLV) here because its particle structure resembles that of ASGV[42,76,77] (Figure 2) and because the size of the coat protein (Table 4) is similar to that of ASGV and PVT rather than to the viruses in subgroups I and II. None of the subgroup III viruses appears to be aphid transmitted; this may not be significant, given that all three viruses have natural hosts that are vegetatively propagated. The cytopathological data on subgroup III viruses are insufficient to help in their classification (Table 3), but LCLV appears to induce cytopathic effects somewhat different from those of subgroup I viruses.[58]

II. PARTICLE STRUCTURE AND COMPOSITION

A. Particle Structure

The particles of some representative Closteroviruses are shown in Figures 1 and 2. Modal lengths (Tables 1 through 3) are difficult to measure accurately because the particles are

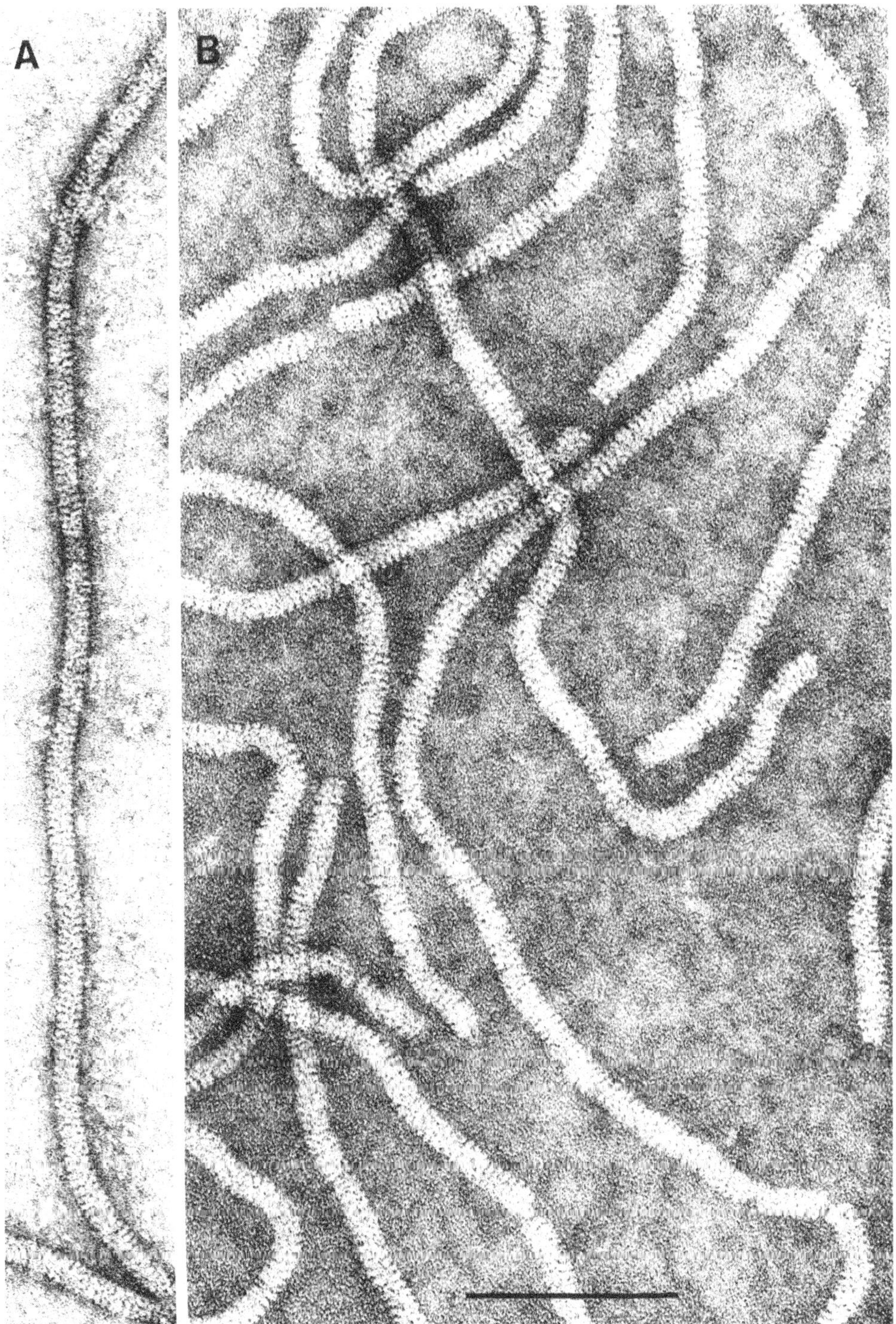

FIGURE 1. A, Purified particles of carnation necrotic fleck virus (CNFV) and B, citrus tristeza virus (CTV), both subgroup I Closteroviruses in uranyl formate stain. In (B) the characteristic open structure is well seen but the particles appear to be slightly damaged. Bar = 100 nm. (Purified virus preparations courtesy of Drs. J. Cohen and M. Bar-Joseph, Institute of Plant Protection, Volcani Center, Bet Dagen, Israel.)

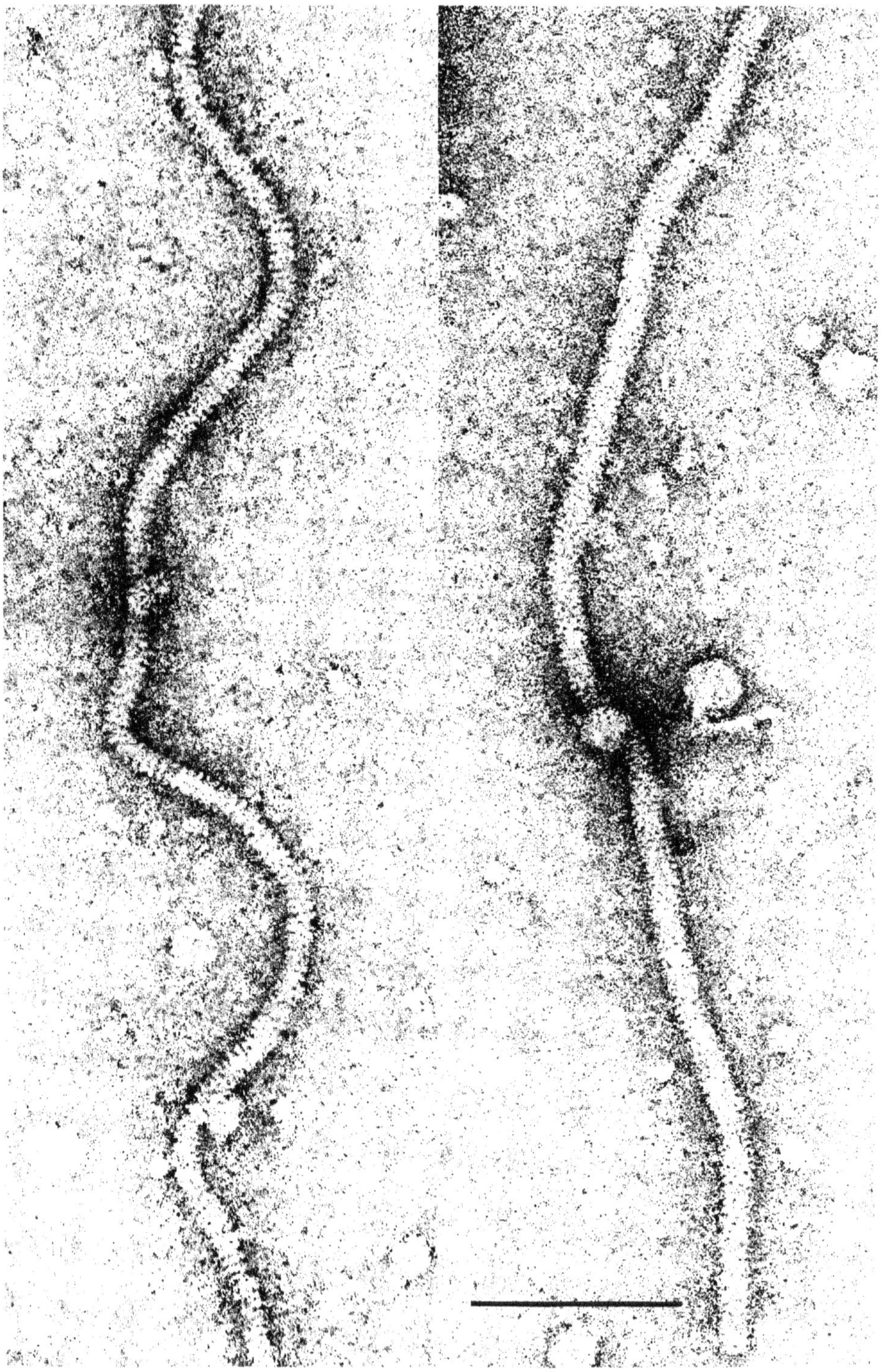

FIGURE 2. The particle on the left is that of grapevine virus A (GVA), a subgroup II Closterovirus, and the one on the right, that of apple stem grooving virus (ASGV), a subgroup III Closterovirus. Both particles were selected from the same micrograph of a mixture of purified preparations of the two viruses stained in uranyl formate. Note that the ASGV particle has a larger diameter and a less open structure. The pitch of the helix is also shorter. Bar = 100 nm.

Table 4
ESTIMATED SIZES OF THE NUCLEIC ACIDS AND COAT PROTEINS OF SOME CLOSTEROVIRUSES

Virus	Nucleic acid ($M_r \times 10^6$)	Coat protein ($M_r \times 10^3$)	Ref.
Subgroup I			
Beet yellows virus (BYV)	4.0—4.7	22.5—23.5	65—67
Carnation necrotic fleck virus (CNFV)	3.95	23.5	16,20
Citrus tristeza virus (CTV)	6.3—6.9	25	22,68,69
Subgroup II			
Apple chlorotic leafspot virus (ACLV)	2.15—2.48	23.5	45,70,71
Grapevine virus A (GVA)	2.55	22	72
Heracleum latent virus (HLV)	2.15—2.52	23.5	45,71
Subgroup III			
Lilac chlorotic leafspot virus (LCLV)	ND[a]	27	59
Potato virus T (PVT)	2.2—2.5	27	55,60

[a] ND, not determined.

very flexuous, often become entangled, easily break, and may assume different dimensions in different negative stains.[2,44,56] Bem and Murant[44] found with HLV that in 2% neutral sodium phosphotungstate the particles were very flexuous with a modal length of 730 nm; in 2% uranyl acetate, pH 3.5, the particles were straighter with a modal length of 770 nm; and in 2% neutral ammonium molybdate they were very straight with a modal length of 610 nm. The width of the particles also varied, being 10.5 nm in ammonium molybdate, 12.5 nm in the phosphotungstate, and 13.7 nm in uranyl acetate. Salazar and Harrison[56] found that PVT was slightly damaged and possibly shortened in neutral phosphotungstate but had the same modal length in uranyl acetate or phosphotungstate at pH 3.3, or after fixation and shadowing. As with HLV, the width of the particles varied from 10.8 to 13.1 nm in different negative stains. In contrast to these results which were obtained in the same laboratory, Conti and colleagues[42] using slightly different techniques, found a constant modal length of 800 nm for GVA, whether the particles were mounted in uranyl acetate or neutral phosphotungstate. Because of the difficulties of measurement, the use of different methods and stains, and the general lack of comparative data, differences of 100 nm in a modal length of, say, 1500 nm are probably not significant, and the figures quoted in Table 1 (and to a lesser extent in Tables 2 and 3) should be accepted with caution.

The pitch of the primary helix has been estimated for several Closteroviruses (Tables 1 through 3) and as with modal length, it is by no means certain that the various values all represent real differences, because closely controlled comparisons have seldom been made. Salazar and Harrison[56] did compare the pitch of PVT and ASGV and found them both to be 3.4 nm, in contrast to the figure of 3.8 nm reported for ASGV by Lister.[54] The most rigorous and perhaps the only really acceptable procedure when comparing the morphology of virus particles by electron microscopy is to mix the particles and co-image them in the same photographic field. Conti and colleagues[42] compared particles of GVA and ASGV co-imaged in uranyl acetate (Figure 2), and found the pitch of ASGV shorter in the ratio 1:0.87, while the ASGV particles were wider in the ratio 1:1.2.

Estimates of the size of the coat proteins of different viruses in subgroups I and II have given rather consistent figures of M_r between 22.5 and 25 $\times$ 10^3 (Table 4). This contrasts with the value of M_r 27 $\times$ 10^3 for the two subgroup III viruses for which data are available (Table 4).

The amino acid compositions of the coat proteins of BYV and CNFV have been determined.[2,62,66] The proteins differed by about 50 amino acids, and tryptophan was absent from both. Absence of tryptophan is also characteristic of apple chlorotic leafspot virus (ACLV)[65] and HLV[45] and possibly GVA.[72] This anomalous amino acid composition of these viruses raised the A260/A280 UV absorbance ratio to around 1.5. The ratio is only 1.18 for ASGV and PVT (subgroup III) indicating that aromatic amino acids are probably present. This trend begins to look interesting until we see that CTV (subgroup I) has the low A260/A280 value of 1.21 to 1.22.[3] There appears to be some conflict regarding LCLV (subgroup III) in that the A260/A280 ratio reported for the type virus is 1.69,[59] whereas that reported for a serologically closely related iris isolate is only 1.16.[76] It is thus difficult at present to draw conclusions on this point, especially as some preparations may contain UV-absorbing contaminants influencing the A260/A280 ratio.[67]

Carpenter and colleagues[66] after analyzing the protein and RNA of BYV and considering published data on the particle structure, concluded that the RNA is embedded at a radius of 3.3 nm. In agreement with Bar-Joseph and Hull,[65] they suggested that the protein subunit is more or less spherical, 3.7 nm in diameter, and that 4 nucleotides are probably associated with each subunit in the helix, there being about 3400 subunits in each particle. It seems that there are likely to be about 10 subunits per turn of the helix.

With PVT, Salazar and his colleagues[56,61] found that the virus varied in appearance according to its history of purification and preparation for electron microscopy. Five morphological aspects were noted. Particles in neutral phosphotungstate had no obvious fine structure, while those in uranyl acetate or uranyl formate presented appearances termed "cross-banding", "criss-cross", and "rope" (of two types). Optical diffraction studies indicated that the helix had a pitch of 3.4 nm and that the different surface patterns observed were probably due to slight distortions of the helix combined with different clustering of the protein monomers. Thus it was suggested that clustering in pairs could give rise to the "criss-cross" pattern, grouping in sevens could generate the principal "rope" pattern and lack of clustering could produce the "cross-banding", which represented the unmodified helix. ASGV, when prepared under similar conditions, gave only the "cross-banding" pattern (Figure 2).[42] The "cross-banding" structure was also noted for the iris strain of LCLV.[76]

It has been pointed out[2,68] that the ratio of RNA mass to modal length in Closterovirus particles ranges between 2821 and 3437 M_r/nm. BYV (3230), CNFV (2821-3160), ACLV (3191), and HLV (3013) form one group with lower values, whereas PVT (3437) has a higher value nearer to that of Potex-, Carla-, and Potyviruses (4038 to 4112).[78]

B. The Viral Genome

The RNA of BYV is infectious[65] and that of CTV has been shown to act directly as a messenger in a rabbit reticulocyte translation system.[79] Thus presumably these viruses and probably all Closteroviruses have positive-sense RNA. The sizes reported for the viral RNAs of the several Closteroviruses studied (Table 4) vary somewhat according to the method of estimation.[2] Electrophoresis in nondenaturing aqueous gels has generally been used, and this gives a value somewhat higher than in denaturing gels containing 8 *M* urea. Bem and Murant[45] estimated the sizes of the RNAs of HLV and ACLV in aqueous gels and urea gels. They found for both viruses a value of 2.3×10^6 M_r in aqueous gels and 2.15×10^6 in urea gels, and suggested that the latter figure was more reliable. More recently, Murant and colleagues[71] have recommended the use of agarose-gel electrophoresis and electron microscopic measurement of RNA denatured in glyoxal for more accurate molecular-weight estimates, especially for RNAs of M_r over 2×10^6. Using this system for ACLV and HLV RNAs, they obtained values of M_r 2.48 and 2.52×10^6, respectively. Thus it seems that most of the values in Table 4 may have to be revised upward.

Little work has yet been done on analysis of Closterovirus genomes. However, Nagel and colleagues[79] obtained translation of CTV RNA in rabbit reticulocyte lysates and detected a product of M_r about 26 × 10^3 which they identified serologically as the coat protein. In addition, a major product and two minor ones of M_r about 50, 65, and 33 × 10^3, respectively, were detected, though roles for these proteins remain undetermined.

Recently, double-stranded RNAs have been isolated from plants infected with ACLV, BYV, and CTV.[69] The sizes of the molecules, estimated by electron microscopy and by gel electrophoresis, were 4.6, 8.4, and 13.3 × 10^6 M_r for the three viruses, respectively, which corresponds to twice the size of the best estimates for their genomic single-stranded RNAs. They are probably the viral replicative forms. Multiple double-stranded RNAs of smaller sizes were also detected; these could be the templates for transcription of subgenomic messenger RNAs.

The genomes of Closteroviruses will need further analysis before the reasons for such large differences in their sizes (and hence in the lengths of the particles) become apparent.

III. CYTOPATHOLOGY

The cytopathology of viruses in the group has been reviewed from time to time.[2,3,80-82] Closteroviruses, in common with many kinds of viruses, induce the range of pathological changes that is generally found in chlorotic and necrotic tissues. Most characteristic are the accumulation of virus particles in large masses, usually in the phloem and phloem parenchyma (Figures 3 and 4) and the proliferation of cytoplasmic membranous vesicles (Figure 5).

With members of subgroup I, though apparently not subgroups II and III, the virus particles may be aligned in ranks to give cross-banded inclusions visible in the light as well as the electron microscope.[2,11,13,15,80,83]

Care has not always been taken to distinguish clearly between virus particles and P-protein, which can occur in several morphological forms depending on the species and the developmental stage of the cell[51,80,84,85] (Figure 6). Where possible, both P-protein and virus particles should be clearly demonstrated in the same cell section.[80]

Many subgroup I Closteroviruses induce what appear in the light microscope as amorphous cytoplasmic inclusions. In thin sections these are seen to consist mainly of membranous vesicles together with virus particles and normal components of the cytoplasm.[11,80,86] The vesicles contain fine fibrils that could be virus-specific RNA (Figure 5), and Esau and Hoefert[11] suggested for BYV that the vesicles might be the sites of virus synthesis. This idea has remained plausible but, so far, without support, except that similar vesicles seem to be a constant feature of infection by subgroup I viruses. These vesicles, referred to as "BYV-type" in Tables 1 and 2, have not been demonstrated with the two subgroup II viruses examined so far, but have been seen in tissues infected with GLR-AV and with CuYV which are considered to be possible members of this subgroup.

Virus like particles have been observed in the nuclei of plants infected with BYV[15,80] (Figure 4) and NSPV.[53] In addition, NSPV forms unique tubular-coil structures in the cytoplasm which are probably composed of virus particles.[53]

In plants infected with LCLV (subgroup III), BYV-type vesicles are not found, but there is a proliferation of the endoplasmic reticulum.[58] The cytopathology of the other subgroup III viruses has not been investigated.

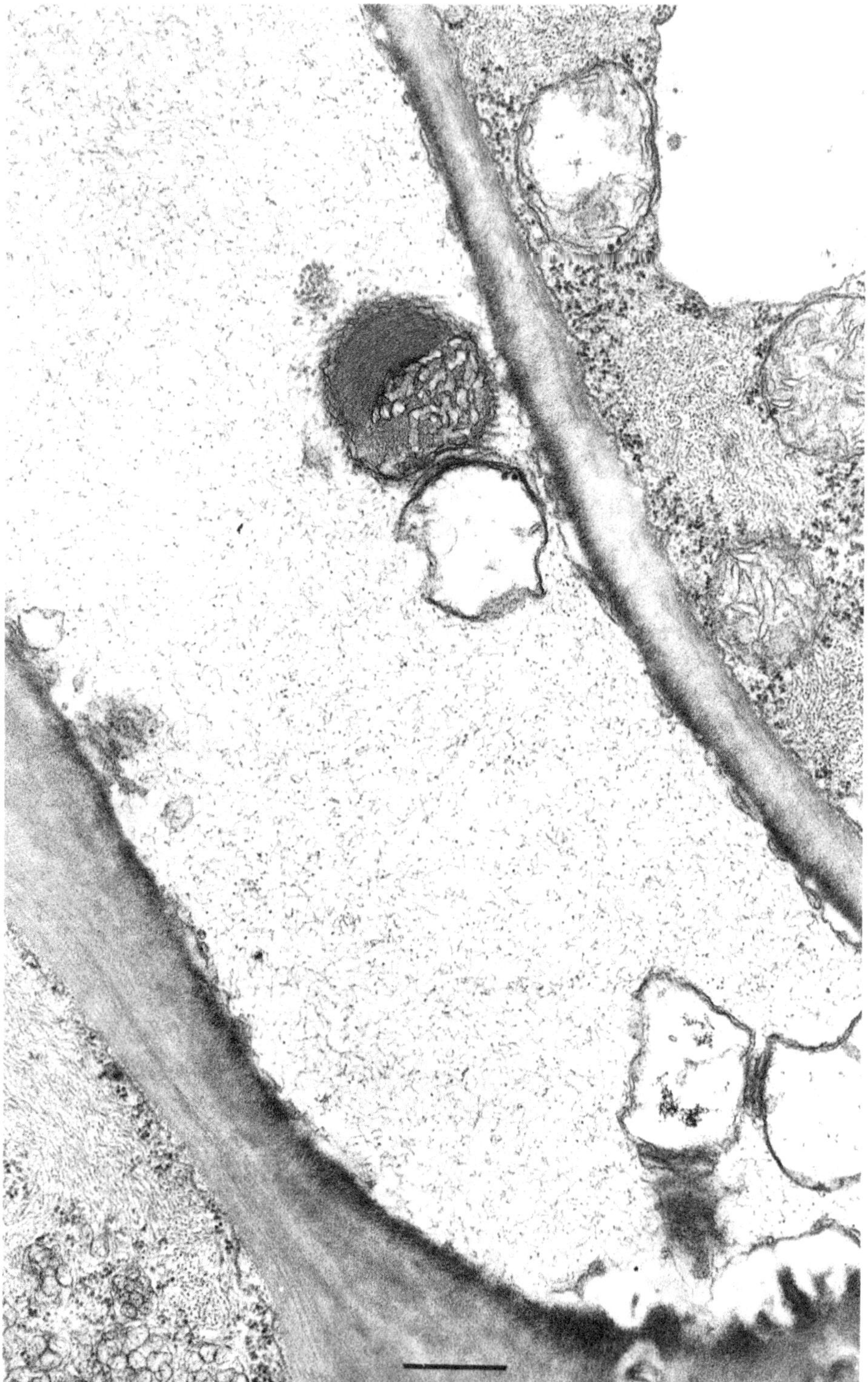

FIGURE 3. Thin section of a sieve tube flanked by two phloem parenchyma cells of a *Beta vulgaris* (sugarbeet) plant infected with beet yellows virus (BYV). Masses of virus particles are seen in all three cells. "BYV-type" vesicles appear in the lower left corner of the figure. Bar = 500 nm. (Courtesy of Dr. K. Esau, Department of Biological Sciences, University of California, Santa Barbara.)

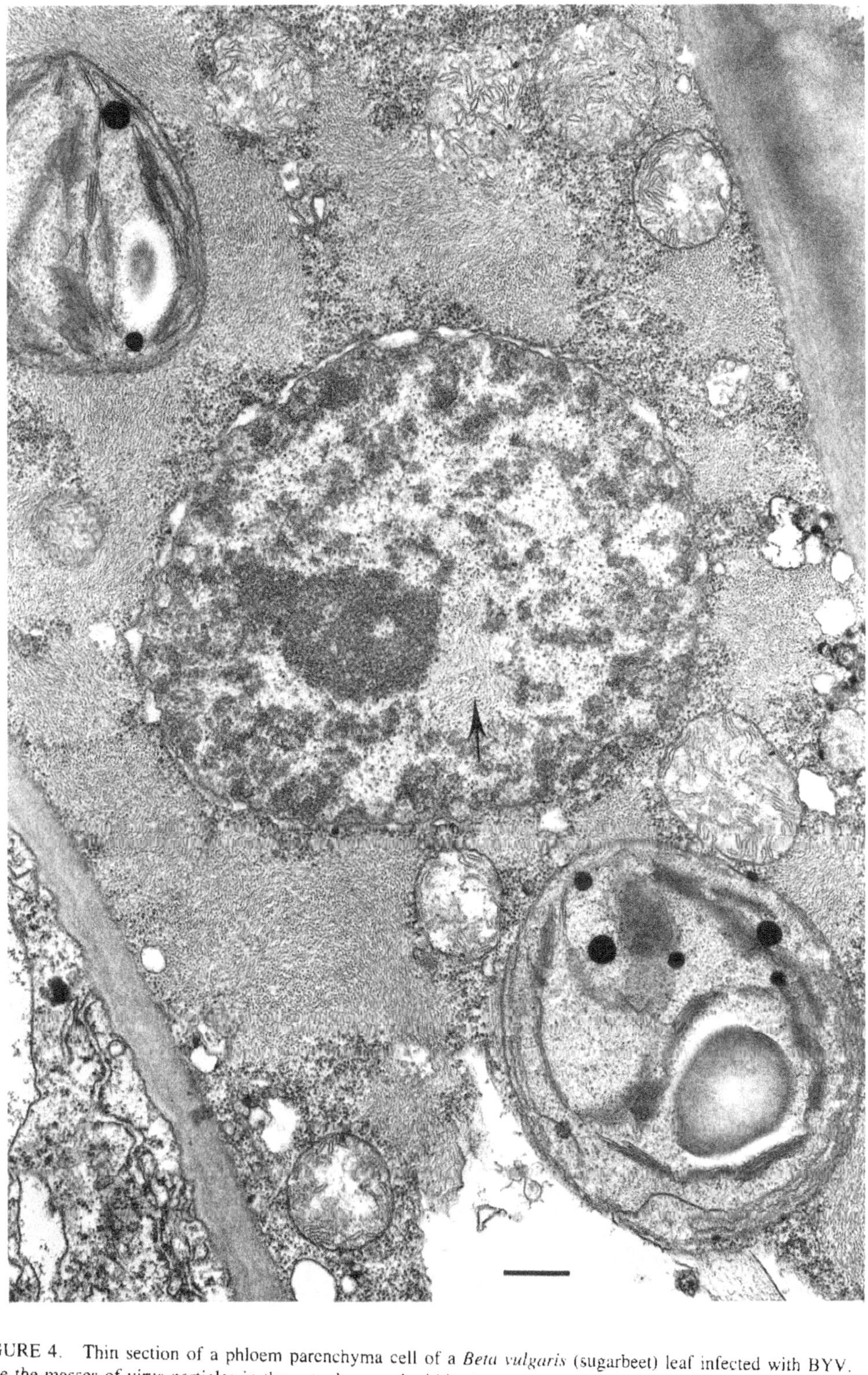

FIGURE 4. Thin section of a phloem parenchyma cell of a *Beta vulgaris* (sugarbeet) leaf infected with BYV. Note the masses of virus particles in the cytoplasm and within the nucleus (arrow). Bar = 500 nm. (Courtesy of Dr. K. Esau, Department of Biological Sciences, University of California, Santa Barbara.)

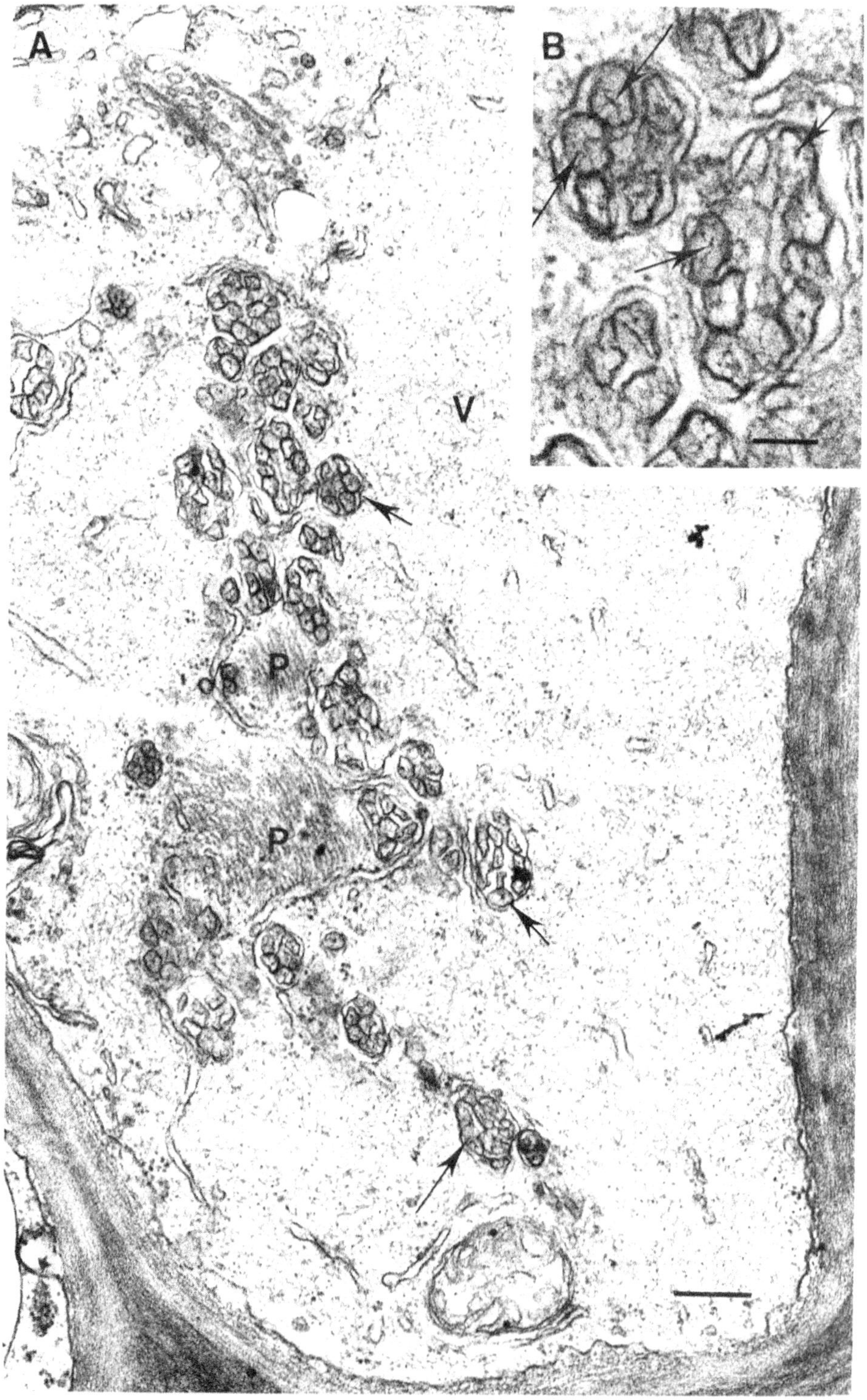

FIGURE 5. Thin section of a maturing sieve tube from a CTV-infected *Citrus* sp. leaf. A, Virus particles (V), tubular P-protein (P), and "BYV-type" vesicles (arrows); B, details of the vesicles containing fibrils (arrows) reminiscent of nucleic acid. Bars = 500 nm in (A) and 100 nm in (B). (Embedded block courtesy of Dr. J. Cohen, Institute of Plant Protection, Volcani Center, Bet Dagen, Israel.)

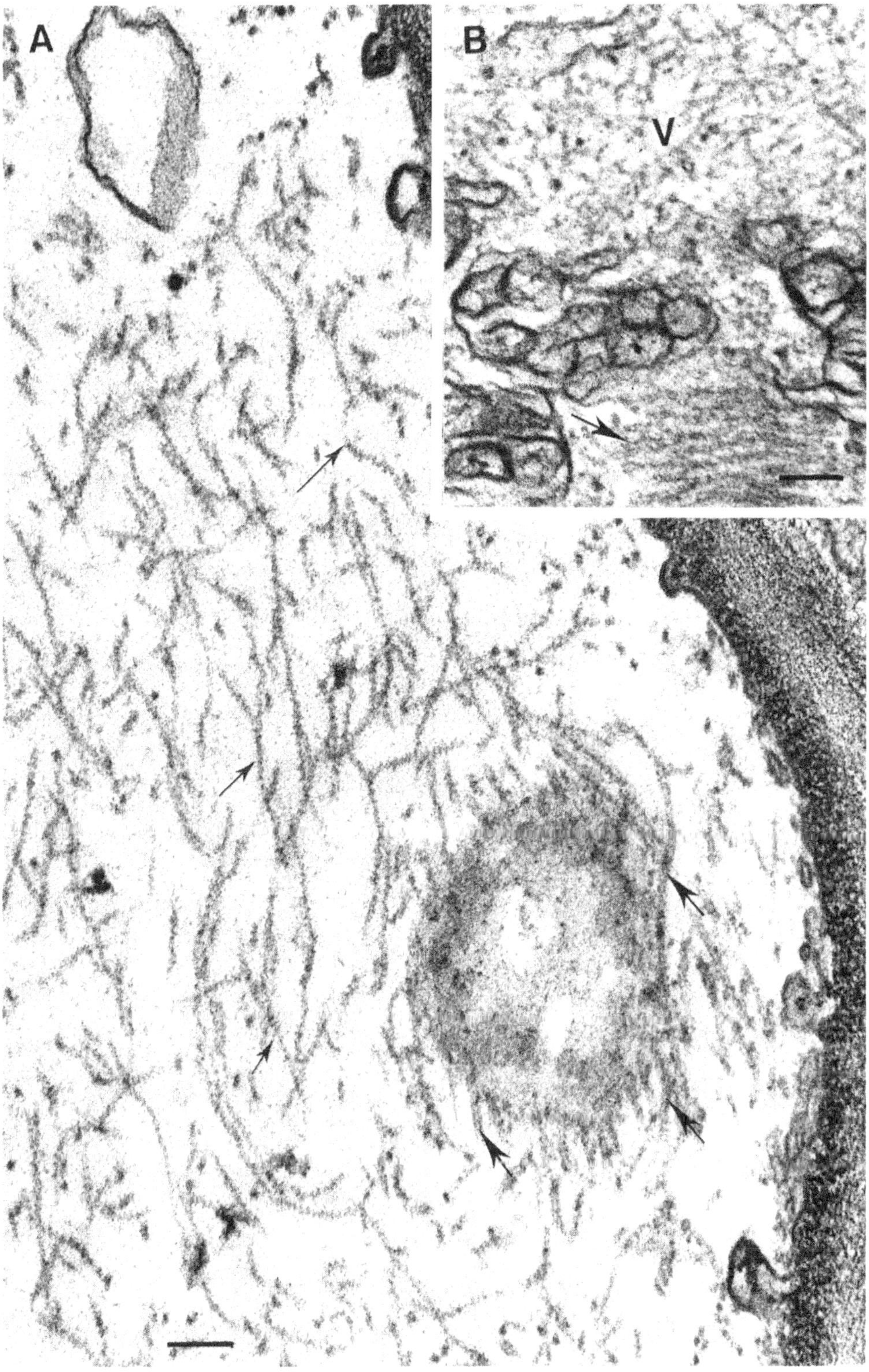

FIGURE 6. Thin sections of sieve tubes from a CTV-infected *Citrus* sp. leaf. A, P-protein appearing as tubules (large arrows) and as beaded filaments (small arrows) but no virus particles; B, tubular P-protein (arrow), a mass of virus particles (V), and some virus-specific vesicles. Bars = 100 nm. (Embedded block courtesy of Dr. J. Cohen, Institute of Plant Protection, Volcani Center, Bet Dagen, Israel.)

REFERENCES

1. **Matthews, R. E. F.,** Classification and nomenclature of viruses. Fourth report of the International Committee on Taxonomy of Viruses, *Intervirology*, 17, 1, 1982.
2. **Bar-Joseph, M., Garnsey, S. M., and Gonsalves, D.,** The Closteroviruses: a distinct group of elongated plant viruses, *Adv. Virus Res.*, 25, 93, 1979.
3. **Lister, R. M. and Bar-Joseph, M.,** Closteroviruses, in *Handbook of Plant Virus Infections and Comparative Diagnosis*, Kurstak, E., Ed., Elsevier, Amsterdam, 1981, 809.
4. **Bar-Joseph, M. and Murant, A. F.,** Closterovirus group, *CMI/AAB Descriptions of Plant Viruses*, No. 260, 1982.
5. **Duffus, J. E.,** The yellowing virus diseases of beet, *Adv. Virus Res.*, 18, 347, 1973.
6. **Rosciglione, B., Castellano, M. A., Martelli, G. P., Savino, V., and Cannizzaro, G.,** Mealybug transmission of grapevine virus A, *Vitis*, 22, 331, 1983.
7. **Yamashita, S., Doi, Y., Yora, K., and Yoshino, M.,** Cucumber yellows virus: its transmission by the greenhouse whitefly, *Trialeurodes vaporariorum* (Westwood), and the yellowing disease of cucumber and muskmelon caused by the virus, *Ann. Phytopathol. Soc. Jpn.*, 45, 484, 1979.
8. **Duffus, J. E.,** Whitefly vectors: virus-vector-plant interactions, transmission mechanisms and virus spread, *4th Int. Congr. Plant Pathol. Melbourne, Abstr. Papers*, No 2, 1983, p. 1.
9. **Varma, A., Gibbs, A. J., Woods, R. D., and Finch, J. T.,** Some observations on the structure of the filamentous particles of several plant viruses, *J. Gen. Virol.*, 2, 107, 1968.
10. **Russell, G. E.,** Beet yellows virus, *CMI/AAB Descriptions of Plant Viruses*, No. 13, 1970.
11. **Esau, K. and Hoefert, L. L.,** Cytology of beet yellows virus infection in *Tetragonia*. I. Parenchyma cells in infected leaf, *Protoplasma*, 72, 255, 1971.
12. **Esau, K. and Hoefert, L. L.,** Cytology of beet yellows virus infection in *Tetragonia*. II. Vascular elements in infected leaf, *Protoplasma*, 72, 459, 1971.
13. **Esau, K. and Hoefert, L. L.,** Cytology of beet yellows virus infection in *Tetragonia*. III. Conformation of virus in infected cells, *Protoplasma*, 73, 51, 1971.
14. **Polák, J.,** Physical properties and serological relationships of beet yellows virus strains, *Phytopathol. Z.*, 72, 235, 1971.
15. **Plaskitt, K. A. and Bar-Joseph, M.,** Electron microscopy of beet yellows virus infection in *Chenopodium hybridum L.*, *Micron*, 9, 109, 1978.
16. **Inouye, T.,** Carnation necrotic fleck virus, *CMI/AAB Descriptions of Plant Viruses*, No. 136, 1974.
17. **Poupet, A., Cardin, L., Marais, A., and Cadilhac, B.,** La bigarrure de l'oeillet: isolement et propriétés d'un virus filamenteux, *Ann. Phytopathol.*, 7, 277, 1975.
18. **Cadilhac, B., Poupet, A., Cardin, L., and Marais, A.,** La bigarrure de l'oeillet (carnation streak virus): cytologie de l'infection, *C. R. Acad. Sci. Paris*, 281, 639, 1975.
19. **Bar-Joseph, M. and Smookler, M.,** Purification, properties, and serology of carnation yellow fleck virus, *Phytopathology*, 66, 835, 1976.
20. **Bar-Joseph, M., Inouye, T., and Sutton, J.,** Serological relationships among thread-like viruses infecting carnations from Japan, Israel, and Australia, *Plant Dis. Rep.*, 60, 851, 1976.
21. **Price, W. C.,** Citrus tristeza virus, *CMI/AAB Descriptions of Plant Viruses*, No. 33, 1970.
22. **Bar-Joseph, M., Loebenstein, G., and Cohen, J.,** Further purification and characterization of threadlike particles associated with the citrus tristeza disease, *Virology*, 50, 821, 1972.
23. **Gonsalves, D., Purcifull, D. E., and Garnsey, S. M.,** Purification and serology of citrus tristeza virus, *Phytopathology*, 68, 553, 1978.
24. **Tsuchizaki, T., Sasaki, A., and Saito, Y.,** Purification of citrus tristeza virus from diseased citrus fruits and the detection of the virus in citrus tissues by fluorescent antibody techniques, *Phytopathology*, 68, 139, 1978.
25. **Hoefert, L. L., Esau, K., and Duffus, J. E.,** Electron microscopy of Beta leaves infected with beet yellow stunt virus, *Virology*, 42, 814, 1970.
26. **Duffus, J. E.,** Beet yellow stunt virus, *CMI/AAB Descriptions of Plant Viruses*, No. 207, 1979.
27. **Esau, K.,** Beet yellow stunt virus in cells of *Sonchus oleraceus* L. and its relation to host mitochondria, *Virology*, 98, 1, 1979.
28. **Esau, K. and Hoefert, L. L.,** Beet yellow stunt virus in the phloem of *Sonchus oleraceus* L., *J. Ultrastruct. Res.*, 75, 326, 1981.
29. **Nakano, M. and Inouye, T.,** Host range and cytopathology of a filamentous virus (Bd-F) detected in burdock, *Ann. Phytopathol. Soc. Jpn.*, 44, 96, 1978.
30. **Nakano, M. and Inouye, T.,** Burdock yellows virus, a closterovirus from *Arctium lappa* L., *Ann. Phytopathol. Soc. Jpn.*, 46, 7, 1980.
31. **Yamashita, S., Ohki, S. T., Doi, Y., and Yora, K.,** Two yellows-type viruses detected from carrot, *Ann. Phytopathol. Soc. Jpn.*, 42, 382, 1976.

32. **Ohki, S. T., Doi, Y., and Yora, K.,** Clover yellows virus, *Ann. Phytopathol. Soc. Jpn.*, 42, 313, 1976.
33. **Inouye, T.,** Wheat yellow leaf virus, *CMI/AAB Descriptions of Plant Viruses*, No. 157, 1976.
34. **Inouye, T.,** personal communication.
35. **Hill, H. R. and Zettler, F. W.,** A virus-like stunting disease of alligatorweed from Florida, *Phytopathology*, 63 (Abstr.), 443, 1973.
36. **Lesemann, D. E.,** Long, filamentous virus-like particles associated with vein necrosis of *Dendrobium phalaenopsis*, *Phytopathol. Z.*, 89, 330, 1977.
37. **Schmidt, H. B., Richter, J., Hertzsch, W., and Klinkowski, M.,** Untersuchungen über eine virusbedingte Nekrose an Futtergräsern, *Phytopathol. Z.*, 47, 66, 1963.
38. **Bem, F. and Murant, A. F.,** Transmission and differentiation of six viruses infecting Hogweed (*Heracleum sphondylium*) in Scotland, *Ann. Appl. Biol.*, 92, 237, 1979.
39. **Murant, A. F.,** Dependence of heracleum latent virus on a fellow closterovirus for transmission by the aphid *Cavariella theobaldi*, *4th Int. Congr. Plant Pathol. Melbourne, Abstr. Papers*, No. 480, 1983, p. 121.
40. **Lister, R. M.,** Apple chlorotic leaf spot virus, *CMI/AAB Descriptions of Plant Viruses*, No. 30, 1970.
41. **Lister, R. M. and Hadidi, A. F.,** Some properties of apple chlorotic leaf spot virus and their relation to purification problems, *Virology*, 45, 240, 1971.
42. **Conti, M., Milne, R. G., Luisoni, E., and Boccardo, G.,** A Closterovirus from a stem-pitting-diseased grapevine, *Phytopathology*, 70, 394, 1980.
43. **Milne, R. G., Conti, M., Lesemann, D. E., Tanne, E., and Cohen, J., Stellmach, G.,** Closterovirus-like particles of two types associated with diseased grapevines, *Phytopathol. Z.*, in press, 1984.
44. **Bem, F. and Murant, A. F.,** Host range, purification and serological properties of heracleum latent virus, *Ann. Appl. Biol.*, 92, 243, 1979.
45. **Bem, F. and Murant, A. F.,** Comparison of particle properties of heracleum latent and apple chlorotic leaf spot viruses, *J. Gen. Virol.*, 44, 817, 1979.
46. **Bem, F. and Murant, A. F.,** Heracleum latent virus, *CMI/AAB Descriptions of Plant Viruses*, No. 228, 1980.
47. **Murant, A. F., Taylor, M., and Duncan, G. H.,** Estimates of molecular weight of virus RNA by electrophoresis and by electron microscopy, *Scott. Hortic. Res. Inst. Ann. Rep. for 1978*, p. 98, 1978.
48. **Semancik, J. S. and Weathers, L. G.,** Partial purification of a mechanically transmissible virus associated with tatter leaf of citrus, *Phytopathology*, 55, 1354, 1965.
49. **Inoue, N., Maeda, T., and Mitsuhata, K.,** Citrus tatter leaf virus isolated from lily, *Ann. Phytopathol. Soc. Jpn.*, 45, 712, 1979.
50. **Namba, S., Yamashita, S., Doi, Y., Yora, K., Terai, Y., and Yano, R.,** Grapevine leafroll virus, a possible member of Closteroviruses, *Ann. Phytopathol. Soc. Jpn.*, 45, 497, 1979.
51. **Faoro, F., Tornaghi, R., Fortusini, A., and Belli, G.,** Association of a possible Closterovirus with grapevine leafroll in Northern Italy, *Riv. Patol. Veg.*, 17, 183, 1981.
52. **Yora, K., Saito, Y., Doi, Y., Inouye, T., and Tomaru, K., Eds.,** *Handbook of Plant Viruses*, Asakura Shoten, Tokyo, 1983 (in Japanese), 632 pp.
53. **Ahmed, N. A., Christie, S. R., and Zettler, F. W.,** Identification and partial characterization of a Closterovirus infecting *Nandina domestica*, *Phytopathology*, 73, 470, 1983.
54. **Lister, R. M.,** Apple stem grooving virus, *CMI/AAB Descriptions of Plant Viruses*, No. 31, 1970.
55. **Van der Meer, F. A.,** Observation on apple stem grooving virus, *Acta Hortic.*, 67, 293, 1976.
56. **Salazar, L. F. and Harrison, B. D.,** Host range, purification and properties of potato virus T, *Ann. Appl. Biol.*, 89, 223, 1978.
57. **Brunt, A. A.,** The occurrence, hosts and properties of lilac chlorotic leafspot virus, a newly recognized virus from *Syringa vulgaris*, *Ann. Appl. Biol.*, 88, 383, 1978.
58. **Brunt, A. A. and Stace-Smith, R.,** The intracellular location of lilac chlorotic leafspot virus, *J. Gen. Virol.*, 39, 63, 1978.
59. **Brunt, A. A.,** Lilac chlorotic leafspot virus, *CMI/AAB Descriptions of Plant Viruses*, No. 202, 1979.
60. **Salazar, L. F. and Harrison, B. D.,** Potato virus T, *CMI/AAB Descriptions of Plant Viruses*, No. 187, 1978.
61. **Salazar, L. F., Hutcheson, A. M., Tollin, P., and Wilson, H. R.,** Optical diffraction studies of particles of potato virus T, *J. Gen. Virol.*, 39, 333, 1978.
62. **Short, M. N., Hull, R., Bar-Joseph, M., and Rees, M. W.,** Biochemical and serological comparisons between carnation yellow fleck virus and sugar beet yellows virus protein subunits, *Virology*, 77, 408, 1977.
63. **Smookler, M. and Loebenstein, G.,** Carnation yellow fleck virus, *Phytopathology*, 64, 979, 1974.
64. **Duffus, J. E.,** Beet yellow stunt, a potentially destructive virus disease of sugar beet and lettuce, *Phytopathology*, 62, 161, 1972.
65. **Bar-Joseph, M. and Hull, R.,** Purification and partial characterization of sugar beet yellows virus, *Virology*, 62, 552, 1974.

66. **Carpenter, J. M., Kassanis, B., and White, R. F.,** The protein and nucleic acid of beet yellows virus, *Virology*, 77, 101, 1977.
67. **Kassanis, B., Carpenter, J. M., White, R. F., and Woods, R. D.,** Purification and some properties of beet yellows virus, *Virology*, 77, 95, 1977.
68. **Bar-Joseph, M., Loebenstein, G., and Cohen, J.,** Comparison of particle characteristics and cytopathology of citrus tristeza virus with other morphologically similar viruses, *Proc. 7th Conf. Int. Org. Citrus Virol.*, Calavan, E. C., Ed., 1976, 39.
69. **Dodds, J. A. and Bar-Joseph, M.,** Double-stranded RNA from plants infected with closteroviruses, *Phytopathology*, 73, 419, 1983.
70. **Bar-Joseph, M., Hull, R., and Lane, L. C.,** Biophysical and biochemical characterization of apple chlorotic leafspot virus, *Virology*, 62, 563, 1974.
71. **Murant, A. F., Taylor, M., Duncan, G. H., and Raschké, J. H.,** Improved estimates of molecular weight of plant virus RNA by agarose gel electrophoresis and electron microscopy after denaturation with glyoxal, *J. Gen. Virol.*, 53, 321, 1981.
72. **Boccardo, G. and d'Aquilio, M.,** The protein and nucleic acid of a Closterovirus isolated from a grapevine with stem-pitting symptoms, *J. Gen. Virol.*, 53, 179, 1981.
73. **Mendgen, K.,** Untersuchungen über eine Vergilbungskrankheit der Reben an Rhein, Mosel und Saar, *Weinberg Keller*, 18, 345, 1971.
74. **Mossop, D. W.,** personal communication, 1983.
75. **Hatta, T.,** unpublished observations, 1980.
76. **Lisa, V.,** Two viruses from rhizomatous iris, *Acta Hortic.*, 110, 39, 1980.
77. **Milne, R. G., and Masenga, V.,** unpublished observations.
78. **Veerisetty, V.,** Relationships among structural parameters of virions of helical symmetry, *Virology*, 84, 523, 1978.
79. **Nagel, J., Hiebert, E., and Lee, R. F.,** Citrus tristeza virus RNA translated with a rabbit reticulocyte lysate: capsid protein identified as one of the products, *Phytopathology*, 72 (Abstr.), 953, 1982.
80. **Esau, K.,** *Viruses in Plant Hosts; Form, Distribution and Pathologic Effects*, University of Wisconsin Press, Madison, 1968, 225 pp.
81. **Christie, R. G. and Edwardson, J. R.,** Light and electron microscopy of plant virus inclusions, *Fla. Agr. Exp. Sta. Monogr. Ser.*, No. 9, 1977, 150 pp.
82. **Martelli, G. P. and Russo, M.,** Plant virus inclusion bodies, *Adv. Virus Res.*, 21, 175, 1977.
83. **Bar-Joseph, M., Josephs, R., and Cohen, J.,** Carnation yellow fleck virus particles "in vivo": a structural analysis, *Virology*, 81, 144, 1977.
84. **Cronshaw, J.,** P-proteins, in *Phloem Transport*, Aronoff, S., Dainty, J., Gorham, P. R., Srivastava, L. M., and Swanson, C. A., Eds., Plenum Press, New York, 1975, 79.
85. **Evert, R. F.,** Phloem structure and histochemistry, *Annu. Rev. Plant Physiol.*, 28, 199, 1977.
86. **Cronshaw, J., Hoefert, L., and Esau, K.,** Ultrastructural features of *Beta* leaves infected with beet yellows virus, *J. Cell Biol.*, 31, 429, 1966.

Chapter 16

SOME UNCLASSIFIED VIRUSES

In this chapter we discuss some viruses that have not so far been assigned to any group recognized by the ICTV.[1] A number of the viruses do in fact fall into recognizable clusters that may well receive official recognition in the future as new virus groups. Others still defy classification, and several of these have already been mentioned in some of the preceding chapters. For convenience, we treat these viruses in eight separate categories.

I. OAT BLUE DWARF VIRUS AND POSSIBLE RELATIVES

Oat blue dwarf virus (OBDV) has small polyhedral particles about 30 nm in diameter which sediment as a single component of 119S, and each particle contains a single molecule of RNA which appears to be single-stranded and of M_r about 2.1×10^6.[2] The virus is restricted to the phloem of infected plants.[2] These properties suggest that the virus could be a Luteovirus, but OBDV differs from the Luteoviruses in its vector transmission. It is transmitted by a leafhopper in which it also appears to multiply,[2,3] whereas Luteoviruses are transmitted circulatively in aphids but apparently without multiplication (see Volume 1, Chapter 9).

It has been suggested that maize rayado fino virus (MRFV) has affinities with OBDV.[4] Both viruses do replicate in their leafhopper vectors,[3,5,6] but they are not related serologically; moreover, their physical properties differ significantly.[2,4,7,8] The physical and chemical properties of MRFV show some remarkable similarities to the Tymoviruses which have been discussed in Chapter 8 of Volume 1.

The isometric particle component of rice tungro disease (rice tungro isometric virus; RTIV), which is also called rice waika and penyakit habang, is another virus with small isometric particles containing single-stranded RNA and is transmitted by leafhoppers. However, RTIV has a reported S value of 175 and is semipersistently but nonpropagatively transmitted.[9-13] It is listed by the ICTV as a possible member of the maize chlorotic dwarf virus group[1] and has been discussed in Chapter 7 of the first volume.

II. PHLEUM MOTTLE AND SIMILAR VIRUSES

In 1977 Hull[14] suggested that phleum mottle virus (PhMV) and a number of other isolates formed a cluster of viruses with similar properties which were good candidates for a new taxonomic group. We list these and some other similar viruses in Table 1. Four of the viruses are so closely related serologically to PhMV[15] that they may be best considered as strains of the same virus (Table 1).

Panicum mosaic virus (PanMV) is serologically related to PhMV,[15] but the relationship is so remote that the two viruses can be safely considered as distinct. However, it seems that St. Augustine decline and molinia streak viruses are, serologically and in other respects, sufficiently similar to PanMV to be considered as strains of that virus. In addition to these two clusters of serologically related virus isolates, lolium mottle and maize chlorotic mottle viruses have properties so similar to PhMV and PanMV that they warrant inclusion in the same group (Table 1). Purified particles of some of these viruses are shown in Figure 1. Viruses of this type do not appear to induce any distinctive cytopathic effects.[17]

Cocksfoot mottle virus (CoMV) and cynosurus mottle virus (CyMoV), although infecting grasses and having similar particles to PhMV and PanMV, do not fit into the same group.

Table 1
SOME PROPERTIES OF VIRUSES RESEMBLING PHLEUM MOTTLE VIRUS

Virus[a]	Particle diameter (nm)	S_{20w}	M_r of RNA ($\times 10^6$)	M_r of coat protein ($\times 10^3$)	Tissue location	Vectors	Ref.
Phleum mottle virus (PhMoV)	~28	106	1.48 1.5 (+0.5)[b]	25 26.5 28.5	Cytoplasm of parenchyma cells	Beetles	14—17
Brome stem-leaf mottle							14, 15
Cocksfoot mild mosaic							14, 15, 18, 19
Festuca mottle							14, 20
Holcus transitory mottle							15
Panicum mosaic virus (PanMV)	25—30	102—109	ND[c]	25	Cytoplasm of parenchyma cells	None found	15, 21
St. Augustine decline							15
Molinia streak							14, 15
Lolium mottle virus (LolMV)	~30	113	1.38	24	Cytoplasm of parenchyma cells	None found	14, 22
Maize chlorotic mottle virus (MCMoV)	~30	ND	ND	24	Cytoplasm of parenchyma cells	Beetles	15, 23, 24

[a] Virus strains are indicated by indenting.
[b] Possibly a second genomic RNA.
[c] ND, not determined.

They appear to have more similarities with viruses such as rice yellow mottle virus (RYMV)[25,26] and members of the Sobemovirus group with which they are discussed in Volume 1, Chapter 10.

III. CARNATION MOTTLE AND SOME SIMILAR VIRUSES

Hull[14] pointed out the similarities between carnation mottle virus (CarMV), pelargonium flower break virus (PFBV), elderberry latent virus (EBLV), and narcissus tip necrosis virus (NTNV) and suggested that this cluster of viruses was a suitable candidate for a new taxonomic group. These and five other similar viruses are listed in Table 2. All these viruses have polyhedral particles with a granular appearance (Figure 2) and a diameter of about 30 nm. They sediment as one component with S values between 118 and 125. The single-stranded RNA, where determined, is of M_r between 1.4 and 1.6 $\times$ 10^6 and the coat protein is near 40 $\times$ 10^3. Essentially, these viruses would be distinguished from Sobemoviruses by having a coat protein of M_r abound 40 rather than 30 $\times$ 10^3. They differ from Tombusviruses in being slightly smaller, without the clear subunit structure conferred on the particles by dimer clustering of the coat protein (see Volume 1, Chapter 12, Figures 1A, 2, and 3) and their S values are lower than those of Tombusviruses. Numerous virus particles can be seen in the cytoplasm of infected cells (Figure 3) but no multivesicular bodies have been observed. None of the viruses have been shown to be serologically interrelated, and it is probably premature to suggest the formation of an official virus group. Nevertheless, the list (Table 2) may have value in helping to direct future comparative work.

Saguaro cactus virus (SCV) has been considered as a possible member of the Tombusvirus group by the ICTV in its last report.[1] However, the relatively low sedimentation rate of SCV particles[42] and the absence of characteristic multivesicular bodies in infected cells[43] argues against its classification as a Tombusvirus. Its properties should be carefully compared to viruses such as CarMV with which it may have affinities.

IV. VIRUSES WITH SMALL POLYHEDRAL PARTICLES AND SINGLE-STRANDED RNA GENOMES, BUT NO CLEAR TAXONOMIC AFFINITIES

Table 3 lists what appear to be five different viruses which have stable particles and which occur in high concentrations in their hosts, but whose taxonomic position is at present obscure. More information about these viruses is needed before their taxonomic affinities can be assessed. If possible, they should be compared in the same laboratory with Sobemoviruses, Tombusviruses, and viruses such as PhMV and CarMV, at least with respect to morphology, stability, under various conditions, sedimentation coefficient, buoyant density, and the sizes of their RNAs and coat proteins. The presence of multivesicular bodies in infected cells seems to be diagnostic for Tombusviruses (Volume 1, Chapter 12) and this also should be investigated. None of the viruses in Table 3 appear to be serologically related to any virus tested. The cytopathology of most of the viruses appears not to have been examined but it seems that maize white line mosaic virus (MWLMV) does not induce multivesicular bodies in its hosts.[44]

In addition to the viruses listed in Table 3, there are some others with small polyhedral particles and single-stranded RNA genomes whose taxonomic affinities are obscure. A number of these, including galinsoga mosaic virus (GaMV), hibiscus chlorotic ringspot virus (HCRSV), and cucumber necrosis virus (CuNV), have been discussed with the Tombusviruses in Volume 1, Chapter 12. Nevertheless, some of their properties differ significantly from those of the true Tombusviruses and only further work can clarify their true taxonomic affinities.

V. CACAO SWOLLEN SHOOT AND STRUCTURALLY SIMILAR VIRUSES

There are a number of viruses with relatively long bacilliform particles (Table 4), but none have been well characterized as they are difficult to purify. However, their distinct morphology (Figure 4) seems to place them in a cluster that may well deserve recognition as a virus group. The only possible confusion is with the so-called "naked rhabdoviruses" (Volume 1, Chapter 5, Figures 17 through 19). However, the latter are different in having rhabdovirus-like nucleocapsids about 40 nm in diameter with clearly visible cross-striations, and in always being associated with the inner nuclear membrane in infected cells. The cacao swollen shoot virus (CSSV) cluster have particles only 23 to 31 nm in diameter without any clearly visible substructure, and the particles are found in the cytoplasm, not in close association with membranes. Colocasia bacilliform virus (CoBV) was reported to possess a striate structure in negative stain[32] but the particles illustrated are almost certainly the nucleocapsids of a contaminant rhabdovirus. There are no data regarding the nucleic acid and protein composition of these viruses.

VI. RICE STRIPE AND SIMILAR VIRUSES

Shikata and Gálvez[61] first described fine flexuous thread-like particles in cells of plant and insect hosts infected with the rice hoja blanca agent. Later, other workers found similar particles in rice stripe[62] (Figure 5A) and maize stripe-infected plants.[63] Rice grassy stunt (Figure 5B) is now the fourth disease of Gramineae found to be caused by such viruses (Table 5), and no doubt a virus group will soon be formally constituted to include them.

Rice stripe virus (RStV) and maize stripe virus (MStV) have been shown to be serologically closely related,[68] and rice grassy stunt virus (RGStV) is distantly related to RStV.[69,70] Rice hoja blanca virus (RHBV) seems to be serologically unrelated to RStV.[65,66] RStV and MStV have differing host ranges within the Gramineae and different vectors: *Laodelphax* and *Peregrinus*, respectively. It is not yet clear whether they should be treated as separate viruses or as strains of one virus.[68]

Table 2
SOME PROPERTIES OF VIRUSES RESEMBLING CARNATION MOTTLE VIRUS

Virus	Particle diameter (nm)	S_{20w}	Buoyant density in CsCl (g/cm^3)	M_r of RNA (× 10^6)	M_r of coat protein (× 10^3)	Vectors	Ref.
Carnation mottle virus (CarMV)	~28	122	1.34	1.4	38	None found	14, 27—31
Blackgram mottle virus (BMoV)	~28	122	1.36	1.4	38	Beetles	32
Cowpea mottle virus (CpMoV)	~30	122	1.35	1.4	44	Beetles	33
Cucumber soilborne virus (CSBV)	~31	120	1.34	1.5	41	Soil transmitted	34
Elderberry latent virus (EBLV)	~30	112	1.36	1.55	40	None found	14, 35
Narcissus tip necrosis virus (NTNV)	~30	123	1.36	1.6	42 (39)[a]	None found	14, 36
Pelargonium flower break virus (PFBV)	~30	125	ND[b]	ND	41	None found	14, 37, 38
Pelargonium line pattern virus (PLPV)	~30	ND	ND	ND	40	None found	39—41
Saguaro cactus virus (SCV)	~32	118	ND	1.4	39	None found	42, 43

[a] Value in brackets indicates a minor protein component detected by polyacrylamide-gel electrophoresis.
[b] ND, not determined.

Table 3
SOME PROPERTIES OF SEVERAL VIRUSES WITH SMALL ISOMETRIC PARTICLES AND SINGLE-STRANDED RNA GENOMES SEDIMENTING AS ONE COMPONENT BUT WITH NO CLEAR AFFINITIES

Virus	Particle diameter (nm)	S_{20w}	Buoyant density in CsCl (g/cm^3)	M_r of RNA ($\times 10^6$)	M_r of coat protein ($\times 10^3$)	Vectors	Ref.
Arracacha virus B (AVB)	~26	126	1.48 + 1.49	ND[a]	26 (+ 20—22)[b]	None found	45
Bean mild mosaic virus (BMMV)	~28	127	ND	1.27	—	Beetles	46
Maize white line mosaic virus (MWLMV)	~35	117	1.35	1.25	32—35	None found	44, 47—49
Parsley latent virus (ParLV)	~27	127	1.449	ND	22	None found	50
Parsnip yellow fleck virus (PYFV)	~30	148	ND	3.7	30 (+27 +22)	Aphids (semi-persistent)	51

[a] ND, not determined.

[b] Minor polypeptides detected in coat protein preparations analyzed by polyacrylamide-gel electrophoresis.

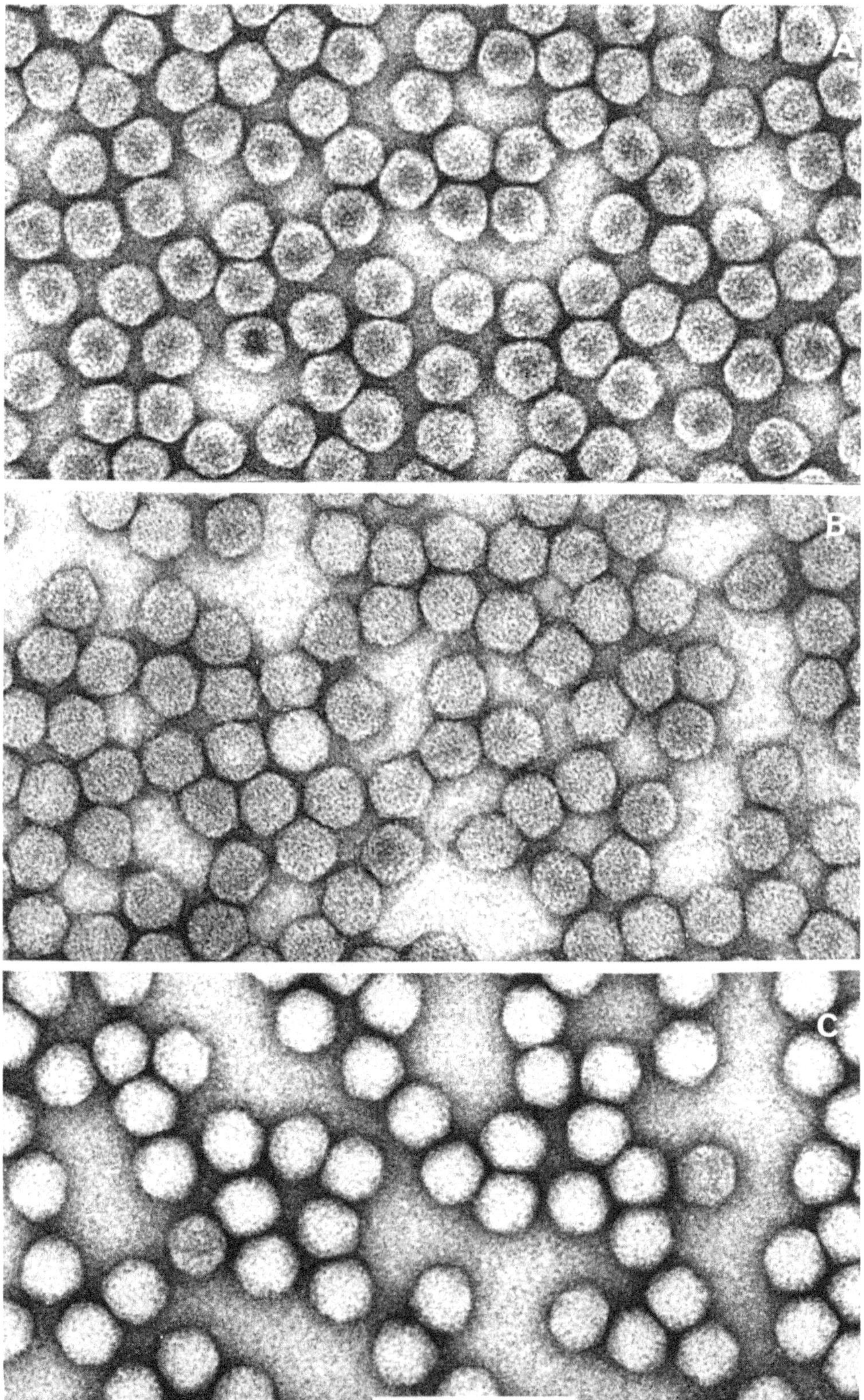

FIGURE 1.

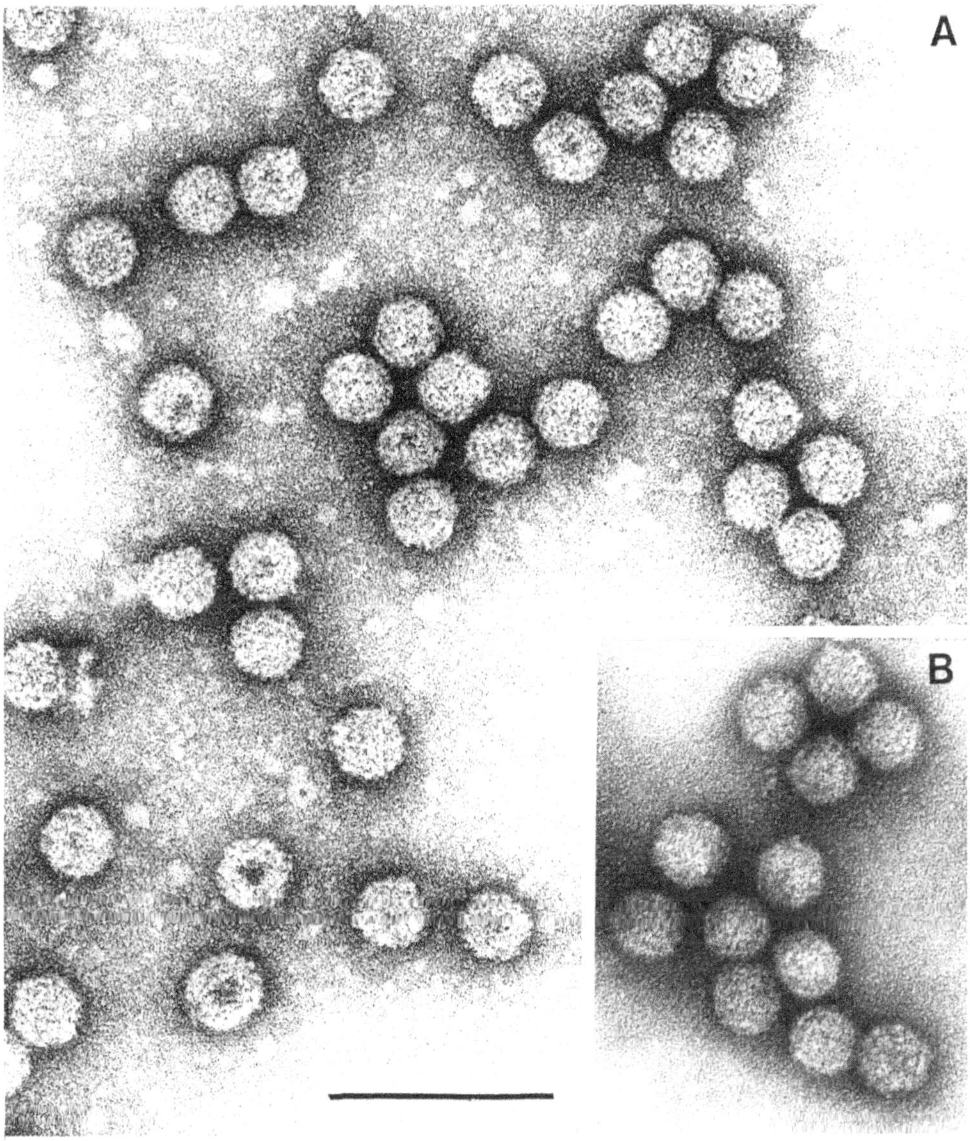

FIGURE 2. Purified preparations of carnation mottle virus (CarMV). A and B, Micrographs of preparations from two different laboratories, both stained in uranyl acetate and photographed at the same magnification. Bar = 100 nm.

FIGURE 1. Purified preparations of A, holcus transitory mottle virus (closely related to phleum mottle virus, PhMoV); B, molinia streak virus (closely related to panicum mosaic virus, PanMV); C, maize chlorotic mottle virus (MCMoV). All preparations stained in uranyl acetate and photographed at the same magnification. Bar = 100 nm. (Courtesy of Dr. D. -E. Lesemann, Institute of Plant Virus Research, Braunschweig, West Germany.)

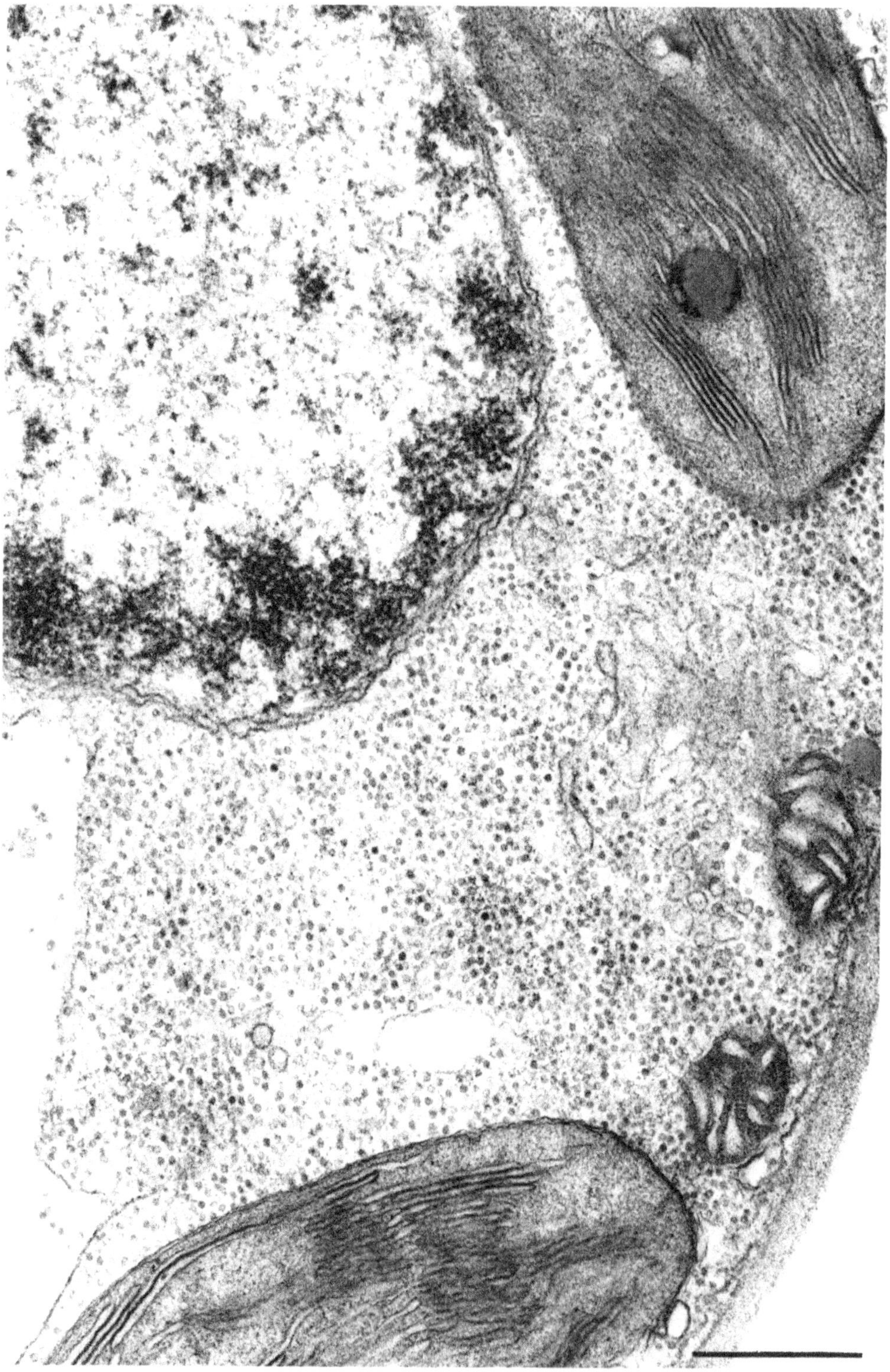

FIGURE 3. Thin section of a *Chenopodium quinoa* leaf cell infected with CarMV showing masses of virus particles in the cytoplasm and some scattered in the nucleus and vacuole. (Tissue cleared of ribosomes by RNase treatment.) Bar = 500 nm.

Table 4
SOME PROPERTIES OF VIRUSES RESEMBLING CACAO SWOLLEN SHOOT VIRUS

Virus	Dimensions of bacilliform particles (nm)	S_{20w}	Tissue location	Vectors	Ref.
Cacao swollen shoot virus (CSSV)	27 × 142	218	Cytoplasm of parenchyma cells	Mealybugs	53, 54
Aucuba ringspot virus (Au RSV)	30 × 180	ND	Cytoplasm	Not found	55
Canna yellow mottle virus (CaYMV)	23 × 120—130	ND	Cytoplasm	Not found	55
Colocasia bacilliform virus (CoBV)	28 × 125	285	Cytoplasm of phloem cells	Not found	52
Rice tungro bacilliform virus (RTBV)	25 × 140	ND	Cytoplasm of parenchyma and phloem cells	Leafhopper (semipersistent, dependent on RTIV)	11, 13, 56, 57
Rubus yellow net virus (RYNV)	25—31 × 80—150		Initially cytoplasm of phloem cells, then cytoplasm of other cells	Aphid (semipersistent)	58, 59
Yam internal brown spot virus (YIBSV)	ND	ND	Cytoplasm	Not found	60

[a] ND, not determined.

The precise particle structure of these viruses is not yet clear, but there seem to be four single-stranded RNAs and one coat protein (Table 5). When isolated, the RNAs of RStV are linear molecules. The nucleoprotein particles appear as thin kinky filaments 3 nm wide or supercoiled pseudobranched particles 6 to 8 nm wide (Table 5 and Figure 5A). After treatment with NaCl, RStV particles have been seen as open circles.[67]

RStV has been shown to multiply in its vectors, and the other viruses in the cluster are persistent and, in all likelihood, also propagative.

It has been suggested[62] that the European wheat striate mosaic disease (not to be confused with American wheat striate mosaic, caused by a Rhabdovirus) may be caused by a virus of the RStV type. The disease is transmitted persistently by the planthopper *Javesella*, but no virus-like particles were detected in thin sections of infected plants or infective planthoppers[73] or in purified infectious preparations.[74,75] The question needs reexamination.

VII. CARROT MOTTLE VIRUS AND SIMILAR VIRUSES

Carrot mottle virus (CMotV) is listed by Matthews[1] as a possible member of the family Togaviridae, which otherwise consists of viruses infecting vertebrates and transmitted by arthropods, in which they also multiply.[1] The suggestion that CMotV is a Togavirus is based on two findings: the detection in thin sections of membrane-bound particles about 50 nm in diameter attached to the tonoplasts of infected cells, and the presence of irregular dense structures of similar diameter in partially purified preparations stained in uranyl acetate.[76] It is not clear whether the structures found were virus particles. However, the partially purified preparations contained infectious material which sedimented at about 270S and had a buoyant density of 1.15 g/cm^3 in CsCl. The infectivity was sensitive to chloroform, ether, deoxycholate, and ribonuclease. CMotV is mechanically transmissible and also circulatively transmitted by the aphid *Cavariella aegopodii*, though only in the presence of carrot red leaf virus, a Luteovirus. There is no evidence that either virus multiplies in the aphid.[77,78]

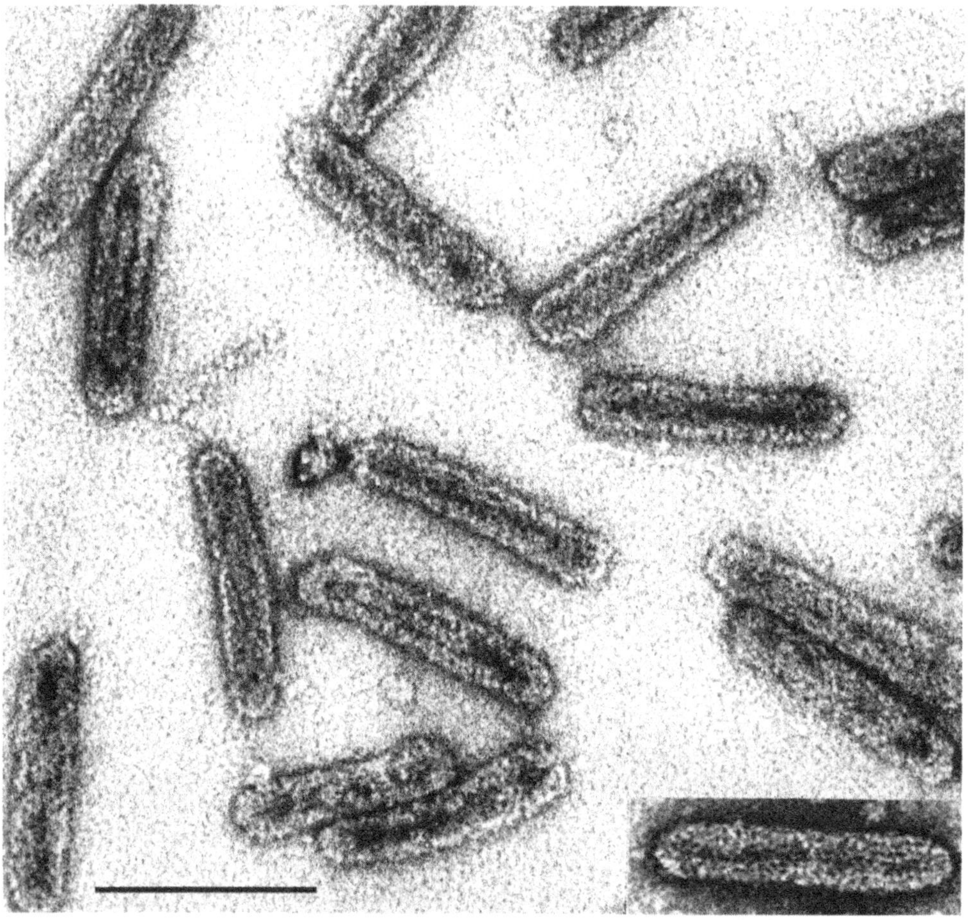

FIGURE 4. Purified preparation of cacao swollen shoot virus (CSSV) stained in uranyl acetate. Bar = 100 nm. (Courtesy of Dr. D. -E. Lesemann, Institute of Plant Virus Research, Braunschweig, West Germany.)

Lettuce speckles mottle virus (LSMV)[79,80] and bean yellow vein banding virus (BYVBV)[81,82] appear to have properties similar to those of CMotV. Cells infected by both these viruses contain similar tonoplast vesicles and aphid transmission of the viruses is dependent on coinfection with appropriate Luteoviruses. The viruses can be transmitted mechanically, but no conventional virus particles have been purified. In regard to the tonoplast vesicles, it is interesting to note that similar structures have been detected in cells infected by well characterized viruses such as tobacco necrosis virus and members of the Cucumovirus, Tobamovirus, and Potexvirus groups (see Volume 1, Chapter 11 and Chapters 6, 9, and 12 in this volume). Much more work is required before the true nature of these viruses is understood.

VIII. CRYPTIC VIRUSES

The name "cryptic virus" was first used for a small isometric virus-like particle in sugar beet that caused no symptoms and was not mechanically or graft transmissible but was transmitted to a high degree through the seed.[83] However, the nucleic acid of this beet cryptic virus (BCV) has not yet been characterized, nor has that of the second cryptic virus found, vicia cryptic virus (VCV)[84] (Table 6). Virus-like particles with properties similar to

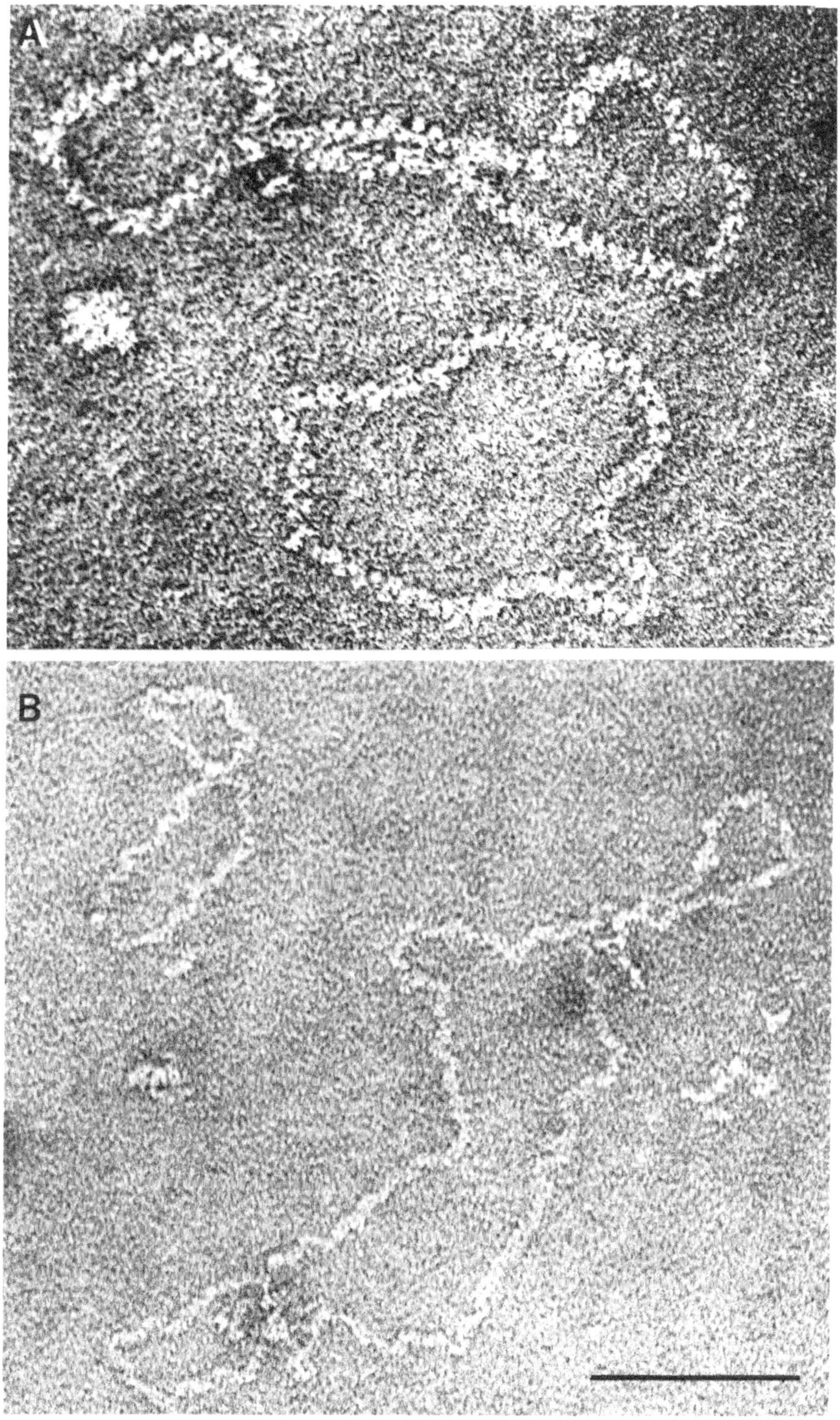

FIGURE 5. Purified preparations of A, rice stripe virus (RStV); B, rice grassy stunt virus (RGStV). Both stained in uranyl acetate and photographed at the same magnification. Bar = 100 nm. (Courtesy of Dr. H. Hibino, International Rice Research Institute, Manila.)

Table 5
SOME PROPERTIES OF VIRUSES RESEMBLING RICE STRIPE VIRUS

Virus	Particle morphology	S_{20w}	Properties of nucleic acid	M_r of coat protein (× 10[3])	Tissue location	Vectors	Ref.
Rice stripe virus (RStV)	Kinky filaments 3 nm diameter (8 nm when supercoiled)	98 + 80 + 65—72	Linear, single-stranded RNA of M_r ~1.9 + 1.4 + 1.0 + 0.9	32	Cytoplasm of mesophyll and phloem cells	*Laodelphax* (planthopper)	64—68
Maize stripe virus (MStV)	Kinky filaments 3 nm diameter	ND[a]	Single-stranded RNA	32.7	ND	*Peregrinus* (planthopper)	63, 68
Rice grassy stunt virus (RGStV)	Filaments 6—8 × 1000—1300 nm	ND	RNA	31	ND	*Nilaparvata* (planthopper)	69, 70
Rice hoja blanca virus (RHBV)	Filaments 7—8 nm in diameter	ND	ND	ND	All tissues	*Sogatodes* and *Peregrinus* (planthoppers)	61, 71, 72

[a] ND, not determined.

Table 6
SOME PROPERTIES OF SELECTED CRYPTIC VIRUSES

Virus	Particle diameter (nm)	S_{20w}	M_r of double-stranded RNA (× 10^6)	M_r of coat protein (× 10^3)	Ref.
Carnation cryptic virus (CarCV)	~29	ND	1.04 + 0.95 + 0.84 (+0.88)[b]	ND	85—88
Alfalfa cryptic virus (ACV)	~30	ND	1.27 + 1.17	ND	88
Alfalfa temperate virus (ATeV)	~30	ND	1.58 + 1.55 + 1.48	ND	89
Beet cryptic virus (BCV)	~30	ND	ND	ND	83, 90, 91
Beet temperate virus (BTeV)	~30	115—122	1.45 + 1.28 + 0.65 + 0.56	ND	89
Festuca cryptic virus (FCV)	~30	ND	1.32 + 1.27 + 1.22 + 0.95 + 0.90	ND	88
Garland chrysanthemum temperate virus (GCTeV)	~30	ND	1.25 + 1.20	ND	92
Hop trefoil cryptic virus (HTCV)	~30	ND	1.30 + 1.07 (+1.22 + 1.17)	53	88
Radish yellow edge virus (RYEV)	~30	118	1.30 + 1.25 + 1.21 (+1.14 + 1.09)	63 + 61	92, 93
Red clover cryptic virus (RCCV)	~30	ND	1.56 + 1.51	ND	88
Ryegrass cryptic virus (RCV) (= ryegrass spherical, = ryegrass seedborne)	~30	ND	1.03 + 0.92	ND	88, 94—98
Spinach temperate virus (STeV)	~30	ND	1.31 + 1.21 + 1.10	ND	89
Vicia cryptic (VCV)	~30	ND	ND	ND	84
White clover small cryptic virus (WCSCV)	~30	ND	1.26 + 1.03	53	88, 89
White clover large cryptic virus (WCLCV)	~35	ND	1.59 + 1.47	ND	88,99
White clover temperate virus (WCTeV)	~30	ND	1.50 + 1.25 (+ 4 minor)	ND	92

[a] ND, not determined.
[b] Figures in parentheses indicate minor components.

those of BCV and VCV, but which were shown to contain double-stranded RNA, were later found in both Italy and Japan (Table 6). Those from Italy were called cryptic viruses following the usage of Kassanis and colleagues,[83] whereas those found in Japan were called "temperate" viruses, though evidently they have the same characteristics as cryptic viruses (Table 6 and Figures 6 and 7). Rather than continuing to use two names for the same type of virus, we propose to retain the term cryptic virus for this possible new virus group.

The cryptic viruses have not yet graduated to the full status of plant viruses because they have not been transmitted experimentally and because most of them have not been demonstrated within the cells of plants. They have only been isolated in small amounts as purified and highly concentrated preparations. Thus, there is the possibility that mycoviruses, which resemble cryptic viruses in many aspects,[100-103] may have been associated with the plants in internal or external mycoflora. In at least the majority of cases, however, the mycovirus contamination hypothesis seems extremely unlikely[88] and so we appear to be faced with plant viruses of a novel kind. Nevertheless, careful work is still needed to clear up any remaining doubts.

Persuasive evidence that hop trefoil cryptic virus (HTCV; Table 6) infects *Medicago lupulina* plants, from which it can be purified, comes from electron microscopy of plant tissues cleared of ribosomes by RNase treatment (Figure 8).[99] Groups of particles of the correct morphology to be HTCV particles can be observed in both the cytoplasm (Figures 8A and B) and nuclei (Figure 8C) of the *M. lupulina* leaf cells.

In all cases but one, that of radish yellow edge virus (RYEV), the cryptic viruses listed in Table 6 do not appear to cause any symptoms in plants. They also all occur in low concentrations and appear, so far, not to be transmissible by mechanical means, vectors, or even grafting.[83-99] All the viruses examined have been shown to contain at least two molecules of double-stranded RNA (Table 6) and all that have been investigated appear to be serologically unrelated to each other or to any other virus tested.[88]

In addition to the viruses listed in Table 6, what appear to be cryptic viruses have been reported in carrot (*Daucus carota*) and two cultivars of turnip (*Brassica rapa*).[89,92] It is not yet clear whether the viruses in white clover and alfalfa found in both Italy and Japan are the same, but BCV is not serologically related to beet temperate virus (BTeV).[89] There is also recent evidence for a possible cryptic virus in avocado.[104] A virus named poinsettia cryptic virus (PCrV) has not yet been shown to possess the characters of other cryptic viruses.[105] Plants of the following species have been screened for the presence of cryptic viruses, but with negative results: cowpea, French bean cv. Saxa, soybean, *Hedysarum coronarium, Lotus corniculatus, Melilotus alba, Onobrychis viciaefolia* (= *O. sativa*), *Trigonella foenum-graecum, Trifolium incarnatum, T. resupinatum, T. squarrosum* (all Leguminosae); *Phleum pratense* and *Poa pratensis* (Gramineae); and *Physalis floridana* (Solanaceae).[88]

Experience so far suggests that many plants over a wide range of plant families harbor small isometric double-stranded RNA viruses that are transmitted only vertically (inherited), are present in low concentrations, and cause no symptoms. Those working with conventional viruses, whether isometric or otherwise, will have to be particularly careful that their "healthy" control plants are not cryptic carriers.

IX. CONCLUDING REMARKS

There still remain numerous plant diseases which are thought to be caused by viruses but whose etiologic agents have not been properly identified. Some of these diseases may well turn out to be incited by agents other than viruses, as many have been in the past. For example, many diseases previously thought to have viral etiology now appear to be caused by viroids, mycoplasmas, rickettsias, corynebacteria or insect toxins.[106-108] However, there seems little doubt that many plant diseases of uncertain etiology will be shown to be caused by viruses. Some of these may turn out to belong to established virus taxa discussed in Chapters 1 through 15 of this volume as well as in Volume 1, Chapters 2 through 12; others will have inscrutable taxonomic affinities, like some of those mentioned in this chapter; yet others may be viruses unlike any that have been encountered to date.

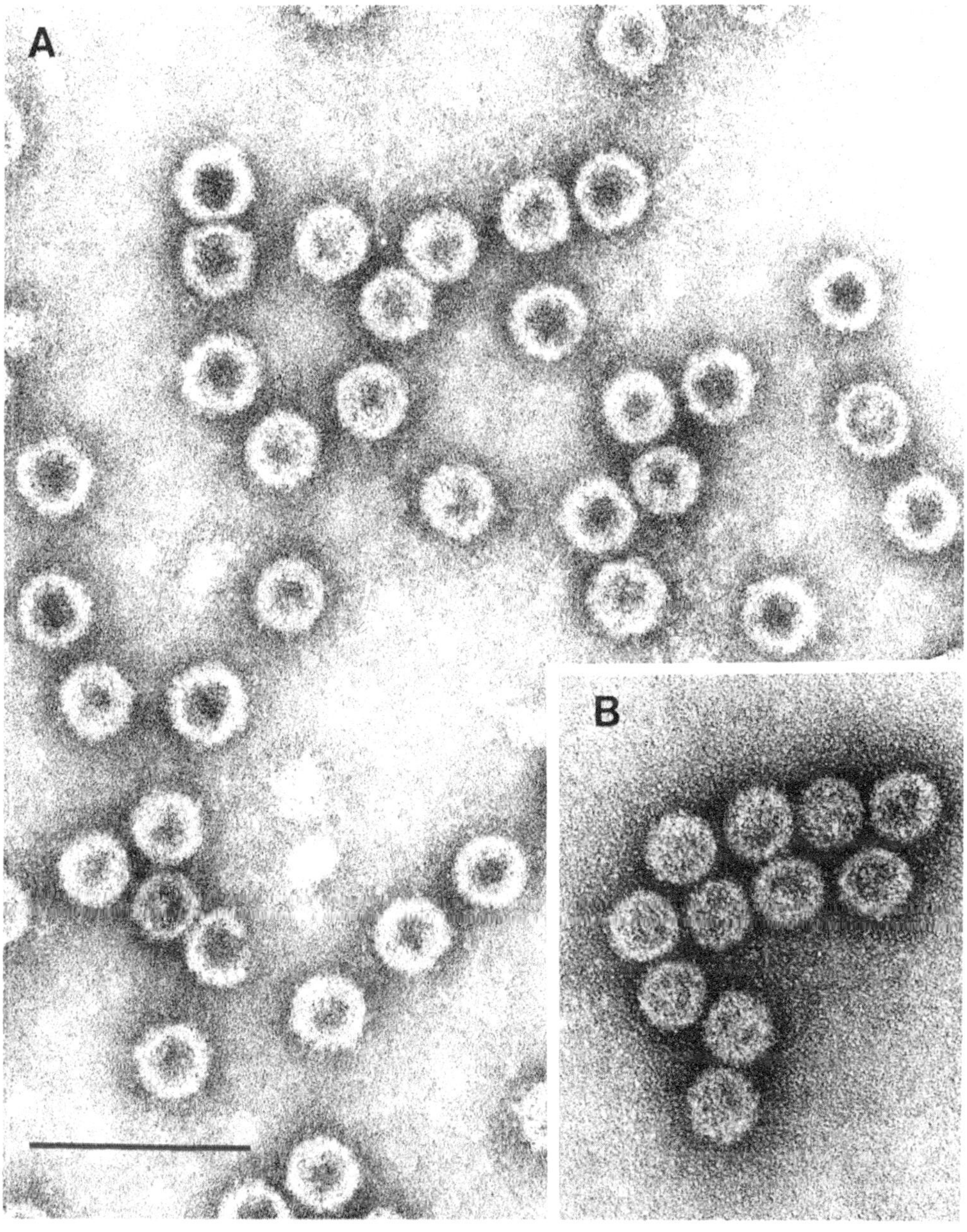

FIGURE 6. Purified preparations of A. Ryegrass cryptic virus (RCV); and B. carnation cryptic virus (CarCV). Both stained in uranyl acetate and photographed at the same magnification. Bar = 100 nm.

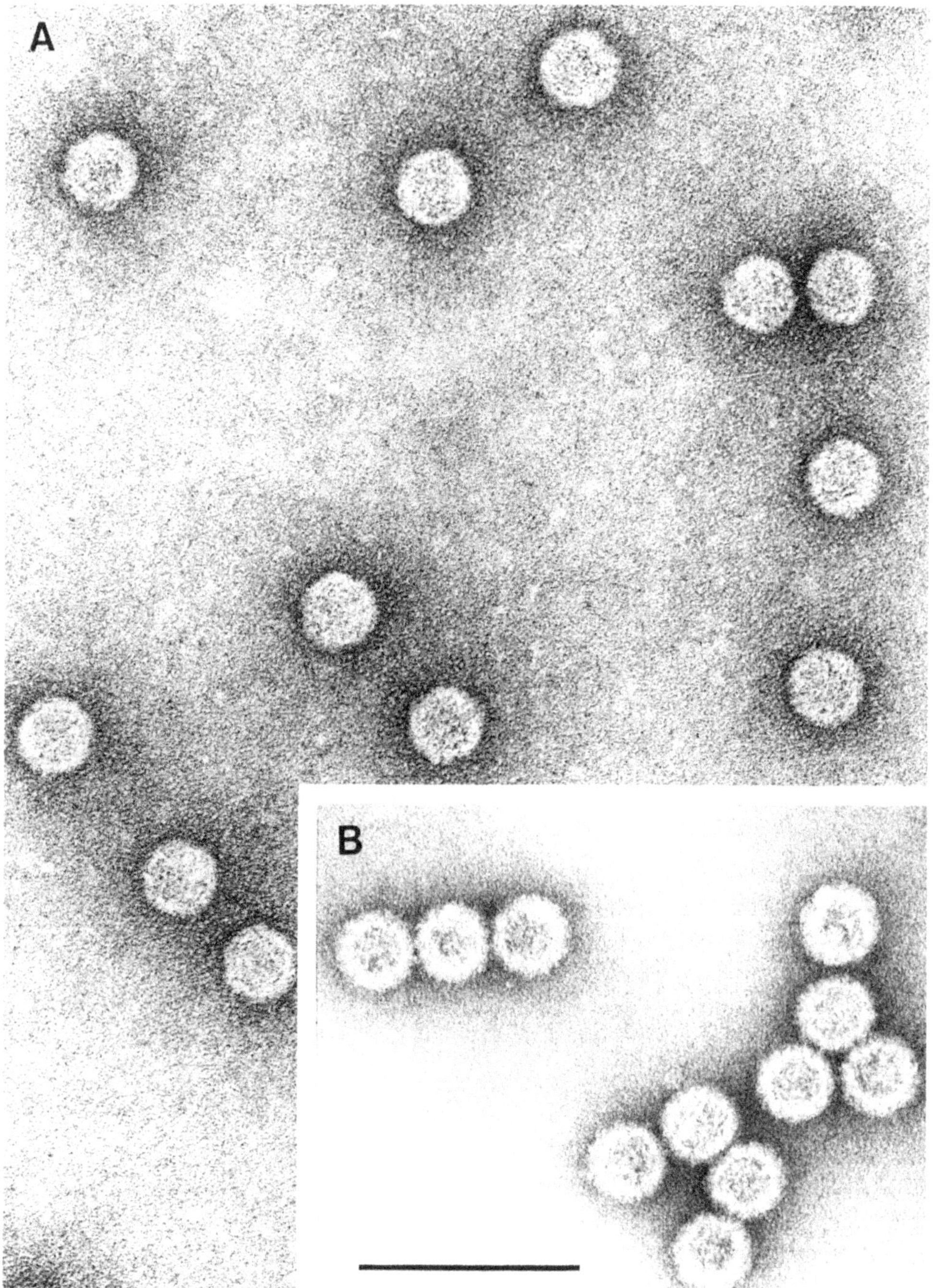

FIGURE 7. Purified preparations of A, White clover small cryptic virus (WCSCV); and B, white clover large cryptic virus (WCLCV). Both stained in uranyl acetate and photographed at the same magnification. Bar = 100 nm.

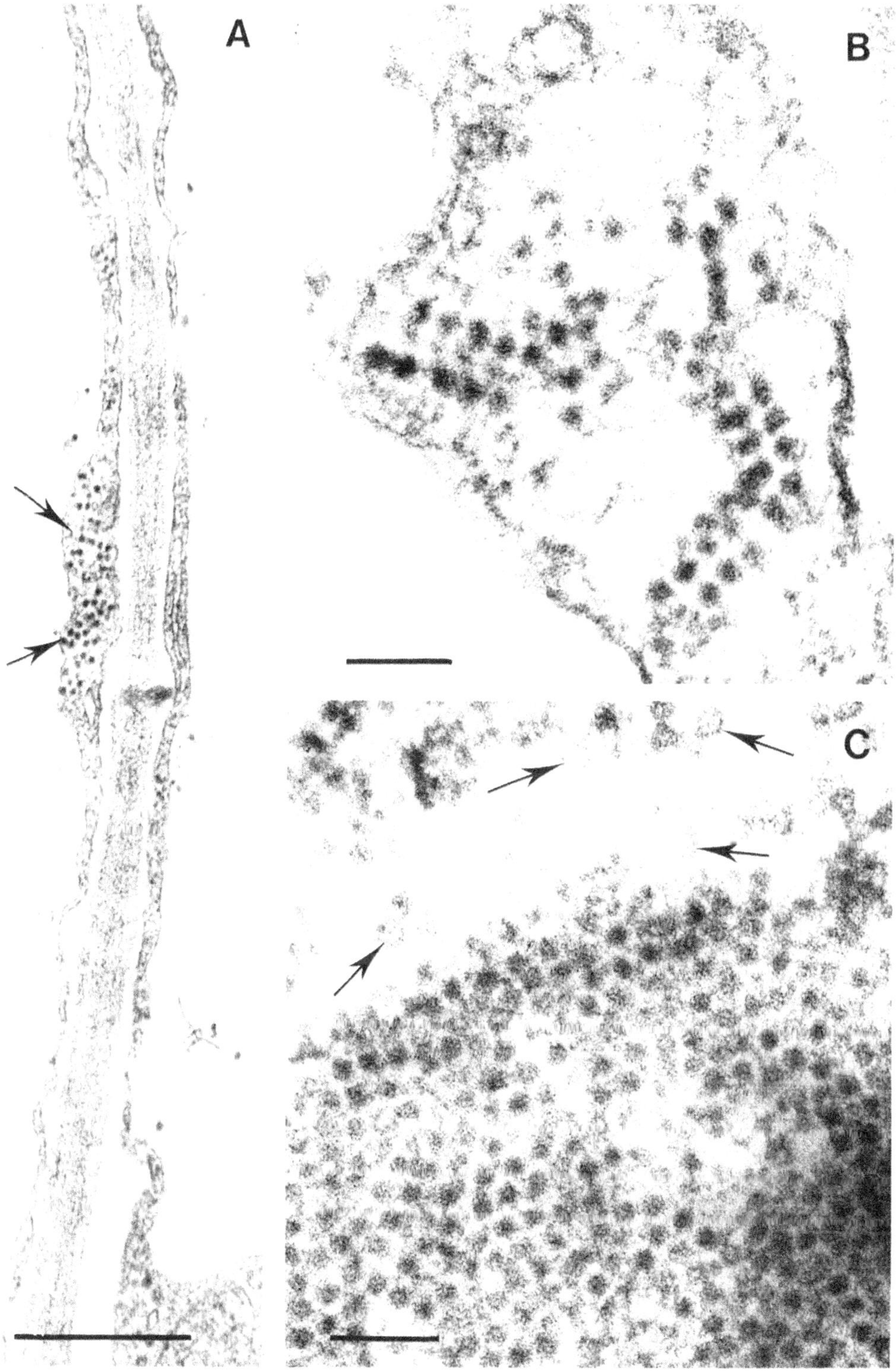

FIGURE 8. Thin sections of *Medicago lupulina* leaf cells carrying hop trefoil cryptic virus (HTCV). A, A small group of particles (arrows) in the cytoplasm of an otherwise typically healthy leaf cell; bar = 500 nm. B, Virus-like particles in the cytoplasm at high magnification. C, A larger group of particles some of which appear to be empty (arrows), in the nucleoplasm. (Tissues cleared of ribosomes by RNase treatment.) Bars = 100 nm.

REFERENCES

1. **Matthews, R. E. F.,** Classification and nomenclature of viruses. Fourth report of the International Committee on Taxonomy of Viruses. *Intervirology*, 17, 1, 1982.
2. **Banttari, E. E. and Zeyen, R. J.,** Oat blue dwarf virus, *CMI/AAB Descriptions of Plant Viruses*, No. 123, 1973.
3. **Banttari, E. E. and Zeyen, R. J.,** Multiplication of the oat blue dwarf virus in the aster leafhopper, *Phytopathology*, 66, 896, 1976.
4. **León, P. and Gámez, R.,** Some physicochemical properties of maize rayado fino virus, *J. Gen. Virol.*, 56, 67, 1981.
5. **Rivera, C., Kozuka, Y., and Gámez, R.,** Rayado fino virus; detection in salivary glands and evidence of increase in virus titer in the leafhopper vector *Dalbulus maidis*, *Turrialba*, 31, 77, 1981.
6. **Kitajima, E. W., and Gámez, R.,** Electron microscopy of maize rayado fino virus in the internal organs of its leafhopper vector, *Intervirology*, 19, 129, 1983.
7. **Gámez, R.,** Maize rayado fino virus, *CMI/AAB Descriptions of Plant Viruses*, No. 220, 1980.
8. **Gingery, R. E., Gordon, D. T., and Nault, L. R.,** Purification and properties of an isolate of maize rayado fino virus from the United States, *Phytopathology*, 72, 1313, 1982.
9. **Gálvez, G. E.,** Rice tungro virus, *CMI/AAB Descriptions of Plant Viruses*, No. 67, 1971.
10. **Yamashita, S., Doi, Y., and Yora, K.,** Some properties and intracellular appearance of rice waika virus, *Ann. Phytopathol. Soc. Jpn.*, 43, 278, 1977.
11. **Hibino, H., Roechan, M., and Sudarisman, S.,** Association of two types of virus particles with penyakit habang (tungro disease) of rice in Indonesia, *Phytopathology*, 68, 1412, 1978.
12. **Hibino, H.,** Transmission of two rice tungro-associated viruses and rice waika virus from doubly or singly infected source plants by leafhopper vectors, *Plant Dis.*, 67, 774, 1983.
13. **Omura, T., Saito, Y., Usugi, T., and Hibino, H.,** Purification and serology of rice tungro spherical and rice tungro bacilliform viruses, *Ann. Phytopathol. Soc. Jpn.*, 49, 73, 1983.
14. **Hull, R.,** The grouping of small spherical plant viruses with single RNA components, *J. Gen. Virol.*, 36, 289, 1977.
15. **Paul, H. L., Querfurth, G., and Huth, W.,** Serological studies on the relationships of some isometric viruses of *Gramineae*, *J. Gen. Virol.*, 47, 67, 1980.
16. **Catherall, P. L. and Chamberlain, J. A.,** Relationships, host-ranges and symptoms of some isolates of phleum mottle virus, *Ann. Appl. Biol.*, 87, 147, 1977.
17. **Chamberlain, J. A. and Catherall, P. L.,** Electron microscopy of some grasses and cereals infected with cocksfoot mottle, phleum mottle and cocksfoot mild mosaic viruses, *J. Gen. Virol.*, 30, 41, 1976.
18. **Huth, W. and Paul, H. L.,** Cocksfoot mild mosaic virus, *CMI/AAB Descriptions of Plant Viruses*, No. 107, 1972.
19. **Torrance, L. and Harrison, B. D.,** Properties of Scottish Isolates of cocksfoot mild mosaic virus and their comparison with others, *Ann. Appl. Biol.*, 97, 285, 1981.
20. **Carpenter, J. M. and Gibson, R. W.,** Festuca mottle virus (FMV), *Rothamsted Exp. Sta. Rep. for 1974*, Part 1, p. 214, 1975.
21. **Niblett, C. L., Paulsen, A. Q., and Toler, R. W.,** Panicum mosaic virus, *CMI/AAB Descriptions of Plant Viruses*, No. 177, 1977.
22. **A'Brook, J.,** Lolium mottle virus, *Plant Pathol.*, 21, 118, 1972.
23. **Nault, L. R., Styer, W. E., Coffey, M. E., Gordon, D. T., Negi, L. S., and Niblett, C. L.,** Transmission of maize chlorotic mottle virus by chrysomelid beetles, *Phytopathology*, 68, 1071, 1978.
24. **Uyemoto, J. K.,** Biology and control of maize chlorotic mottle virus, *Plant Dis.*, 67, 7, 1983.
25. **Mohamed, N. A. and Mossop, D. W.,** Cynosurus and cocksfoot mottle viruses: a comparison, *J. Gen. Virol.*, 55, 63, 1981.
26. **Bakker, W.,** Rice yellow mottle virus, *CMI/AAB Descriptions of Plant Viruses*, No. 149, 1975.
27. **Waterworth, H. E. and Kaper, J. M.,** Purification and properties of carnation mottle virus and its ribonucleic acid, *Phytopathology*, 62, 959, 1972.
28. **Boccardo, G., Redolfi, P., Cantisani, A., d'Aquilio, M., and Pennazio, S.,** Heat inactivation of carnation mottle virus in purified suspensions, *Phytopathol. Mediterr.*, 13, 36, 1974.
29. **Hollings, M. and Stone, O. M.,** Carnation mottle virus, *CMI/AAB Descriptions of Plant Viruses*, No. 7, 1970.
30. **Tremaine, J. H.,** Physical, chemical, and serological studies on carnation mottle virus, *Virology*, 42, 611, 1970.
31. **Robleda, S. C.,** Ultrastructure of cells infected with carnation mottle virus, *Phytopathol. Z.*, 78, 134, 1973.
32. **Scott, H. A. and Hoy, J. W.,** Blackgram mottle virus, *CMI/AAB Descriptions of Plant Viruses*, No. 237, 1981.

33. **Bozarth, R. F. and Shoyinka, S. A.,** Cowpea mottle virus, *CMI/AAB Descriptions of Plant Viruses*, No. 212, 1979.
34. **Koenig, R., Lesemann, D. -E., Huth, W., and Makkouk, K. M.,** Comparison of a new soilborne virus from cucumber with tombus-, diantho-, and other similar viruses, *Phytopathology*, 73, 515, 1983.
35. **Jones, A. T.,** Elderberry latent virus, *CMI/AAB Descriptions of Plant Viruses*, No. 127, 1974.
36. **Mowat, W. P., Asjes, C. J., and Brunt, A. A.,** Narcissus tip necrosis virus, *CMI/AAB Descriptions of Plant Viruses*, No. 166, 1976.
37. **Stone, O. M. and Hollings, M.,** Some properties of pelargonium flower-break virus, *Ann. Appl. Biol.*, 75, 15, 1973.
38. **Hollings, M. and Stone, O. M.,** Pelargonium flower break virus, *CMI/AAB Descriptions of Plant Viruses*, No. 130, 1974.
39. **Stefanac, Z., Pleše, N., and Wrischer, M.,** Intracellular changes provoked by pelargonium line pattern virus, *Phytopathol. Z.*, 105, 288, 1982.
40. **Stone, O. M.,** Nine viruses isolated from pelargonium in the United Kingdom, *Acta Hortic.*, 110, 177, 1980.
41. **Pleše, N. and Štefanac, Z.,** Some properties of a distinctive isometric virus from pelargonium, *Acta Hortic.*, 110, 183, 1980.
42. **Nelson, M. R., Yoshimura, M. A., and Tremaine, J. H.,** Saguaro cactus virus, *CMI/AAB Descriptions of Plant Viruses*, No. 148, 1975.
43. **Russo, M. and Martelli, G. P.,** Ultrastructure of turnip crinkle- and saguaro cactus virus-infected tissues, *Virology*, 118, 109, 1982.
44. **de Zoeten, G. A., Arny, D. C., Grau, C. R., Saad, S. M., and Gaard, G.,** Properties of the nucleoprotein associated with maize white line mosaic in Wisconsin, *Phytopathology*, 70, 1019, 1980.
45. **Jones, R. A. C. and Kenten, R. H.,** Arracacha virus B, *CMI/AAB Descriptions of Plant Viruses*, No. 270, 1983.
46. **Waterworth, H.,** Bean mild mosaic virus, *CMI/AAB Descriptions of Plant Viruses*, No. 231, 1981.
47. **Boothroyd, C. W. and Israel, H. W.,** A new mosaic disease of corn, *Plant Dis.*, 64, 218, 1980.
48. **Conti, M.,** Maize viruses and virus diseases in Italy and other Mediterranean countries, in *Proc. Int. Maize Virus Dis. Colloq. 1982, Wooster, Ohio*, Gordon, D. T., Knoke, J. K., Nault, L. R., and Ritter, R. M., 1983, 103.
49. **Louie, R., Gordon, D. T., Knoke, J. K., Gingery, R. E., Bradfute, O. E., and Lipps, P. E.,** Maize white line mosaic virus in Ohio, *Plant Dis.*, 66, 167, 1982.
50. **Bos, L., Huttinga, H., and Maat, D. Z.,** Parsley latent virus, a new and prevalent seed-transmitted, but possibly harmless virus of *Petroselinum crispum*, *Neth. J. Plant Pathol.*, 85, 125, 1979.
51. **Murant, A. F.,** Parsnip yellow fleck virus, *CMI/AAB Descriptions of Plant Viruses*, No. 129, 1974.
52. **James, M., Kenten, R. H., and Woods, R. D.,** Virus-like particles associated with two diseases of *Colocasia esculenta* (L.) Schott in the Solomon Islands, *J. Gen. Virol.*, 21, 145, 1973.
53. **Brunt, A. A.,** Cacao swollen shoot virus, *CMI/AAB Descriptions of Plant Viruses*, No. 10, 1970.
54. **Adomako, D., Lesemann, D. -E., and Paul, H. L.,** Improved methods for the purification and detection of cocoa swollen shoot virus, *Ann. Appl. Biol.*, 103, 109, 1983.
55. **Yora, K., Saito, Y., Doi, Y., Inouye, T., and Tomaru, K., Eds.,** *Handbook of Plant Viruses* Asakura Shoten, Tokyo, 1983, 632 pp. (in Japanese).
56. **Saito, Y., Roechan, M., Tantera, D. M., and Iwaki, M.,** Small bacilliform particles associated with penyakit habang (tungro-like) disease of rice in Indonesia, *Phytopathology*, 65, 793, 1975.
57. **Milne, R. G., Boccardo, G., and Ling, K. C.,** Electron microscopy of particles of rice tungro virus complex, *Int. Rice Res. Newslett.*, 6(5), 13, 1981.
58. **Stace Smith, R. and Jones, A. T.,** Rubus yellow net virus, *CMI/AAB Descriptions of Plant Viruses*, No. 188, 1978.
59. **Jones, A. T., Roberts, I. M., and Murant, A. F.,** Association of different kinds of bacilliform particle with vein chlorosis and mosaic diseases of raspberry (*Rubus idaeus*), *Ann. Appl. Biol.*, 77, 283, 1974.
60. **Harrison, B. D. and Roberts, I. M.,** Association of virus like particles with internal brown spot of yam *(Dioscorea alata)*, *Trop. Agric. (Trinidad)*, 50, 335, 1973.
61. **Shikata, E. and Gálvez, G. E.,** Fine flexuous threadlike particles in cells of plants and insect hosts infected with rice hoja blanca virus, *Virology*, 39, 635, 1969.
62. **Koganezawa, H., Doi, Y., and Yora, K.,** Purification of rice stripe virus, *Ann. Phytopathol. Soc. Jpn.*, 41, 148, 1975.
63. **Gingery, R. E., Nault, L. R., and Bradfute, O. E.,** Maize stripe virus: characteristics of a member of a new virus class, *Virology*, 112, 99, 1981.
64. **Toriyama, S.,** Characterization of rice stripe virus: a heavy component carrying infectivity, *J. Gen. Virol.*, 61, 187, 1982.
65. **Toriyama, S.,** Three ribonucleic acids associated with rice stripe virus, *Ann. Phytopathol. Soc. Jpn.*, 48, 482, 1982.

66. **Toriyama, S.,** Rice stripe virus, *CMI/AAB Descriptions of Plant Viruses,* No. 269, 1983.
67. **Horita, M., Tshushima, S., Uyeda, I., and Shikata, E.,** Serology of rice stripe virus, *Mem. Fac. Agric. Hokkaido Univ.,* 13, 551, 1983.
68. **Gingery, R. E., Nault, L. R., and Yamashita, S.,** Relationship between maize stripe virus and rice stripe virus, *J. Gen. Virol.,* 64, 1765, 1983.
69. **Hibino, H., Usugi, T., Omura, T., and Shohara, K.,** Morphology and serological relationship of grassy stunt-associated filamentous nucleoprotein and rice stripe virus, *Int. Rice Res. Newslett.,* 8(1), 9, 1983.
70. **Hibino, H., Usugi, T., Tsuchizaki, T., Iwasaki, M., and Izumi, S.,** Purification and properties of grassy stunt-associated filamentous particles, *Int. Rice Res. Newslett.,* 8(2), 11, 1983.
71. **Morales, F. J. and Niessen, A. I.,** Association of spiral filamentous virus like particles with rice hoja blanca, *Phytopathology,* 73, 971, 1983.
72. **Varón de agudelo, F. and Martinez-López, G.,** Maize hoja blanca: a complex of viruses transmitted by *Peregrinus maidis, Phytopathology,* 73, 125, 1983 (abstr.).
73. **Ammar, E. D., Milne, R. G., and Watson, M. A.,** unpublished results, 1969.
74. **Slykhuis, J. T. and Watson, M. A.,** Striate mosaic of cereals in Europe and its transmission by *Delphacodes pellucida* (Fab.), *Ann. Appl. Biol.,* 46, 542, 1958.
75. **Serjeant, E. P.,** The transmission of European wheat striate mosaic virus by *Javesella pellucida* (Fabr.) injected with extracts of plants and plant-hoppers, *Ann. Appl. Biol.,* 59, 39, 1967.
76. **Murant, A. F.,** Carrot mottle virus, *CMI/AAB Descriptions of Plant Viruses,* No. 137, 1974.
77. **Elnagar, S. and Murant, A. F.,** Relations of carrot red leaf and carrot mottle viruses with their aphid vector, *Cavariella aegopodii, Ann. Appl. Biol.,* 89, 237, 1978.
78. **Elnagar, S. and Murant, A. F.,** Aphid-injection experiments with carrot mottle virus and its helper virus, carrot red leaf, *Ann. Appl. Biol.,* 89, 245, 1978.
79. **Falk, B. W., Duffus, J. E., and Morris, T. J.,** Transmission, host range, and serological properties of the viruses that cause lettuce speckles disease, *Phytopathology,* 69, 612, 1979.
80. **Falk, B. W., Morris, T. J., and Duffus, J. E.,** Unstable infectivity and sedimentable ds-RNA associated with lettuce speckles mottle virus, *Virology,* 96, 239, 1979.
81. **Cockbain, A. J. and Jones, P.,** Bean yellow vein-banding virus. *Rothamsted Exp. Sta. Rep. for 1980,* Part 1, p. 189, 1981.
82. **Cockbain, A. J., and Jones, P.,** Bean yellow veinbanding and its helper viruses, *8th Mtg. Working Group Legume Viruses* (abstr.), Versailles, 1981.
83. **Kassanis, B., White, R. F., and Woods, R. D.,** Beet cryptic virus, *Phytopathol. Z.,* 90, 350, 1977.
84. **Kenten, R. H., Cockbain, A. J., and Woods, R. D.,** Vicia cryptic virus (VCV), *Rothamsted Exp. Sta. Rep. for 1979,* Part 1, p. 176, 1980.
85. **Lisa, V., Luisoni, E., and Milne, R. G.,** Carnation cryptic virus, *Acta Hortic.,* 110, 175, 1980.
86. **Lisa, V., Luisoni, E., and Milne, R. G.,** A possible virus cryptic in carnation, *Ann. Appl. Biol.,* 98, 431, 1981.
87. **Lisa, V., Boccardo, G., and Milne, R. G.,** Double-stranded ribonucleic acid from carnation cryptic virus, *Virology,* 115, 410, 1981.
88. **Boccardo, G., Lisa, V., and Milne, R. G.,** Cryptic viruses in plants, in *Double-Stranded RNA Viruses,* Compans, R. W. and Bishop, D. H. L., Eds., Elsevier, Amsterdam, 1983, 425.
89. **Natsuaki, T., Yamashita, S., Doi, Y., Okuda, S., and Teranaka, M.,** Two seed-borne double-stranded RNA viruses, beet temperate virus and spinach temperate virus, *Ann. Phytopathol. Soc. Jpn.,* 49, 709, 1983.
90. **Kassanis, B., Russell, G. E., and White, R. F.,** Seed and pollen transmission of beet cryptic virus in sugar beet plants, *Phytopathol. Z.,* 91, 76, 1978.
91. **White, R. F. and Woods, R. D.,** Beet cryptic virus in some varieties of *Beta vulgaris, Phytopathol. Z.,* 91, 91, 1978.
92. **Natsuaki, T., Yamashita, S., Doi, Y., Okuda, S., and Teranaka, M.,** Radish yellow edge virus, a seed-borne virus with double-stranded RNA, of a possible new group, *Ann. Phytopathol. Soc. Jpn.,* 49, 593, 1983.
93. **Natsuaki, T., Yamashita, S., Doi, Y., and Yora, K.,** Radish yellow edge virus, *Ann. Phytopathol. Soc. Jpn.,* 45, 313, 1979.
94. **Plumb, R. T.,** Ryegrass spherical virus (RSV), *Rothamsted Exp. Sta. Rep. for 1972,* Part 1, p. 125, 1973.
95. **Milne, R. G.,** Electron microscopy of thin sections of Italian ryegrass infected with both ryegrass cryptic virus and oat sterile dwarf virus, *Microbiologica,* 3, 333, 1980.
96. **Luisoni, E. and Milne, R. G.,** Some properties of ryegrass spherical virus, *Proc. 3rd. Conf. Virus Diseases of Gramineae in Europe,* 1981, p. 73.
97. **Plumb, R. T. and Lennon, E. A.,** Seed transmission and effects of ryegrass spherical virus, *Proc. 3rd. Conf. Virus Diseases of Gramineae in Europe,* 1981, p. 75.
98. **Plumb, R. T. and Lennon, E. A.,** Ryegrass seed-borne virus, *Rothamsted Exp. Sta. Rep. for 1980,* Part 1, p. 187, 1981.

99. **Boccardo, G., Lisa, V., Luisoni, E., and Milne, R. G.,** unpublished results, 1983.
100. **Bruenn, J. A.,** Virus-like particles of yeast, *Ann. Rev. Microbiol.*, 34, 49, 1980.
101. **Bozarth, R. F., Koltin, Y., Weissman, M. B., Parker, R. L., Dalton, R. E., and Steinlauf, R.,** The molecular weight and packaging of dsRNAs in the mycovirus from *Ustilago maydis* killer strains, *Virology*, 113, 492, 1981.
102. **Buck, K. W., Almond, M. R., McFadden, J. J. P., Romanos, M. A., and Rawlinson, C. J.,** Properties of thirteen viruses and virus variants obtained from eight isolates of the wheat take-all fungus, *Gaeumannomyces graminis* var. *tritici*, *J. Gen. Virol.*, 53, 235, 1981.
103. **Hollings, M.,** Mycoviruses and plant pathology, *Plant Dis.* 66, 1106, 1982.
104. **Jordan, R. L., Dodds, J. A., and Ohr, H. D.,** Evidence for virus-like agents in avocado, *Phytopathology*, 73, 1130, 1983.
105. **Koenig, R. and Lesemann, D. -E.,** Two isometric viruses in poinsettias, *Plant Dis.*, 64, 782, 1980.
106. **Diener, T. O.,** *Viroids and Viroid Diseases*, John Wiley & Sons, New York, 1979, 252 pp. 1979.
107. **Maramorosch, K.,** Mycoplasmas and rickettsiae in relation to plant diseases, *Ann. Rev. Microbiol.*, 28, 301, 1974.
108. **Ofori, F. A. and Francki, R. I. B.,** Evidence that maize wallaby ear disease is caused by an insect toxin, *Ann. Appl. Biol.*, 103, 185, 1983.

VIRUS ABBREVIATIONS AND TAXONOMIC GROUPS

Virus	Abbreviation	Group*
Agropyron mosaic virus	AgMV	Potyvirus
Alfalfa cryptic virus	ACV	Unclassified
Alfalfa mosaic virus	AMV	Alfalfa mosaic virus group
Alfalfa temperate virus	ATeV	Unclassified
Alligatorweed stunting virus	AWSV	Closterovirus I?
American hop latent virus	AHLV	Carlavirus
American plum line pattern virus	APLPV	Ilarvirus
American wheat striate mosiac virus	AWSMV	Rhabdoviridae
Andean potato latent virus	APLV	Tymovirus
Andean potato mottle virus	APMV	Comovirus
Anthoxanthum latent blanching virus	ALBV	Hordeivirus?
Apple chlorotic leafspot virus	ACLV	Closterovirus II
Apple mosaic virus	ApMV	Ilarvirus
Apple stem grooving virus	ASGV	Closterovirus III
Arabis mosaic virus	ArMV	Nepovirus
Arracacha virus A	AVA	Nepovirus
Arracacha virus B	AVB	Unclassified
Arrhenatherum blue dwarf virus	ABDV	Reoviridae *(Fijivirus)*
Artichoke Italian latent virus	AILV	Nepovirus
Artichoke mottled crinkle virus	AMCV	Tombusvirus
Artichoke vein banding virus	AVBV	Nepovirus
Artichoke yellow ringspot virus	AYRV	Nepovirus
Asparagus stunt virus	ASV	Ilarvirus
Asparagus virus II	AVII	Ilarvirus
Bamboo mosaic virus	BaMV	Potexvirus
Barley stripe mosaic virus	BSMV	Hordeivirus
Barley yellow mosaic virus	BaYMV	Potyvirus?
Barley yellow dwarf virus	BYDV	Luteovirus
Barley yellow striate mosaic virus	BYSMV	Rhabdoviridae
Barrel cactus virus	BaCV	Potexvirus
Bean common mosaic virus	BCMV	Potyvirus
Bean curly dwarf mosaic virus	BCDMV	Comovirus
Bean golden mosaic virus	BGMV	Geminivirus
Bean mild mosaic virus	BMMV	Unclassified
Bean pod mottle virus	BPMV	Comovirus
Bean rugose mosaic virus	BRMV	Comovirus
Bean summer death virus	BSDV	Geminivirus
Bean yellow mosaic virus	BYMV	Potyvirus
Bean yellow vein banding virus	BYVBV	Unclassified
Bearded iris mosaic virus	BIMV	Potyvirus
Beet cryptic virus	BCV	Unclassified
Beet curly top virus	BCTV	Geminivirus
Beet leaf curl virus	BLCV	Rhabdoviridae

* Group or family names are those approved by the ICTV. In the family Reoviridae, the genus into which a virus is classified is also indicated in parentheses. Question marks indicate uncertainty of the taxonomic status of the viruses. In the case of the Closterovirus group, we have divided the members into subgroups I to III (see Chapter 15, Volume II) which are also indicated here.

Beet mild yellowing virus	BMYV	Luteovirus
Beet mosaic virus	BtMV	Potyvirus
Beet necrotic yellow vein virus	BNYVV	Tobamovirus?
Beet temperate virus	BTeV	Unclassified
Beet western yellows virus	BWYV	Luteovirus
Beet yellow stunt virus	BYSV	Closterovirus I
Beet yellows virus	BYV	Closterovirus I
Belladonna mottle virus	BeMV	Tymovirus
Bidens mottle virus	BiMV	Potyvirus
Black raspberry latent virus	BRLV	Ilarvirus
Blackeye cowpea mosaic virus	BlCMV	Potyvirus
Blackgram mottle virus	BMoV	Unclassified
Blueberry red ringspot virus	BRRV	Caulimovirus
Blueberry shoestring virus	BSSV	Sobemovirus?
Boletus virus	BoV	Potexvirus?
Broad bean mottle virus	BBMV	Bromovirus
Broad bean necrosis virus	BBNV	Tobamovirus?
Broad bean stain virus	BBSV	Comovirus
Broad bean true mosaic virus	BBTMV	Comovirus
Broad bean wilt virus	BBWV	Comovirus?
Broccoli necrotic yellows virus	BNYV	Rhabdoviridae
Brome mosaic virus	BMV	Bromovirus
Burdock yellows virus	BuYV	Closterovirus I
Cacao necrosis virus	CNV	Nepovirus
Cacao swollen shoot virus	CSSV	Unclassified
Cacao yellow mosaic virus	CYMV	Tymovirus
Cactus virus 2	CV2	Carlavirus
Cactus virus X	CVX	Potexvirus
Canna yellow mottle virus	CaYMV	Unclassified
Carnation cryptic virus	CarCV	Unclassified
Carnation etched ring virus	CERV	Caulimovirus
Carnation Italian ringspot virus	CIRSV	Tombusvirus
Carnation latent virus	CLV	Carlavirus
Carnation mottle virus	CarMV	Unclassified
Carnation necrotic fleck virus	CNFV	Closterovirus I
Carnation ringspot virus	CRSV	Dianthovirus
Carnation vein mottle virus	CVMV	Potyvirus
Carnation yellow stripe virus	CYSV	Tobacco necrosis virus group?
Carrot latent virus	CaLV	Rhabdoviridae
Carrot mottle virus	CMotV	Unclassified
Carrot red leaf virus	CaRLV	Luteovirus
Carrot thin leaf virus	CTLV	Potyvirus
Carrot yellow leaf virus	CYLV	Closterovirus I
Cassava common mosaic virus	CsCMV	Potexvirus
Cassava latent virus	CaLV	Geminivirus
Cassava vein mosaic virus	CaVMV	Caulimovirus
Cauliflower mosaic virus	CaMV	Caulimovirus
Celery mosaic virus	CeMV	Potyvirus
Cereal chlorotic mottle virus	CeCMV	Rhabdoviridae
Cereal tillering disease virus	CTDV	Reoviridae *(Fijivirus)*
Chara corallina virus	CCV	Tobamovirus?

Cherry leafroll virus	CLRV	Nepovirus
Cherry rasp leaf virus	CRLV	Nepovirus?
Chicory yellow mottle virus	ChYMV	Nepovirus
Chloris striate mosaic virus	CSMV	Geminivirus
Chrysanthemum virus B	CVB	Carlavirus
Citrus crinkly leaf virus	CiCLV	Ilarvirus
Citrus leaf rugose virus	CiLRV	Ilarvirus
Citrus tatter leaf virus	CiTLV	Closterovirus II?
Citrus tristeza virus	CTV	Closterovirus I
Citrus variegation virus	CVV	Ilarvirus
Clitoria yellow vein virus	CYVV	Tymovirus
Clover blotch virus	CBV	Cucumovirus
Clover primary leaf necrosis virus	CPLNV	Dianthovirus
Clover yellow mosaic virus	ClYMV	Potexvirus
Clover yellows virus	CYV	Closterovirus I
Clover yellow vein virus	ClYVV	Potyvirus
Cocksfoot mottle virus	CoMV	Sobemovirus?
Cocksfoot streak virus	CSV	Potyvirus
Coffee ringspot virus	CRV	Rhabdoviridae
Colocasia bacilliform virus	CoBV	Unclassified
Colocasia bobone disease virus	CBDV	Rhabdoviridae
Commelina virus X	ComVX	Potexvirus
Cow parsnip mosaic virus	CoPMV	Rhabdoviridae
Cowpea aphid-borne mosaic virus	CABMV	Potyvirus
Cowpea chlorotic mottle virus	CCMV	Bromovirus
Cowpea mild mottle virus	CPMMV	Carlavirus?
Cowpea mosaic virus	CPMV	Comovirus
Cowpea mottle virus	CpMoV	Unclassified
Cowpea ringspot virus	CPRSV	Cucumovirus?
Cowpea severe mosaic virus	CPSMV	Comovirus
Crimson clover latent virus	CCLV	Nepovirus
Cucumber green mottle mosaic virus	CGMMV	Tobamovirus
Cucumber mosaic virus	CMV	Cucumovirus
Cucumber necrosis virus	CuMV	Tobacco necrosis virus group?
Cucumber soilborne virus	CSBV	Unclassified
Cucumber virus 4	CV4	Tobamovirus
Cucumber yellows virus	CuYV	Closterovirus II?
Cymbidium mosaic virus	CybMV	Potexvirus
Cymbidium ringspot virus	CyRSV	Tombusvirus
Cynara virus	CV	Rhabdoviridae
Cynosurus mottle virus	CyMoV	Sobemovirus?
Dahlia mosaic virus	DMV	Caulimovirus
Dandelion latent virus	DLV	Carlavirus
Daphne virus X	DVX	Potexvirus
Dasheen mosaic virus	DaMV	Potyvirus
Dendrobium vein necrosis virus	DVNV	Closterovirus I?
Desmodium yellow mottle virus	DYMV	Tymovirus
Digitaria striate virus	DSV	Rhabdoviridae
Dulcamara mottle virus	DuMV	Tymovirus
Eggplant mosaic virus	EMV	Tymovirus
Eggplant mottled crinkle virus	EMCV	Tombusvirus

Eggplant mottled dwarf virus	EMDV	Rhabdoviridae
Elderberry carlavirus	ECV	Carlavirus
Elderberry latent virus	EBLV	Unclassified
Elm mottle virus	EMoV	Ilarvirus
Erysimum latent virus	ErLV	Tymovirus
Euphorbia mosaic virus	EuMV	Geminivirus
Festuca cryptic virus	FCV	Unclassified
Festuca necrosis virus	FNV	Closterovirus I?
Fiji disease virus	FDV	Reoviridae *(Fijivirus)*
Figwort mosaic virus	FiMV	Caulimovirus
Finger millet mosaic virus	FMMV	Rhabdoviridae
Foxtail mosaic virus	FoMV	Potexvirus
Frangipani mosaic virus	FMV	Tobamovirus
Galinsoga mosaic virus	GaMV	Tombusvirus?
Garland chrysanthemum temperate virus	GCTeV	Unclassified
Glycine mosaic virus	GMV	Comovirus
Glycine mottle virus	GMoV	Tombusvirus
Gomphrena virus	GV	Rhabdoviridae
Grapevine Bulgarian latent virus	GBLV	Nepovirus
Grapevine chrome mosaic virus	GCMV	Nepovirus
Grapevine fan leaf virus	GFLV	Nepovirus
Grapevine leafroll-associated virus	GLR-AV	Closterovirus II
Grapevine virus A	GVA	Closterovirus II
Grapevine virus B	GVB	Closterovirus II?
Helenium virus S	HeVS	Carlavirus
Henbane mosaic virus	HMV	Potyvirus
Heracleum latent virus	HLV	Closterovirus II
Heracleum virus 6	HV6	Closterovirus I?
Hibiscus chlorotic ringspot virus	HCRSV	Tombusvirus?
Hibiscus latent ringspot virus	HLRV	Nepovirus
Hippeastrum mosaic virus	HiMV	Potyvirus
Hop latent virus	HoLV	Carlavirus
Hop mosaic virus	HoMV	Carlavirus
Hop trefoil cryptic virus	HTCV	Unclassified
Hordeum mosaic virus	HorMV	Potyvirus?
Horseradish latent virus	HLaV	Caulimovirus
Hydrangea ringspot virus	HRSV	Potexvirus
Hypochoeris mosaic virus	HyMV	Tobamovirus?
Impatiens latent virus	ILV	Potexvirus?
Indian peanut clump virus	IPCV	Tobamovirus?
Iris mild mosaic virus	IMMV	Potyvirus
Ivy vein clearing virus	IVCV	Rhabdoviridae
Kalanchoe virus 1	KV1	Carlavirus
Kennedya yellow mosaic virus	KYMV	Tymovirus
Lamium mild mosaic virus	LMMV	Comovirus?
Legume yellows virus	LYV	Luteovirus
Leek yellow stripe virus	LYSV	Potyvirus
Lettuce mosaic virus	LMV	Potyvirus
Lettuce necrotic yellows virus	LNYV	Rhabdoviridae
Lettuce speckles mottle virus	LSMV	Unclassified
Lilac chlorotic leafspot virus	LCLV	Closterovirus III
Lilac mottle virus	LiMV	Carlavirus

Lilac ring mottle virus	LRMV	Ilarvirus
Lily symptomless virus	LSV	Carlavirus
Lily virus X	LVX	Potexvirus
Lolium enation virus	LEV	Reoviridae *(Fijivirus)*
Lolium mottle virus	LoMV	Unclassified
Lonicera latent virus	LLV	Carlavirus
Lucerne Australian latent virus	LALV	Nepovirus
Lucerne enation virus	LuEV	Rhabdoviridae
Lucerne transient streak virus	LTSV	Sobemovirus?
Lychnis ringspot virus	LRSV	Hordeivirus
Maclura mosaic virus	MaMV	Potyvirus?
Maize chlorotic dwarf virus	MCDV	Maize chlorotic dwarf virus group
Maize chlorotic mottle virus	MCMoV	Unclassified
Maize mosaic virus	MMV	Rhabdoviridae
Maize rayado fino virus	MRFV	Unclassified
Maize rough dwarf virus	MRDV	Reoviridae *(Fijivirus)*
Maize sterile stunt virus	MSSV	Rhabdoviridae
Maize streak virus	MSV	Geminivirus
Maize stripe virus	MStV	Unclassified
Maize white line mosaic virus	MWLMV	Unclassified
Malva yellows virus	MYV	Luteovirus
Melandrium yellow fleck virus	MYFV	Bromovirus
Melilotus latent virus	MeLV	Rhabdoviridae
Mirabilis mosaic virus	MiMV	Caulimovirus
Mulberry latent virus	MLV	Carlavirus
Mulberry ringspot virus	MRV	Nepovirus
Mungbean yellow mosaic virus	MYMV	Geminivirus
Muskmelon vein necrosis virus	MkVNV	Carlavirus
Myrobalan latent ringspot virus	MLRV	Nepovirus
Nandina mosaic virus	NanMV	Potexvirus
Nandina stem-pitting virus	NSPV	Closterovirus II?
Narcissus latent virus	NLV	Carlavirus
Narcissus mosaic virus	NaMV	Potexvirus
Narcissus tip necrosis virus	NTNV	Unclassified
Narcissus yellow stripe virus	NYSV	Potyvirus
Nerine latent virus	NeLV	Carlavirus
Nerine virus X	NeVX	Potexvirus
Nicotiana velutina mosaic virus	NVMV	Tobamovirus?
Northern cereal mosaic virus	NCMV	Rhabdoviridae
Oat blue dwarf virus	OBDV	Unclassified
Oat mosaic virus	OMV	Potyvirus?
Oat sterile dwarf virus	OSDV	Reoviridae *(Fijivirus)*
Oat striate mosaic virus	OSMV	Rhabdoviridae
Odontoglossum ringspot virus	ORSV	Tobamovirus
Okra mosaic virus	OkMV	Tymovirus
Olive latent ringspot virus	OLRV	Nepovirus
Onion yellow dwarf virus	OYDV	Potyvirus
Ononis yellow mosaic virus	OYMV	Tymovirus
Orchid fleck virus	OFV	Rhabdoviridae?
Pangola stunt virus	PaSV	Reoviridae *(Fijivirus)*
Panicum mosaic virus	PanMV	Unclassified

Papaya mosaic virus	PapMV	Potexvirus
Papaya ringspot virus	PRSV	Potyvirus
Parsley latent virus	ParLV	Unclassified
Parsley rhabdovirus	PRV	Rhabdoviridae
Parsley virus 5	PaIV5	Potexvirus?
Parsnip mosaic virus	ParMV	Potyvirus
Parsnip virus 3	ParV3	Potexvirus?
Parsnip yellow fleck virus	PYFV	Unclassified
Paspalum striate mosaic virus	PSMV	Geminivirus
Passiflora latent virus	PaLV	Carlavirus
Passionfruit woodiness virus	PWV	Potyvirus
Peach rosette mosaic virus	PRMV	Nepovirus
Pea early-browning virus	PEBV	Tobravirus
Pea enation mosaic virus	PEMV	Pea enation mosaic virus group
Pea green mottle virus	PGMV	Comovirus
Pea leafroll virus	PeLRV	Luteovirus
Pea mild mosaic virus	PMMV	Comovirus
Pea seed-borne mosaic virus	PSMV	Potyvirus
Pea streak virus	PeSV	Carlavirus
Peanut clump virus	PCV	Tobamovirus?
Peanut mottle virus	PeMoV	Potyvirus
Peanut stunt virus	PSV	Cucumovirus
Peanut yellow mottle virus	PYMV	Tymovirus
Pelargonium flower break virus	PFBV	Unclassified
Pelargonium leaf curl virus	PLCV	Tombusvirus
Pelargonium line pattern virus	PLPV	Unclassified
Pelargonium vein clearing virus	PVCV	Rhabdoviridae
Pelargonium zonate spot virus	PZSV	Ilarvirus?
Pepino latent virus	PeLV	Carlavirus
Pepino mosaic virus	PepMV	Potexvirus
Pepper mottle virus	PepMoV	Potyvirus
Pepper veinal mottle virus	PVMV	Potyvirus
Petunia asteroid mosaic virus	PAMV	Tombusvirus
Petunia vein clearing virus	PeVCV	Caulimovirus
Phleum mottle virus	PhMoV	Unclassified
Physalis mosaic virus	PhMV	Tymovirus
Pisum virus	PV	Rhabdoviridae
Pittosporum vein yellowing virus	PVYV	Rhabdoviridae
Plantago mottle virus	PlMV	Tymovirus
Plantago severe mottle virus	PlSMV	Potexvirus
Plantago virus 4	PlV4	Caulimovirus
Plantain virus X	PlVX	Potexvirus
Plum pox virus	PPV	Potyvirus
Poa semilatent virus	PSLV	Hordeivirus
Poinsettia cryptic virus	PCrV	Unclassified
Pokeweed mosaic virus	PokMV	Potyvirus
Poplar mosaic virus	PopMV	Carlavirus
Potato aucuba mosaic virus	PAuMV	Potexvirus?
Potato black ringspot virus	PBRV	Nepovirus
Potato leafroll virus	PLRV	Luteovirus
Potato mop-top virus	PMTV	Tobamovirus?

Potato virus A	PVA	Potyvirus
Potato virus M	PVM	Carlavirus
Potato virus S	PVS	Carlavirus
Potato virus T	PVT	Closterovirus III
Potato virus X	PVX	Potexvirus
Potato virus Y	PVY	Potyvirus
Potato yellow dwarf virus	PYDV	Rhabdoviridae
Prune dwarf virus	PDV	Ilarvirus
Prunus necrotic ringspot virus	PNRSV	Ilarvirus
Quail pea mosaic virus	QPMV	Comovirus
Radish mosaic virus	RaMV	Comovirus
Radish yellow edge virus	RYEV	Unclassified
Raphanus virus	RV	Rhabdoviridae
Raspberry bushy dwarf virus	RBDV	Ilarvirus?
Raspberry ringspot virus	RRV	Nepovirus
Raspberry vein chlorosis virus	RVCV	Rhabdoviridae
Red clover cryptic virus	RCCV	Unclassified
Red clover mottle virus	RCMV	Comovirus
Red clover necrotic mosaic virus	RCNMV	Dianthovirus
Red clover vein mosaic virus	RCVMV	Carlavirus
Ribgrass mosaic virus	RMV	Tobamovirus
Rice black-streaked dwarf virus	RBSDV	Reoviridae *(Fijivirus)*
Rice dwarf virus	RDV	Reoviridae *(Phytoreovirus)*
Rice gall dwarf virus	RGDV	Reoviridae *(Phytoreovirus)*
Rice grassy stunt virus	RGStV	Unclassified
Rice hoja blanca virus	RHBV	Unclassified
Rice necrosis mosaic virus	RNMV	Potyvirus?
Rice ragged stunt virus	RRSV	Reoviridae?
Rice stripe virus	RStV	Unclassified
Rice transitory yellowing virus	RTYV	Rhabdoviridae
Rice tungro isometric virus	RTIV	Unclassified
Rice tungro bacilliform virus	RTBV	Unclassified
Rice yellow mottle virus	RYMV	Sobemovirus?
Robinia mosaic virus	RobMV	Cucumovirus
Rubus yellow net virus	RYNV	Unclassified
Ryegrass mosaic virus	RGMV	Potyvirus?
Ryegrass cryptic virus	RCV	Unclassified
Saguaro cactus virus	SCV	Unclassified
Sammons' opuntia virus	SOV	Tobamovirus
Satsuma dwarf virus	SDV	Nepovirus
Scrophularia mottle virus	ScrMV	Tymovirus
Shallot latent virus	SLV	Carlavirus
Soilborne wheat mosaic virus	SBWMV	Tobamovirus?
Solanum nodiflorum mottle virus	SNMV	Sobemovirus?
Solanum yellows virus	SYV	Luteovirus
Sonchus virus	SonV	Rhabdoviridae
Sonchus yellow net virus	SYNV	Rhabdoviridae
Southern bean mosaic virus	SBMV	Sobemovirus
Sowbane mosaic virus	SoMV	Sobemovirus?
Sowthistle yellow vein virus	SYVV	Rhabdoviridae
Soybean dwarf virus	SoyDV	Luteovirus
Soybean mosaic virus	SoyMV	Potyvirus

Spinach latent virus	SPLV	Ilarvirus
Spinach temperate virus	STeV	Unclassified
Squash mosaic virus	SqMV	Comovirus
Strawberry crinkle virus	StCV	Rhabdoviridae
Strawberry latent ringspot virus	SLRV	Nepovirus?
Strawberry vein banding virus	SVBV	Caulimovirus
Subterranean clover mottle virus	SCMoV	Sobemovirus?
Subterranean clover red leaf virus	SCRLV	Luteovirus
Sugarcane mosaic virus	SCMV	Potyvirus
Sunn-hemp mosaic virus	SHMV	Tobamovirus
Sweet clover necrotic mosaic virus	SCNMV	Dianthovirus
Sweet potato mild mottle virus	SPMMV	Potyvirus?
Tephrosia symptomless virus	TeSV	Tombusvirus?
Thistle mottle virus	ThMoV	Caulimovirus
Tobacco etch virus	TEV	Potyvirus
Tobacco leaf curl virus	TLCV	Geminivirus
Tobacco mosaic virus	TMV	Tobamovirus
Tobacco mosaic U_2 virus	T_2MV	Tobamovirus
Tobacco necrosis virus	TNV	Tobacco necrosis virus group
Tobacco necrotic dwarf virus	TNDV	Luteovirus
Tobacco rattle virus	TRV	Tobravirus
Tobacco ringspot virus	TRSV	Nepovirus
Tobacco streak virus	TSV	Ilarvirus
Tobacco vein mottling virus	TVMV	Potyvirus
Tobacco yellow dwarf virus	TYDV	Geminivirus
Tomato aspermy virus	TAV	Cucumovirus
Tomato black ring virus	TBRV	Nepovirus
Tomato bushy stunt virus	TBSV	Tombusvirus
Tomato golden mosaic virus	ToGMV	Geminivirus
Tomato mosaic virus	ToMV	Tobamovirus
Tomato ringspot virus	ToRSV	Nepovirus
Tomato spotted wilt virus	TSWV	Tomato spotted wilt virus group
Tomato top necrosis virus	TTNV	Nepovirus?
Tomato yellow dwarf virus	ToYDV	Geminivirus
Tomato yellow leaf curl virus	ToYLV	Geminivirus
Tomato yellow mosaic virus	ToYMV	Geminivirus
Tomato yellow top virus	TYTV	Luteovirus
Tulare apple mosaic virus	TAMV	Ilarvirus
Tulip breaking virus	TBV	Potyvirus
Tulip virus X	TuVX	Potexvirus
Turnip crinkle virus	TCV	Tombusvirus?
Turnip mosaic virus	TuMV	Potyvirus
Turnip rosette virus	TRoV	Sobemovirus
Turnip yellow mosaic virus	TYMV	Tymovirus
Turnip yellows virus	TYV	Luteovirus
Ullucus virus C	UVC	Comovirus
Velvet tobacco mottle virus	VTMoV	Sobemovirus?
Vicia cryptic virus	VCV	Unclassified
Viola mottle virus	VMV	Potexvirus
Watermelon mosaic virus	WMV	Potyvirus

Wheat chlorotic streak virus	WCSV	Rhabdoviridae
Wheat dwarf virus	WDV	Geminivirus
Wheat rosette stunt virus	WRSV	Rhabdoviridae
Wheat spindle streak mosaic virus	WSSMV	Potyvirus?
Wheat streak mosaic virus	WSMV	Potyvirus?
Wheat yellow leaf virus	WYLV	Closterovirus I
Wheat yellow mosaic virus	WYMV	Potyvirus?
White clover large cryptic virus	WCLCV	Unclassified
White clover mosaic virus	WClMV	Potexvirus
White clover small cryptic virus	WCSCV	Unclassified
White clover temperate virus	WCTeV	Unclassified
Wild cucumber mosaic virus	WCMV	Tymovirus
Winter wheat (Russian) mosaic virus	WWMV	Rhabdoviridae
Wound tumor virus	WTV	Reoviridae *(Phytoreovirus)*
Yam internal brown spot virus	YIBSV	Unclassified
Youcai mosaic virus	YMV	Tobamovirus
Zucchini yellow fleck virus	ZYFV	Potyvirus
Zygocactus virus X	ZVX	Potexvirus

INDEX

A

B

D

E

I

J

K

L

M

N

O

P

Q

R

S

T

U

V

W

X

Y

Z